Teaching Learning Based Optimization Algorithm

R. Venkata Rao

Teaching Learning Based Optimization Algorithm

And Its Engineering Applications

Springer

R. Venkata Rao
Department of Mechanical Engineering
Sardar Vallabhbhai National Institute of
Technology
Surat, Gujarat
India

ISBN 978-3-319-37130-6 ISBN 978-3-319-22732-0 (eBook)
DOI 10.1007/978-3-319-22732-0

Springer Cham Heidelberg New York Dordrecht London

Softcover re-print of the Hardcover 1st edition 2016

Printed on acid-free paper

Springer International Publishing AG Switzerland is part of Springer Science+Business Media
(www.springer.com)

Dedicated to my parents (Lakshmi Narayana and Jayamma), dearest wife (Sujatha Rao) and beloved daughter (Jaya Lakshmi)

Foreword

Recent worldwide advances in manufacturing technologies have brought about a metamorphosis in industry. Fast-changing technologies have created a need for an equally fast response from manufacturing industries. The trends are towards:

- globalization of international trading and production
- international cooperation
- customers' special needs and requirements
- new products with new design, shape, functionality, built-in intelligence and low-energy consumption
- environment-friendly products
- increased share of small series and custom order oriented production.

These trends bring higher competition among the companies and struggle for survival on the market. It is very important for the companies to be innovative, to respond to the changes quickly and efficiently not only in traditional ways (e.g., by time and cost reduction), but also by development of new innovative production methods and products. All these approaches lead to introduction of totally integrated, flexible and intelligent production systems in which advanced modeling and optimization techniques play a crucial role. Researchers are pushed to develop new, intelligent optimization algorithms and methods to comply with the requirements given by the engineers from the industry.

Nowadays, several intelligent optimization algorithms are already in use depending on the nature of phenomenon simulated by the algorithms. In this book, Dr. R. Venkata Rao describes a new optimization algorithm named as "Teaching-Learning-Based Optimization (TLBO)" in a clear and lucid style. The TLBO algorithm is used for solving the continuous and discrete optimization problems involving single objective or multiobjectives. The algorithm works on the principle of teaching and learning, where teachers influence the quality of results of learners. The learners also learn from interaction among themselves. The elitist version of TLBO algorithm, named as ETLBO, is also described in this book. Applications of TLBO algorithm in the fields of electrical engineering, mechanical design, thermal engineering, manufacturing engineering, civil engineering, structural engineering, computer engineering, electronics engineering, physics, and biotechnology are presented in the book.

The TLBO algorithm has already gained a large reputation among the optimization research community. I believe that the TLBO algorithm and its elitist version will be very much useful to the scientists, engineers, and practitioners involved in development and use of advanced optimization algorithms. I hope the book will be a delight to the readers.

Joze Balic
Faculty of Mechanical Engineering
Institute for Production Engineering
Head of Laboratory for Intelligent Manufacturing Systems
University of Maribor, Slovenia

Preface

The advanced optimization algorithms may be classified into different groups depending on the criterion being considered such as population based, iterative based, stochastic, deterministic, etc. Depending on the nature of phenomenon simulated by the algorithms, the population-based heuristic algorithms have two important groups: evolutionary algorithms (EA) and swarm intelligence algorithms. Some of the recognized evolutionary algorithms are: genetic algorithms (GA), differential evolution (DE), evolutionary strategy (ES), evolutionary programming (EP), and artificial immune algorithm (AIA). Among all, GA is a widely used algorithm for various applications. GA works on the principle of the Darwinian theory of the survival of the fittest and the theory of evolution of the living beings. DE is similar to GA with specialized crossover and selection operator. ES is based on the hypothesis that during the biological evolution the laws of heredity have been developed for fastest phylogenetic adaptation. ES imitates, in contrast to the GA, the effects of genetic procedures on the phenotype. EP also simulates the phenomenon of natural evolution at phenotype level and AIA works like the immune system of the human being. Some of the well-known swarm intelligence based algorithms are: particle swarm optimization (PSO), which works on the principle of foraging behavior of the swarm of birds, ant colony optimization (ACO) which works on the principle of foraging behavior of the ant for the food, shuffled frog leaping (SFL) algorithm which works on the principle of communication among the frogs, and artificial bee colony (ABC) algorithm which works on the principle of foraging behavior of a honey bee. Besides these evolutionary and swarm intelligence algorithms, there are some other algorithms which work on the principles of different natural phenomena. Some of them are: harmony search (HS) algorithm which works on the principle of music improvisation in a music player, gravitational search Algorithm (GSA) which works on the principle of gravitational force acting between the bodies, biogeography-based optimization (BBO) which works on the principle of immigration and emigration of the species from one place to the other and league championship algorithm (LCA) which mimics the sporting competition in a sport league.

All the above-mentioned algorithms are population-based optimization methods and have some limitations in one or the other aspect. The main limitation of all the algorithms is that different parameters (i.e., algorithm-specific parameters) are required for proper working of these algorithms. Proper tuning of these parameters is essential for the searching of the optimum solution by these algorithms. A change in the algorithm-specific parameters changes the effectiveness of the algorithm. Most commonly used evolutionary optimization algorithm is the genetic algorithm (GA). However, GA provides a near optimal solution for a complex problem having large number of variables and constraints. This is mainly due to the difficulty in determining the optimum controlling parameters such as crossover probability, mutation probability, selection operator, etc. The same is the case with PSO which uses inertia weight and social and cognitive parameters. Similarly, ABC requires optimum controlling parameters of number of bees (employed, scout, and onlookers), limit, etc. HS requires harmony memory consideration rate, pitch adjusting rate, and the number of improvisations. Sometimes, the difficulty in the selection of algorithm-specific parameters increases with modifications and hybridization. Therefore, the efforts must be continued to develop an optimization algorithm which is free from the algorithm-specific parameters.

An optimization algorithm named as "Teaching-Learning-Based Optimization (TLBO)" is presented in this book to obtain global solutions for continuous as well as discrete optimization problems with less computational effort and high consistency. The TLBO algorithm does not require any algorithm-specific parameters. The TLBO algorithm is based on the effect of the influence of a teacher on the output of learners in a class. Here, output is considered in terms of results or grades. The teacher is generally considered as a highly learned person who shares his or her knowledge with the learners. The quality of a teacher affects the outcome of learners. It is obvious that a good teacher trains learners such that they can have better results in terms of their marks or grades. Moreover, learners also learn from interaction among themselves which also helps in their results. The TLBO algorithm is developed with this philosophy. Furthermore, an elitist version of TLBO algorithm (named as ETLBO) and a non-dominated sorting version (named as NSTLBO) for multiobjective optimization are also presented in this book.

The TLBO algorithm is developed by my team and it is gaining wide acceptance in the optimization research community. After its introduction in 2011, the TLBO algorithm is finding a large number of applications in different fields of science and engineering. The major applications, as of April 2015, are found in the fields of electrical engineering, mechanical design, thermal engineering, manufacturing engineering, civil engineering, structural engineering, computer engineering, electronics engineering, physics, chemistry, biotechnology, and economics. Many research papers have been published in various reputed international journals of Elsevier, Springer-Verlag, Taylor & Francis, and IEEE Transactions, in addition to those published in the proceedings of international conferences. The number of research papers is continuously increasing at a faster rate. The algorithm has carved a niche for itself in the field of advanced optimization and many more researchers may find this as a potential optimization algorithm.

This book provides a detailed understanding of the TLBO algorithm and its versions such as elitist TLBO algorithm and non-dominated sorting TLBO algorithm for multiobjective optimization. Also it provides the applications of TLBO algorithm and its versions in different fields of engineering. The computer codes of TLBO and ETLBO algorithm are also included in the book and these will be useful to the readers. The book is expected to be useful to various engineering professionals as it presents the powerful TLBO algorithm to make their tasks easier, logical, efficient, and effective. The book is intended for engineers, practitioners, managers, institutes involved in the optimization related projects, applied research workers, academics and graduate students in mechanical, manufacturing, electrical, computer, civil and structural engineering. As such, this book is expected to become a valuable reference for those wishing to do research by making use of advanced optimization techniques for solving single objective or multiobjective combinatorial design optimization problems.

I am grateful to Anthony Doyle and his team of Springer-Verlag, London, for their support and help in producing this book. I wish to thank various researchers and the publishers of international journals for giving me the permission to reproduce certain portions of their published research works. I gratefully acknowledge the support of my past and present M.Tech. and Ph.D. students (particularly, P.J. Pawar, G.G. Waghmare, Dhiraj Rai, Kiran More, and Vinay Kumar). My special thanks are due to Ms. Jaya Panvalker (Chairperson, Board of Governors of my institute), Director and my colleagues at S.V. National Institute of Technology, Surat, India.

While every attempt has been made to ensure that no errors (printing or otherwise) enter the book, the possibility of these creeping into the book is always there. I will be grateful to the readers if these errors are pointed out. Suggestions for further improvement of the book will be thankfully acknowledged.

Bangkok R. Venkata Rao

Contents

Chapter 1
Introduction to Optimization

Abstract This chapter presents an introduction to the single objective and multi-objective optimization problems and the methods to solve the same. The merits and demerits of the classical and the advanced optimization methods are presented and the need for an algorithm-specific parameter-less algorithm is emphasized.

1.1 Optimization of Single Objective and Multiobjective Problems

Optimization can be defined as finding solution of a problem where it is necessary to maximize or minimize a single or set of objective functions within a domain which contains the acceptable values of variables while some restrictions are to be satisfied. There might be a large number of sets of variables in the domain that maximize or minimize the objective function(s) while satisfying the described restrictions. They are called as the acceptable solutions and the solution which is the best among them is called the optimum solution of the problem. An objective function expresses the main aim of the model which is either to be minimized or maximized. For example, in a manufacturing process, the aim may be to maximize the profit or minimize the cost. In designing a structure, the aim may be to maximize the strength or minimize the deflection or a combination of many objectives.

A set of variables control the value of the objective function and these variables are essential for the optimization problems. We cannot define the objective function and the constraints without the variables. A set of constraints are those which allow the variables to take on certain values but exclude others. The constraints are not essential and their presence depends on the requirements of the optimization problem. The optimization problem is to find the values of the variables that minimize or maximize the objective function while satisfying the constraints.

R.V. Rao, *Teaching Learning Based Optimization Algorithm*,
DOI 10.1007/978-3-319-22732-0_1

The generalized statement of an optimization problem for maximization can be written as

$$\text{To find } X = \begin{bmatrix} x_1 \\ x_2 \\ \vdots \\ x_n \end{bmatrix} \text{ which maximizes } f(X) \tag{1.1}$$

Subject to the constraints:

$$g_i(X) \leq 0, \quad i = 1, 2, \ldots, m \tag{1.2}$$

$$l_j(X) = 0, \quad j = 1, 2, \ldots, p \tag{1.3}$$

where X is an n-dimensional vector called the design vector, $f(X)$ is called the objective function, and $g_i(X)$ and $l_j(X)$ are known as inequality and equality constraints, respectively. The number of variables n and the number of constraints m and p need not be related in any way. This type of problem is called a constrained optimization problem.

The optimization problem may contain only a single objective function or a number of objective functions. A problem containing only one objective function is called the single objective optimization problem. A problem containing more than one objective function is called the multiple or multiobjective optimization problem. An example of a constrained single objective optimization problem is given below.

Minimize cost,

$$Y_1 = f(x_1, x_2, x_3, x_4) = 2x_1 + 3x_2 + 1.5x_3 + 0.8x_4 \tag{1.4}$$

Subject to the constraints:

$$x_1 + 2x_2 + x_3 \leq 150 \tag{1.5}$$

$$x_1 + x_2 + x_3 + 0.3x_4 \leq 170 \tag{1.6}$$

$$3x_1 + 1.5x_2 \leq 128 \tag{1.7}$$

The ranges of the variables:

$$0 \leq x_1 \leq 50 \tag{1.8}$$

$$0 \leq x_2 \leq 32 \tag{1.9}$$

$$0 \le x_3 \le 44 \tag{1.10}$$

$$0 \le x_4 \le 18 \tag{1.11}$$

In this optimization problem, the objective function and the constraints are expressed as linear equations. However, in practical applications, these may be in the form of nonlinear equations. From the considered ranges of the variables given by Eqs. (1.8)–(1.11), it can be observed that the variables can assume any value within the given ranges. For example, x_1 can assume any value between 0 and 50 (including 0 and 50), x_2 can assume any value between 0 and 32 (including 0 and 32), x_3 can assume any value between 0 and 44 (including 0 and 44), and x_4 can assume any value between 0 and 18 (including 0 and 18). The problem is to find the best combination of x_1, x_2, x_3, and x_4 to obtain the minimum value of Y_1. We can solve the above single objective optimization problem by using traditional (such as simplex method, dynamic programming, separable programming, etc.) and advanced optimization methods (such as genetic algorithm (GA), simulated annealing (SA), particle swarm optimization (PSO), ant colony optimization (ACO), artificial bee colony algorithm (ABC), etc. Then, which method will give us the best solution (i.e., minimum value of Y_1)? We do not know! We can know only after applying these methods to the *same* problem and whichever method gives the best solution is called the best optimization method *for the given problem*.

The multiobjective optimization problems require the simultaneous optimization of multiple (often conflicting) objectives over a given space of candidate solutions. These problems occur in many practical applications, rather often as bi-objective problems, with typical pairs of objectives as quality versus cost, strength versus weight, or accuracy versus complexity. Suppose we include another objective function of maximizing the profit to the above single objective optimization problem.

Maximize Profit,

$$Y_2 = f(x_1, x_2, x_3, x_4) = 0.75x_1 + 8x_2 + x_3 + 1.2x_4 \tag{1.12}$$

Then the problem is to find the best combination of x_1, x_2, x_3, and x_4 to obtain the minimum value of Y_1 and maximum value of Y_2.

Now let us assume a set of values (0, 0, 0, 0) of x_1, x_2, x_3 and x_4. These values satisfy the constraints and hence substituting these values in the objective function Y_1 leads to the value of cost of 0. This value is highly appreciable but at the same time if the same set of values (0, 0, 0, 0) is substituted in the second objective function Y_2 then it leads to the value of profit of 0. This is highly undesirable and hence we can say that the set of values (0, 0, 0, 0) of x_1, x_2, x_3, and x_4 do not give the optimum solution. Now let us assume another set of values (0, 5, 40, 10) of x_1, x_2, x_3, and x_4. These values also satisfy the constraints and hence substituting these values in the objective functions Y_1 and Y_2 lead to the values of 83 and 342 respectively. The set of values (0, 5, 40, and 10) of x_1, x_2, x_3, and x_4 can be said as a feasible solution but not the optimum solution. Let us further assume another set of values

(15, 18, 7, and 24) of x_1, x_2, x_3, and x_4. These values also satisfy the constraints and hence substituting these values in the objective functions Y_1 and Y_2 lead to the values of 113.7 and 191.05 respectively. Thus, the set of values (15, 18, 7, and 24) of x_1, x_2, x_3, and x_4 can be said as another feasible solution but not the optimum solution. Thus, how many feasible solutions are possible for the considered problem? Various combinations of x_1, x, x_3, and x_4 are possible and hence we can say that a large number (or almost infinite number) of feasible solutions may be possible for the considered problem. The two objectives are of conflicting type and optimal solution of one objective does not meet the optimal solution of the other and there exist a large number (or almost infinite number) of solutions (as variables can take any value within their bounds). In general, the multiobjective optimization problems have decision variable values which are determined in a continuous or integer domain with either an infinite or a large number of solutions, the best of which should satisfy the designer or decision-maker's constraints and preference priorities.

Here also we can apply different optimization methods to solve the *same* bi-objective optimization problem and whichever method gives the best combination of solution (i.e., most minimum value of Y_1 and most maximum value of Y_2) is called the best optimization method *for the given problem*. In fact, the solution to the multiobjective optimization problem involves finding not one, but a set of solutions that represent the best possible trade-offs among the objective functions being optimized. Such trade-offs constitute the so-called Pareto optimal set, and their corresponding objective function values form the so-called Pareto front.

The multiobjective optimization problem may contain any number of objectives more than one. The example described by Eqs. (1.4)–(1.12) is *only for giving an idea* to the readers about the concepts of single objective and multiobjective optimization problems.

1.2 Merits and Demerits of the Classical and Advanced Optimization Techniques

Engineering design can be characterized as a goal-oriented and constrained decision-making process to create products that satisfy well-defined human needs. Design optimization consists of certain goals (objective functions), a search space (feasible solutions) and a search process (optimization method). The feasible solutions are the set of all designs characterized by all possible values of the design parameters (i.e., design variables). Optimization problems are of high importance both for the industrial world as well as for the scientific world. The optimization algorithms are becoming increasingly popular in engineering design. They are extensively used in those engineering design problems where the emphasis is on maximizing or minimizing certain goal(s). Traditional or classical optimization methods can be used for finding the optimum solution of continuous and differentiable functions. They are comparatively easier to apply and require comparatively few iterations. Some of them are particularly suitable to find the optimum solution of differentiable functions.

The optimization method searches for the optimal design from all available feasible designs. For example, mechanical design includes an optimization process in which designers always consider certain objectives such as strength, deflection, weight, wear, corrosion, etc. In real design problems, the number of design variables can be very large and their influence on the objective function to be optimized can be very complicated with a nonlinear character. The objective function may have many local optima whereas the designer is interested in the global optimum. Such problems may not be handled by the classical methods that may only compute local optima. So there remains a need for efficient and effective optimization methods for mechanical design problems.

It is a very well-known fact that the traditional or classical optimization techniques impose some limitations on solving complex optimization problems. These limitations are mainly interrelated to their inherent search mechanisms. Search strategies of these classical techniques are generally depended on the type of objective and constraint functions (linear, nonlinear, etc.) and the type of variables used in the problem modeling (integer, binary, continuous, etc.), their efficiency is also very much dependent on the size of the solution space, number of variables and constraints used in the problem modeling and the structure of the solution space (convex, non-convex, etc.). They also do not provide generic solution approaches that can be used to solve problems where different types of variables, objective and constraint functions are used. For example, simplex algorithm can be used to solve problems with linear objective and constraint functions; geometric programming can be used to solve nonlinear models with a polynomial objective function, etc. On the other hand, most of the real-life problems including design optimization problems require several types of variables, objective functions, and constraint functions simultaneously in their formulation. Mainly for these reasons, classical optimization algorithms are usually not suitable or difficult to use for making the effective solution.

In order to overcome some of the well-known deficiencies of the classical optimization procedures, metaheuristic optimization techniques (also called the advanced optimization techniques) mainly originated from artificial intelligence research have been developed by researchers. These algorithms are problem- and model-independent and most of them are efficient and flexible. Research on these techniques is very active and many new metaheuristics and improved versions of the older ones are continually appearing in the scientific literature. Some of the advantages of the metaheuristic algorithms are listed below.

- Robust to dynamic changes: Classical methods of optimization are not robust to dynamic changes in the environment and they require a complete restart for providing a solution. In contrary, metaheuristic algorithms can be used to adapt solutions to changing circumstances.

- Broad applicability: Metaheuristic algorithms can be applied to any problems that can be formulated as function optimization problems.
- Hybridization with other methods: Metaheuristic algorithms can be combined with many classical optimization techniques.
- Ability to solve problems that have no solutions: The advantage of evolutionary algorithms includes the ability to address problems for which there is no human expertise. Even though human expertise should be used when it is needed and available; it often proves less adequate for automated problem-solving routines.

In the real world, there are many problems in which it is desirable to optimize two or more objective functions at the same time. These are known as multiobjective optimization problems and continuous research is being conducted in this field and nature inspired heuristic optimization methods are proving to be better than the classical deterministic methods and thus are widely used. The optimization methods include genetic algorithm (GA), ant colony optimization (ACO) algorithm, particle swarm optimization (PSO) algorithm; differential Evolution (DE) algorithm, artificial bee colony (ABC) algorithm, shuffled frog leaping (SFL) algorithm, harmony search (HS) algorithm, etc. These algorithms can be used for single objective and multiobjective optimization problems. These algorithms have been applied to many engineering optimization problems and proved effective for solving some specific kinds of problems. However, the parameter setting of these algorithms is a serious problem which influences the performance of the optimization problem. For example, the GA requires the crossover probability, mutation probability, and selection operator; ACO algorithm requires exponent parameters, pheromone evaporation rate and the reward factor; PSO algorithm requires learning factors, inertia weight and the maximum value of velocity; DE algorithm requires crossover probability and differential weight; ABC algorithm requires number of bees (scout, onlooker, and employed) and the limit; SFL algorithm requires number of memeplexes and iteration per memeplexes; HS algorithm requires harmony memory consideration rate, pitch adjusting rate and number of improvisations. Finding the optimum values of these algorithm-specific parameters is an optimization problem itself! Proper tuning of the algorithm-specific parameters is very crucial factor which affects the performance of the algorithms. The improper tuning of algorithm-specific parameters either increases the computational efforts or yields the local optimum solution. In addition to the tuning of algorithm-specific parameters, the common control parameters such as population size, number of iterations or generations, elite size, etc., need to be tuned which further enhances the effort.

Keeping in view of the burden on the designer or decision-maker for tuning of the algorithm-specific parameters, Rao et al. (2011, 2012a, b) developed a new optimization algorithm known as "Teaching-Leaning-Based Optimization (TLBO)" algorithm which does not require any algorithm-specific parameters. Unlike other optimization techniques, the TLBO algorithm does not require any algorithm-specific parameters to be tuned, thus making the implementation of TLBO algorithm simpler. The TLBO algorithm requires only the common control parameters like population size and number of generations for its working. If elitism is considered then the elite size becomes another common control parameter of the algorithm.

1.3 Organization of the Book

This book presents the performance of the TLBO algorithm and its versions on the following topics.

- Complicated multimodal composite benchmark functions
- Multiobjective constrained and unconstrained benchmark functions
- Benchmark engineering design problems
- Design optimization of a spur gear train
- Design optimization of a plate-fin heat sink
- Optimization of a multiple chiller systems
- Design optimization of a shell and tube condenser
- Design optimization of a smooth flat plate solar air heater
- Design optimization of a robot manipulator
- Multiobjective optimization of design and manufacturing tolerances
- Single objective and multiobjective optimization of traditional and advanced machining processes

Chapter 2 of the book presents the fundamentals of the TLBO, elitist TLBO and non-dominated sorting TLBO algorithms. The working of the TLBO algorithm for solving the constrained and unconstrained optimization problems is presented step-by-step by means of two examples and the readers can easily understand the working of the TLBO algorithm. Another example is also included to demonstrate the working of the non-dominated sorting TLBO algorithm. Chapters 3–14 present the performance of the TLBO algorithm and its versions on the topics listed above. Chapter 15 presents an overview of the applications of the TLBO algorithm and its modifications made by various researchers to solve different single objective and multiobjective optimization problems belonging to the disciplines of electrical engineering, mechanical design, thermal engineering, manufacturing engineering, civil engineering, structural engineering, computer engineering, electronics engineering, physics, chemistry, biotechnology, and economics. The epilog is included after Chap. 15. The codes of the TLBO and ETLBO algorithms are given in the appendix for the benefit of the readers.

The next chapter presents the details of the working of TLBO algorithm and its versions.

References

Rao, R.V., Savsani, V.J., Vakharia, D.P., 2011. Teaching-learning-based optimization: A novel method for constrained mechanical design optimization problems. Computer-Aided Design 43, 303–315.

Rao, R.V., Savsani, V.J., Vakharia, D.P., 2012a. Teaching-learning-based optimization: an optimization method for continuous non-linear large scale problems. Information Sciences 183, 1–15.

Rao, R.V., Savsani, V.J., Balic, J., 2012b. Teaching–learning-based optimization algorithm for unconstrained and constrained real-parameter optimization problems. Engineering Optimization 44(12), 1447–1462.

Chapter 2
Teaching-Learning-Based Optimization Algorithm

Abstract This chapter introduces teaching-learning-based optimization (TLBO) algorithm and its elitist and non-dominated sorting multiobjective versions. Two examples of unconstrained and constrained benchmark functions and an example of a multiobjective constrained problem are presented to demonstrate the procedural steps of the algorithm.

2.1 Teaching-Learning-Based Optimization Algorithm

All evolutionary and swarm intelligence based optimization algorithms require common control parameters like population size, number of generations, elite size, etc. Besides the common control parameters, different algorithms require their own algorithm-specific parameters. For example, GA uses mutation probability and crossover probability and selection operator; PSO uses inertia weight and social and cognitive parameters; ABC algorithm uses number of bees (scout, onlooker, and employed) and limit; and NSGA-II requires crossover probability, mutation probability, and distribution index. Proper tuning of these algorithm-specific parameters is a very crucial factor which affects the performance of the algorithms. The improper tuning of algorithm-specific parameters either increases the computational effort or yields a local optimal solution. In addition to the tuning of algorithm-specific parameters, the common control parameters also need to be tuned which further enhances the effort. Thus, there is a need to develop an algorithm which does not require any algorithm-specific parameters and teaching-learning-based optimization (TLBO) is such an algorithm.

The TLBO algorithm is a teaching-learning process inspired algorithm proposed by Rao et al. (2011, 2012a, b) and Rao and Savsani (2012) based on the effect of influence of a teacher on the output of learners in a class. The algorithm describes two basic modes of the learning: (i) through teacher (known as teacher phase) and (ii) through interaction with the other learners (known as learner phase). In this optimization algorithm, a group of learners is considered as population and different

R.V. Rao, *Teaching Learning Based Optimization Algorithm*,
DOI 10.1007/978-3-319-22732-0_2

subjects offered to the learners are considered as different design variables of the optimization problem and a learner's result is analogous to the 'fitness' value of the optimization problem. The best solution in the entire population is considered as the teacher. The design variables are actually the parameters involved in the objective function of the given optimization problem and the best solution is the best value of the objective function.

The working of TLBO is divided into two parts, 'Teacher phase' and 'Learner phase'. Working of both the phases is explained below.

2.1.1 Teacher Phase

It is the first part of the algorithm where learners learn through the teacher. During this phase, a teacher tries to increase the mean result of the class in the subject taught by him or her depending on his or her capability. At any iteration i, assume that there are 'm' number of subjects (i.e., design variables), 'n' number of learners (i.e., population size, $k = 1, 2,\ldots, n$) and $M_{j,i}$ be the mean result of the learners in a particular subject 'j' ($j = 1, 2,\ldots, m$) The best overall result $X_{\text{total-kbest},i}$ considering all the subjects together obtained in the entire population of learners can be considered as the result of best learner kbest. However, as the teacher is usually considered as a highly learned person who trains learners so that they can have better results, the best learner identified is considered by the algorithm as the teacher. The difference between the existing mean result of each subject and the corresponding result of the teacher for each subject is given by,

$$\text{Difference\_Mean}_{j,k,i} = r_i\left(X_{j,\text{kbest},i} - T_F M_{j,i}\right) \tag{2.1}$$

where, $X_{j,\text{kbest},i}$ is the result of the best learner in subject j. T_F is the teaching factor which decides the value of mean to be changed, and r_i is the random number in the range [0, 1]. Value of T_F can be either 1 or 2. The value of T_F is decided randomly with equal probability as,

$$T_F = \text{round}[1 + \text{rand}(0, 1)\{2 - 1\}] \tag{2.2}$$

T_F is not a parameter of the TLBO algorithm. The value of T_F is not given as an input to the algorithm and its value is randomly decided by the algorithm using Eq. (2.2). After conducting a number of experiments on many benchmark functions it is concluded that the algorithm performs better if the value of T_F is between 1 and 2. However, the algorithm is found to perform much better if the value of TF is either 1 or 2 and hence to simplify the algorithm, the teaching factor is suggested to take either 1 or 2 depending on the rounding up criteria given by Eq. (2.2). Based on the Difference_Mean$_{j,k,i}$, the existing solution is updated in the teacher phase according to the following expression.

$$X'_{j,k,i} = X_{j,k,i} + \text{Difference\_Mean}_{j,k,i} \tag{2.3}$$

where, $X'_{j,k,i}$ is the updated value of $X_{j,k,i}$. $X'_{j,k,i}$ is accepted if it gives better function value. All the accepted function values at the end of the teacher phase are maintained and these values become the input to the learner phase. The learner phase depends upon the teacher phase.

2.1.2 *Learner Phase*

It is the second part of the algorithm where learners increase their knowledge by interacting among themselves. A learner interacts randomly with other learners for enhancing his or her knowledge. A learner learns new things if the other learner has more knowledge than him or her. Considering a population size of '*n*', the learning phenomenon of this phase is explained below.

Randomly select two learners P and Q such that $X'_{\text{total-}P,i} \neq X'_{\text{total-}Q,i}$ (where, $X'_{\text{total-}P,i}$ and $X'_{\text{total-}Q,i}$ are the updated function values of $X_{\text{total-}P,i}$ and $X_{\text{total-}Q,i}$ of P and Q, respectively, at the end of teacher phase)

$$X''_{j,P,i} = X'_{j,P,i} + r_i(X'_{j,P,i} - X'_{j,Q,i}),\ \mathit{If}\ X'_{\text{total}-P,i} < X'_{\text{total}-Q,i} \tag{2.4}$$

$$X''_{j,P,i} = X'_{j,P,i} + r_i(X'_{j,Q,i} - X'_{j,P,i}),\ \mathit{If}\ X'_{\text{total}-Q,I} < X'_{\text{total}-P,i} \tag{2.5}$$

$X''_{j,P,I}$ is accepted if it gives a better function value.

The Eqs. (2.4) and (2.5) are for minimization problems. In the case of maximization problems, the Eqs. (2.6) and (2.7) are used.

$$X''_{j,P,i} = X'_{j,P,i} + r_i(X'_{j,P,i} - X'_{j,Q,i}),\ \mathit{If}\ X'_{\text{total}-Q,i} < X'_{\text{total}-P,i} \tag{2.6}$$

$$X''_{j,P,i} = X'_{j,P,i} + r_i(X'_{j,Q,i} - X'_{j,P,i}),\ \mathit{If}\ X'_{\text{total}-P,i} < X'_{\text{total}-Q,i} \tag{2.7}$$

Teaching-learning-based optimisation (TLBO) is a population-based algorithm which simulates the teaching-learning process of the class room. This algorithm requires only the common control parameters such as the population size and the number of generations and does not require any algorithm-specific control parameters.

2.2 Demonstration of the Working of TLBO Algorithm on Unconstrained Optimization Problems

To demonstrate the working of TLBO algorithm, an unconstrained benchmark function of sphere is considered. The objective function is to find out the values of x_i that minimize the sphere function.

Benchmark function: Sphere
Minimize,

$$f(x_i) = \sum_{i=1}^{n} x_i^2 \tag{2.8}$$

Range of variables: $-100 \leq x_i \leq 100$

The known solution to this benchmark function is 0 for all x_i values of 0. Now to demonstrate the TLBO algorithm, let us assume a population size of 5 (i.e., number of learners), two design variables x_1 and x_2 (i.e., number of subjects) and one iteration as the termination criterion. The initial population is randomly generated within the ranges of the variables and the corresponding values of the objective function are shown in Table 2.1. The mean values of x_1 and x_2 are also shown. As it is a minimization function, the lowest value of $f(x)$ is considered as the best learner (and is considered as equivalent to teacher).

Now the teacher tries to improve the mean result of the class. Assuming random numbers $r_1 = 0.58$ for x_1 and $r_2 = 0.49$ for x_2, and $T_f = 1$, the difference_mean values for x_1 and x_2 are calculated as,

$$\text{difference\_Mean}\,(x_1) = 0.58 * (-18 - (-8.2)) = -5.684$$

$$\text{difference\_Mean}\,(x_2) = 0.49 * (-27 - (-1)) = -12.74$$

The value of difference_mean (x_1) is added to all the values under the x_1 column and the value of difference_mean (x_2) is added to all the values under the x_2 column of Table 2.1. Table 2.2 shows the new values of x_1 and x_2 and the corresponding values of the objective function.

Table 2.1 Initial population

	x_1	x_2	$f(x)$	
	–55	36	4321	
	0	41	1681	
	96	–86	16612	
	–64	31	5057	
	–18	–27	1053	Teacher
Mean	–8.2	–1		

Table 2.2 New values of the variables and the objective function (teacher phase)

x_1	x_2	$f(x)$
–60.684	23.26	4223.575
–5.684	28.26	830.9355
90.316	–98.74	17906.57
–69.684	18.26	5189.287
–23.684	–39.74	2140.199

Table 2.3 Updated values of the variables and the objective function based on fitness comparison (teacher phase)

x_1	x_2	$f(x)$
–60.684	23.26	4223.575
–5.684	28.26	830.9355
96	–86	16612
–64	31	5057
–18	–27	1053

Now, the values of $f(x)$ of Tables 2.1 and 2.2 are compared and the best values of $f(x)$ are considered and placed in Table 2.3. This completes the teacher phase of the TLBO algorithm.

Now, the learner phase (also called student phase) starts and any student can interact with any other student for knowledge transfer. This interaction can be done in random manner. In this example, interactions between learners 1 and 2, 2 and 4, 3 and 5, 4 and 1, and 5 and 3 are considered. It is to be noted that every learner has to interact with any other learner. That is why, in this example, five interactions are considered (i.e., one interaction for each learner). Table 2.4 shows the new values of x_1 and x_2 for the learners after the interactions and considering random numbers $r_1 = 0.81$ for x_1 and $r_2 = 0.92$ for x_2. For example, the new values of x_1 and x_2 for learner 1 are calculated as explained below.

As it is a minimization function, the value of $f(x)$ is better for learner 2 as compared to that of learner 1 and hence the knowledge transfer is from learner 2 to learner 1. Hence the new values of x_1 and x_2 for learner 1 is calculated as,

$$(x_1)\text{ new for learner 1} = -60.684 + 0.81(-5.684 - (-60.684)) = -16.134$$
$$(x_2)\text{ new for learner 1} = 23.26 + 0.92(28.26 - 23.26) = 27.86.$$

Similarly, the new values of x_1 and x_2 for learner 2 are calculated as explained below.

As it is a minimization function, the value of $f(x)$ is better for learner 2 as compared to that of learner 4 and hence the knowledge transfer is from learner 2 to learner 4. Hence the new values of x_1 and x_2 for learner 2 is calculated as,

$$(x_1)\text{ new for learner 2} = -5.684 + 0.81(-5.684 - (-64)) = 41.552$$
$$(x_2)\text{ new for learner 2} = 28.26 + 0.92(28.26 - 31) = 25.7392$$

Table 2.4 New values of the variables and the objective function (learner phase)

x_1	x_2	$f(x)$	Interaction
–16.134	27.86	1036.486	1 and 2
41.552	25.7392	2389.075	2 and 4
3.66	–31.72	1019.554	3 and 5
–61.314	23.879	4329.613	4 and 1
–100(–110.34[a])	27.28	10744.2	5 and 3

[a]This value has crossed the given range of the variable and hence it is assigned the bound value

Now, the values of *f*(*x*) of Tables 2.3 and 2.4 are compared and the best values of *f*(*x*) are considered and placed in Table 2.5. This completes the learner phase and one iteration of the TLBO algorithm.

It can be noted that the minimum value of the objective function in the randomly generated initial population is 1053 and it has been reduced to 830.9355 (shown bold) at the end of first iteration. If we increase the number of iterations then the known value of the objective function (i.e., 0) can be obtained within next few iterations. Also, it is to be noted that in the case of maximization function problems, the best value means the maximum value of the objective function and the calculations are to be proceeded accordingly in the teacher and learner phases.

Now, the same minimization function of sphere is attempted in a different way. Here the interactions in the learner phase are considered in such a manner that each learner interacts with all other learners. For example, learner 1 interacts with the learners 2, 3, 4, and 5; learner 2 interacts with the learners 1, 3, 4, and 5; learner 3 interacts with learners 1, 2, 4, and 5; and so on. Tables 2.6, 2.7, 2.8, 2.9, and 2.10 show these interactions. The best interactions are also shown in the tables.

Table 2.11 shows the updated values of the variables and the objective function based on the best fitness values obtained in Tables 2.6, 2.7, 2.8, 2.9, and 2.10.

Table 2.5 Updated values of the variables and the objective function based on fitness comparison (learner phase)

x_1	x_2	*f*(*x*)
−16.134	27.86	1036.486
−5.684	**28.26**	**830.9355**
3.66	−31.72	1019.554
−61.314	23.879	4329.613
−18	−27	1053

Table 2.6 First learner interacting with every other learner (learner phase)

x_1	x_2	*f*(*x*)	
−60.684	23.26	4223.575	
−16.134	27.86	1036.486	best
−100(−187.598[a])	100(123.779[a])	20000	
−57.998	16.1392	3624.242	
−26.11	-22.9792	1209.776	

[a]These values have crossed the given ranges of the variables and hence they are assigned the bound values

Table 2.7 Second learner interacting with every other learner (learner phase)

x_1	x_2	*f*(*x*)	
38.866	32.86	2590.346	
−5.684	28.26	830.9355	best
−88.048	100(133.379[a])	17752.45	
41.552	25.7392	2389.075	
4.292	79.0992	6275.105	

[a]This value has crossed the given range of the variable and hence it is assigned the bound value

Table 2.8 Third learner interacting with every other learner (learner phase)

x_1	x_2	$f(x)$	
−30.914	14.5192	1166.483	
13.636	19.119	551.4767	best
96	−86	16612	
−33.6	21.64	1597.25	
3.66	−31.72	1019.554	

Table 2.9 Fourth learner interacting with every other learner (learner phase)

x_1	x_2	$f(x)$	
−61.314	23.879	4329.613	
−16.764	28.4792	1092.097	best
−100(−193.6[a])	100(138.64[a])	20000	
−64	31	5057	
−26.74	−22.36	1214.997	

[a]These values have crossed the given ranges of the variables and hence they are assigned the bound values

Table 2.10 Fifth learner interacting with every other learner (learner phase)

x_1	x_2	$f(x)$	
16.574	−73.239	5638.649	
−8.024	23.8392	632.692	best
−100(−110.34[a])	27.28	10744.2	
19.26	−80.36	6828.677	
−18	−27	1053	

[a]This value has crossed the given range of the variable and hence it is assigned the bound value

Table 2.11 updated values of the variables and the objective function based on the best fitness values obtained (learner phase)

x_1	x_2	$f(x)$
−16.134	27.86	1036.486
−5.684	28.26	830.9355
13.636	**19.119**	**551.4767**
−16.764	28.4792	1092.097
−8.024	23.8392	632.692

It can be noted that the minimum value of the objective function in the randomly generated initial population is 1053 and it has been reduced to 551.4767 at the end of first iteration. It can be noted that by following the above approach of every learner interacting with all the remaining learners, the number of calculations get increased but this approach may provide the global optimum value (i.e., 0 in this example) in less number of iterations as compared to the approach of considering random interactions between the learners.

2.3 Demonstration of the Working of TLBO Algorithm on Constrained Optimization Problems

To demonstrate the working of TLBO algorithm, a constrained benchmark function of Himmelblau is considered. The objective function is to find out the values of x_1 and x_2 that minimize the Himmelblau function.

Benchmark function: Himmelblau

Minimize,

$$f(x_i) = (x_1^2 + x_2 - 11)^2 + (x_1 + x_2^2 - 7)^2 \tag{2.9}$$

Constraints:

$$g_1(x) = 26 - (x_1 - 5)^2 - x_2^2 \geq 0 \tag{2.10}$$

$$g_2(x) = 20 - 4x_1 - x_2 \geq 0 \tag{2.11}$$

Ranges of variables: $-5 \leq x_1, x_2 \leq 5$

The known solution to this benchmark function is 0 for $x_1 = 3$ and $x_2 = 2$ and $g_1(x) = 18$ and $g_2(x) = 6$. Now to demonstrate the TLBO algorithm, let us assume a population size of 5 (i.e., number of learners), two design variables x_1 and x_2 (i.e., number of subjects) and one iteration as the termination criterion. The initial population is randomly generated within the ranges of the variables and the corresponding values of the objective function are shown in Table 2.12. The mean values of x_1 and x_2 are also shown. As it is a minimization function, the lowest value of $f(x)$ is considered as the best learner (and is considered as equivalent to teacher). If the constraints are violated then penalties are assigned to the objective function. There are many ways to assign the penalties and in this example the penalty p_1 for violation of $g_1(x)$ is considered as $10 * (g_1(x))^2$ and the penalty p_2 for violation of $g_2(x)$ is considered as $10 * (g_2(x))^2$. As it is a minimization problem, the values of penalties are added to the value of the objective function $f(x)$ and the fitness function is $f'(x) = f(x) + 10 * (g_1(x))^2 + 10 * (g_2(x))^2$. The function $f'(x)$ is called the pseudo-objective function.

Table 2.12 Initial population

	x_1	x_2	$f(x)$	$g_1(x)$	p_1	$g_2(x)$	p_2	$f'(x)$
	3.22	**0.403**	**13.13922**	**22.66919**	**0**	**6.717**	**0**	**13.13922**
	0.191	2.289	77.71054	-2.366	55.979	16.947	0	133.6902
	3.182	0.335	14.02423	22.58265	0	6.937	0	14.02423
	1.66	4.593	261.5732	-6.25125	390.781	8.767	0	652.3543
	2.214	0.867	43.64116	17.48652	0	10.277	0	43.64116
Mean	2.093	1.697						

Table 2.13 New values of the variables, objective function, and the penalties (teacher phase)

x_1	x_2	$f(x)$	$g_1(x)$	p_1	$g_2(x)$	p_2	$f'(x)$
3.5017	−0.794	8.443577	23.12466	0	6.7872	0	8.443577
0.4727	1.092	122.2511	4.311091	0	17.0172	0	122.2511
3.4637	−0.862	7.820563	22.89674	0	7.0072	0	7.820563
1.9417	3.396	56.61739	5.113985	0	8.8372	0	56.61739
2.4957	−0.33	45.34465	19.61958	0	10.3472	0	45.34465

It may be noted that 10 * $(g_1(x))^2$ is used as the penalty function in this example for the violation in constraint $g_1(x)$ and 10 * $(g_2(x))^2$ is used as the penalty function for the violation in constraint $g_2(x)$. Higher penalties are desirable and one may use 50 * $(g_1(x))^2$ or 100 * $(g_1(x))^2$ or 500 * $(g_1(x))^2$ or any such penalty for violation of $g_1(x)$ and 50 * $(g_2(x))^2$ or 100 * $(g_2(x))^2$ or 500 * $(g_2(x))^2$ or any such penalty for violation of $g_2(x)$. The assignment of penalties for violations depends upon the designer/decision-maker/user. Sometimes, the penalty functions assigned may be different for different constraints depending upon the application as decided by the designer/decision-maker/user.

The mean values of x_1 and x_2 are also shown in Table 2.12. As it is a minimization function, the lowest value of $f'(x)$ is considered as the best learner (and is considered as equivalent to teacher).

Now the teacher tries to improve the mean result of the class. Assuming random numbers r_1 = 0.25 for x_1 and r_2 = 0.925 for x_2 and T_f = 1, the difference_mean values for x_1 and x_2 are calculated as,

$$\text{difference\_mean}(x_1) = 0.25 * (3.22 - (2.093)) = 0.2817$$

$$\text{difference\_mean}(x_2) = 0.925 * (0.403 - (1.697)) = -1.197$$

The value of difference_mean (x_1) is added to all the values under the x_1 column and the value of difference_mean (x_2) is added to all the values under the x_2 column of Table 2.12. Table 2.13 shows the new values of x_1 and x_2, penalties and the corresponding values of the objective function.

Now, the values of $f'(x)$ of Tables 2.12 and 2.13 are compared and the best values of $f'(x)$ are considered and placed in Table 2.14. This completes the teacher phase of the TLBO algorithm.

Table 2.14 Updated values of the variables, objective function, and the penalties based on fitness comparison (teacher phase)

x_1	x_2	$f(x)$	$g_1(x)$	p_1	$g_2(x)$	p_2	$f'(x)$
3.5017	−0.794	8.443577	23.12466	0	6.7872	0	8.443577
0.4727	1.092	122.2511	4.311091	0	17.0172	0	122.2511
3.4637	−0.862	7.820563	22.89674	0	7.0072	0	7.820563
1.9417	3.396	56.61739	5.113985	0	8.8372	0	56.61739
2.214	0.867	43.64116	17.48652	0	10.277	0	43.64116

Table 2.15 New values of the variables, objective function and the penalties (learner phase)

x_1	x_2	$f(x)$	$g_1(x)$	p_1	$g_2(x)$	p_2	$f'(x)$	Interaction
5(6.4095[a])	−2.0199	147.8492	21.92	0	2.0199	0	147.8492	1 and 2
1.8829	2.5896	26.19377	9.577659	0	9.8788	0	26.19377	2 and 4
4.6634	−1.9869	79.34047	21.93893	0	3.3333	0	79.34047	3 and 5
3.4393	0.6725	11.91628	23.11196	0	5.5703	0	11.91628	4 and 1
3.8856	0.7207	29.95277	24.2387	0	3.7369	0	29.95277	5 and 2

[a]This value has crossed the given range of the variable and hence it is assigned the bound value

Now, the learner phase (also called student phase) starts and any student can interact with any other student for knowledge transfer. This interaction can be done in random manner. In this example, interactions between learners 1 and 2, 2 and 4, 3 and 5, 4 and 1, and 5 and 2 are considered. Table 2.15 shows the new values of x_1 and x_2 for the learners after the interactions and considering random numbers $r_1 = 0.96$ for x_1 and $r_2 = 0.65$ for x_2. For example, the new values of x_1 and x_2 for learner 1 are calculated as explained below.

As it is a minimization function, the value of $f'(x)$ is better for learner 1 as compared to that of learner 2 and hence the knowledge transfer is from learner 1 to learner 2. Hence the new values of x_1 and x_2 for learner 1 is calculated as,

$$(x_1)\text{ new for learner 1} = 3.5017 + 0.96(3.5017 - (0.4727)) = 6.4095$$
$$(x_2)\text{ new for learner 1} = -0.794 + 0.65(-0.794 - 1.092) = -2.0199$$

Similarly, the new values of x_1 and x_2 for learner 2 are calculated. The value of $f'(x)$ is better for learner 4 as compared to that of learner 2 and hence the knowledge transfer is from learner 4 to learner 2. Hence the new values of x_1 and x_2 for learner 2 is calculated as,

$$(x_1)\text{ new for learner 2} = 0.4727 + 0.96(1.9417 - 0.4727) = 1.8829$$

$$(x_2)\text{ new for learner 2} = 1.092 + 0.65(3.396 - 1.092) = 2.5896$$

Now, the values of $f'(x)$ of Tables 2.14 and 2.15 are compared and the best values of $f'(x)$ are considered and are placed in Table 2.16. This completes the learner phase and one iteration of the TLBO algorithm.

It can be noted that the minimum value of the objective function in the randomly generated initial population is −13.13922 and it has been reduced to 7.820563 at the end of the first iteration. If we increase the number of iterations then the known value of the objective function (i.e., 0) can be obtained within next few iterations.

It is to be noted that in the case of maximization problems, the best value means the maximum value of the objective function and the calculations are to be proceeded accordingly in the teacher and learner phases. The values of penalties are to

Table 2.16 Updated values of the variables, objective function, and the penalties based on fitness comparison (learner phase)

x_1	x_2	$f(x)$	$g_1(x)$	p_1	$g_2(x)$	p_2	$f'(x)$
3.5017	–0.794	8.443577	23.12466	0	6.7872	0	8.443577
1.8829	2.5896	26.19377	9.577659	0	9.8788	0	26.19377
3.4637	–0.862	7.820563	22.89674	0	7.0072	0	7.820563
3.4393	0.6725	11.91628	23.11196	0	5.5703	0	11.91628
3.8856	0.7207	29.95277	24.2387	0	3.7369	0	29.95277

Table 2.17 First learner interacting with every other learner

x_1	x_2	$f(x)$	$g_1(x)$	p_1	$g_2(x)$	p_2	$f'(x)$	
3.5017	–0.794	8.443577	23.12466	0	6.7872	0	8.443577	
5(6.4095[a])	–2.0199	147.8492	21.92	0	2.0199	0	147.8492	
3.465	–0.8382	8.05084	22.9412	0	6.9782	0	8.05084	best
4.999	–3.5175	217.2476	13.62719	0	3.5215	0	217.2476	
4.7378	–1.8737	93.20215	22.4205	0	2.9225	0	93.20215	

[a]This has crossed the given range of the variable and hence is assigned the bound value

be subtracted from the objective function in the case of maximization problems (i.e., $f'(x) = f(x) - 10 * (g_1(x))^2 - 10 * (g_2(x))^2$).

Now, the same minimization function of Himmelblau is attempted in a different way. Here the interactions in the learner phase are considered in such a manner that each learner interacts with all other learners. For example, learner 1 interacts with the learners 2, 3, 4, and 5; learner 2 interacts with the learners 1, 3, 4, and 5; learner 3 interacts with learners 1, 2, 4, and 5; and so on. Tables 2.17, 2.18, 2.19, 2.20, and 2.21 show these interactions. The best interactions are also shown in the tables.

Table 2.22 shows the updated values of the variables, objective function and the penalties based on the best fitness values obtained in Tables 2.17, 2.18, 2.19, 2.20 and 2.21

It can be noted that the minimum value of the objective function in the randomly generated initial population is –13.13922 and it has been reduced to 7.597071 at the end of first iteration. It can be noted that by following the above approach of every learner interacting with all the remaining learners, the number of calculations seem

Table 2.18 Second learner interacting with every other learner

x_1	x_2	$f(x)$	$g_1(x)$	p_1	$g_2(x)$	p_2	$f'(x)$	
3.3805	–0.1339	13.05768	23.35929	0	6.6119	0	13.05768	best
0.4727	1.092	122.2511	4.311091	0	17.0172	0	122.2511	
3.3441	–0.1781	13.13471	23.22628	0	6.8017	0	13.13471	
1.8829	2.5896	26.19377	9.577659	0	9.8788	0	26.19377	
2.1443	0.9457	45.46327	16.95063	0	10.4771	0	45.46327	

Table 2.19 Third learner interacting with every other learner

x_1	x_2	$f(x)$	$g_1(x)$	p_1	$g_2(x)$	p_2	$f'(x)$	
3.4272	–0.9062	7.597071	22.7051	0	7.1974	0	7.597071	best
5(6.3351[a])	–2.1321	147.3284	21.45415	0	2.1321	0	147.3284	
3.4637	–0.862	7.820563	22.89674	0	7.0072	0	7.820563	
4.9248	–3.6297	215.8199	12.81962	0	3.9305	0	215.8199	
4.6634	–1.9859	79.34521	21.9429	0	3.3323	0	79.34521	

[a]This has crossed the given range of the variable and hence is assigned the bound value

Table 2.20 Fourth learner interacting with every other learner

x_1	x_2	$f(x)$	$g_1(x)$	p_1	$g_2(x)$	p_2	$f'(x)$	
3.4393	0.6725	11.91628	23.11196	0	5.5703	0	11.91628	
3.3519	4.8936	438.3633	–0.66355	4.4030	1.6988	0	442.7664	
3.4028	0.6283	11.71331	23.05419	0	5.7605	0	11.71331	best
1.9417	3.396	56.61739	5.113985	0	8.8372	0	56.61739	
2.2031	1.7521	22.29212	15.1075	0	9.4355	0	22.29212	

Table 2.21 Fifth learner interacting with every other learner

x_1	x_2	$f(x)$	$g_1(x)$	p_1	$g_2(x)$	p_2	$f'(x)$	
3.4502	–0.2127	12.75966	23.55288	0	6.4119	0	12.75966	
3.8856	0.7207	29.95277	24.2387	0	3.7369	0	29.95277	
3.4137	–0.2569	12.5497	23.41765	0	6.6021	0	12.5497	best
2.4754	–0.7769	47.28898	19.02282	0	10.8753	0	47.28898	
2.214	0.867	43.64116	17.48652	0	10.277	0	43.64116	

Table 2.22 Updated values of the variables, objective function, and the penalties based on the best fitness values (learner phase)

x_1	x_2	$f(x)$	$g_1(x)$	p_1	$g_2(x)$	p_2	$f'(x)$
3.465	–0.8382	8.05084	22.9412	0	6.9782	0	8.05084
3.3805	–0.1339	13.05768	23.35929	0	6.6119	0	13.05768
3.4272	–0.9062	7.597071	22.7051	0	7.1974	0	7.597071
3.4028	0.6283	11.71331	23.05419	0	5.7605	0	11.71331
3.4137	–0.2569	12.5497	23.41765	0	6.6021	0	12.5497

to be increased but this approach may provide the global optimum value (i.e., 0 in this example) in less number of iterations as compared to the approach of considering random interactions between the learners. The two demonstrations of working of the TLBO algorithm clearly prove its simplicity and effectiveness in solving the unconstrained and constrained optimization problems. The flowchart of the TLBO algorithm is shown in Fig. 2.1.

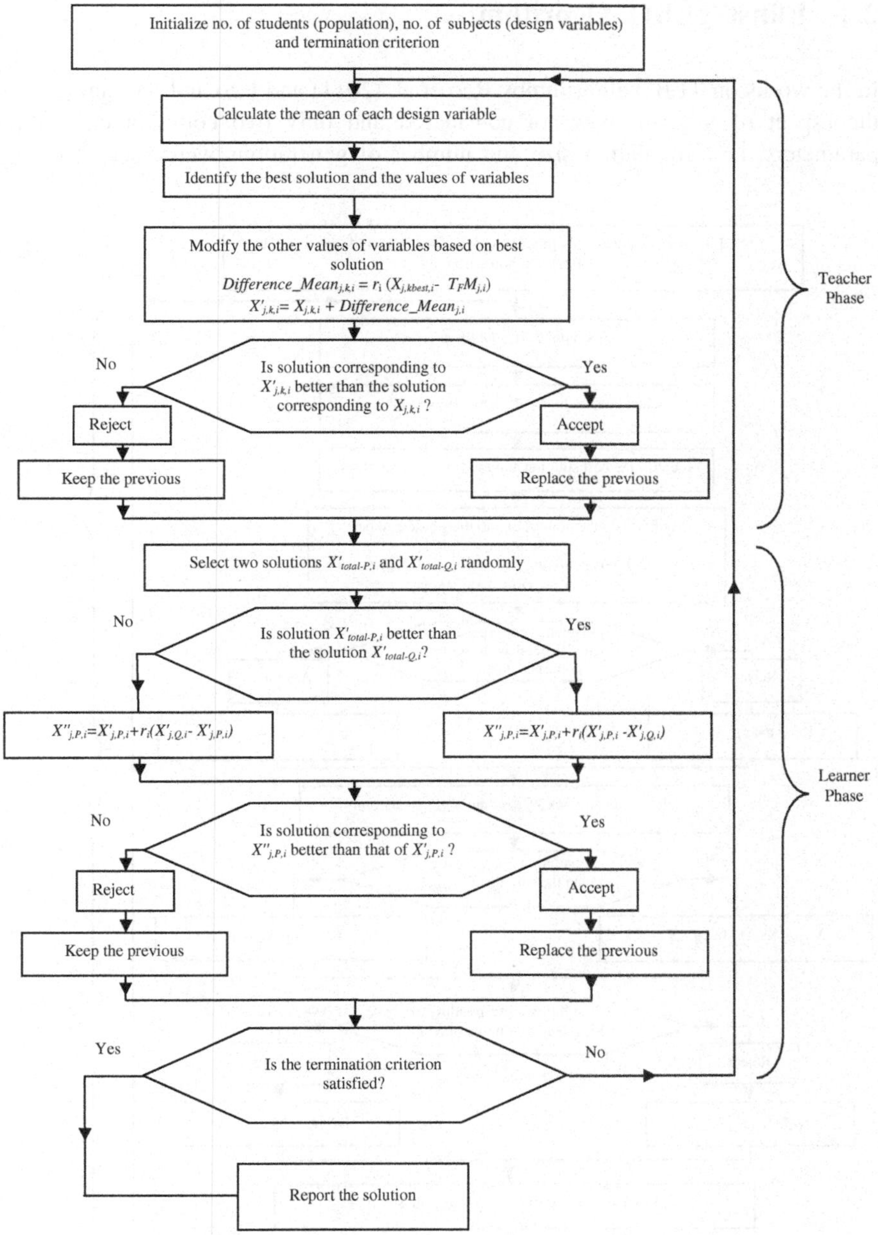

Fig. 2.1 Flowchart of TLBO algorithm

2.4 Elitist TLBO Algorithm

In the works on TLBO algorithmby Rao et al. (2011) and Rao and Savsani (2012), the aspect of 'elitism' was not considered and only two common controlling parameters, i.e., population size and number of generations were used. Rao and

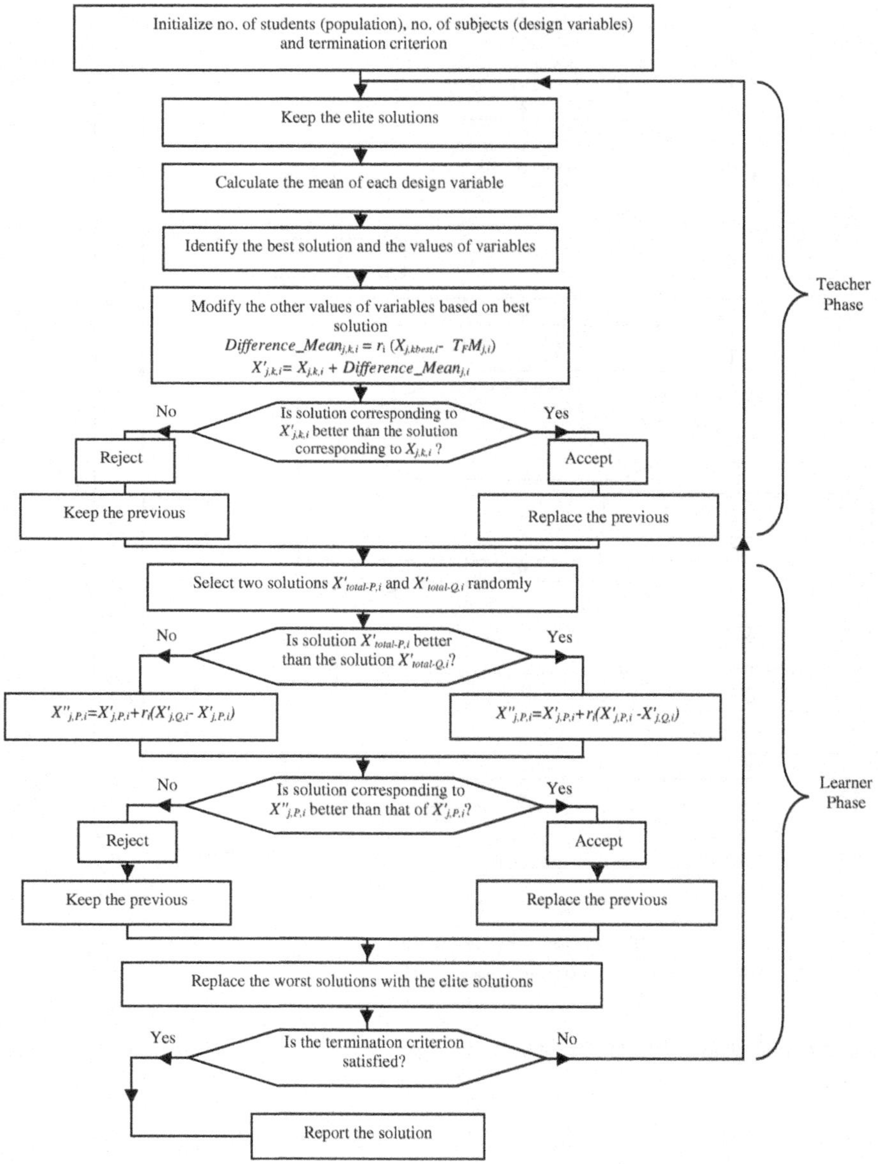

Fig. 2.2 Flowchart of ETLBO algorithm

Patel (2012) introduced ‘elitism’ in the TLBO algorithm to identify its effect on the exploration and exploitation capacities of the algorithm. The concept of elitism is utilized in most of the evolutionary and swarm intelligence algorithms where during every generation the worst solutions are replaced by the elite solutions. The flowchart of the elitist TLBO algorithm is shown in Fig. 2.2.

2.5 Non-dominated Sorting TLBO Algorithm for Multiobjective Optimization

Multiobjective optimization is an area of multiple criteria decision-making that is concerned with mathematical optimization problems involving more than one objective function to be optimized simultaneously. Multiobjective optimization has been applied in many fields of science, engineering, economics, and logistics where optimal decisions need to be taken in the presence of trade-offs between two or more conflicting objectives. Minimizing cost while maximizing comfort while buying a car, and maximizing performance while minimizing fuel consumption, and emission of pollutants of a vehicle are examples of multiobjective optimization problems involving two and three objectives, respectively. In practical problems there can be more than three objectives.

For a multiobjective optimization problem, there does not exist a single solution that simultaneously optimizes each objective. In that case, the objective functions are said to be conflicting and there exist a large number of Pareto optimal solutions. A solution is called non-dominated, Pareto optimal, Pareto efficient, or noninferior, if none of the objective functions can be improved in value without degrading some of the other objective values. Researchers study multiobjective optimization problems from different viewpoints and, thus, there exist different solution philosophies and goals when setting and solving them. The goal may be to find a representative set of Pareto optimal solutions, and/or quantify the trade-offs in satisfying the different objectives, and/or finding a single solution that satisfies the subjective preferences of a decision-maker.

Two important approaches of solving the multiobjective optimization problems are considered in this book and these are (1) a priori approach and (2) a posteriori approach. In the a priori approach, the preferences of the decision-maker are asked and the best solution according to the given preferences is found. The preferences of the decision-maker are in the form of weights assigned to the objective functions. The weights may be assigned through any method like direct assignment, eigen-vector method (Rao 2007), empty method, minimal information method, etc. Once the weights are decided by the decision-maker, the multiple objectives are combined into a scalar objective via the weight vector. However, if the objective functions are simply weighted and added to produce a single fitness, the function with the largest range would dominate the evolution. A poor input value for the objective with the larger range makes the overall value much worse than a poor value for the objective with smaller range. To avoid this, all objective functions are normalized to have same range. For example, if $f_1(x)$ and $f_2(x)$ are the two objective functions to me minimized, then the combined objective function can be written as,

$$\min f(x) = \left\{ w_1 \left[\left(\frac{f_{1(x)}}{f_1^*} \right) \right] + w_2 \left[\left(\frac{f_{2(x)}}{f_2^*} \right) \right] \right\} \tag{2.12}$$

where, $f(x)$ is the combined objective function and f_1^* is the minimum value of the objective function $f_1(x)$ when solved it independently without considering $f_2(x)$ (i.e., solving the multiobjective problem as a single objective problem and considering only $f_1(x)$ and ignoring $f_2(x)$). And f_2^* is the minimum value of the objective function $f_2(x)$ when solved it independently without considering $f_1(x)$ (i.e. solving the multiobjective problem as a single objective problem considering only $f_2(x)$ and ignoring $f_1(x)$). w_1 and w_2 are the weights assigned by the decision-maker to the objective functions $f_1(x)$ and $f_2(x)$ respectively. Suppose $f_1(x)$ and $f_2(x)$ are not of the same type (i.e., minimization or maximization) but one is a minimization function (say $f_1(x)$) and the other is a maximization function (say $f_2(x)$). In that case, the Eq. (2.12) is written as Eq. (2.13) and f_2^* is the maximum value of the objective function $f_2(x)$ when solved it independently without considering $f_1(x)$.

$$\min\ f(x) = \left\{ w_1 \left[\left(\frac{f_{1(x)}}{f_1^*} \right) \right] - w_2 \left[\left(\frac{f_{2(x)}}{f_2^*} \right) \right] \right\} \tag{2.13}$$

In general, the combined objective function can include any number of objectives and the summation of all weights is equal to 1.

A posteriori approach aims to generate all the Pareto optimal solutions or a representative set of Pareto optimal solutions and the decision-maker chooses the best one among them. Evolutionary algorithms are popular approaches for generating the Pareto optimal solutions to a multiobjective optimization problem. Currently, most evolutionary multiobjective optimization algorithms apply Pareto-based ranking schemes. Evolutionary algorithms such as the non-dominated sorting genetic algorithm-II (NSGA-II) and strength Pareto evolutionary algorithm 2 (SPEA-2) have become standard approaches. The main advantage of evolutionary algorithms, when applied to solve multiobjective optimization problems, is the fact that they typically generate sets of solutions, allowing computation of an approximation of the entire Pareto front. The main disadvantage of evolutionary algorithms is their low speed and the Pareto optimality of the solutions cannot be guaranteed. It is only known that none of the generated solutions dominates the others. In this book, a new approach for generating the Pareto optimal solutions to the multiobjective optimization problems is described and it is named as "non-dominated sorting teaching-learning-based optimization algorithm (NSTLBO)."

The NSTLBO algorithm is an extension of the TLBO algorithm. The NSTLBO algorithm is a posteriori approach for solving multiobjective optimization problems and maintains a diverse set of solutions. The NSTLBO algorithm consists of teacher phase and learner phase similar to the TLBO algorithm. However, in order to handle multiple objectives effectively and efficiently, the NSTLBO algorithm is incorporated with non-dominated sorting approach and crowding distance computation mechanism proposed by Deb (2001). The teacher phase and learner phase ensure good exploration and exploitation of the search space while non-dominated sorting approach makes certain that the selection process is always towards the good solutions and the population is pushed towards the Pareto front in each iteration. The crowding distance assignment mechanism assures the selection of teacher from a sparse region of the search space thus averting any chance of premature convergence of the algorithm at local optima.

In the NSTLBO algorithm, the learners are updated according to the teacher phase and the learner phase of the TLBO algorithm. However, in case of single objective optimization it is easy to decide which solution is better than the other based on the objective function value. But in the presence of multiple conflicting objectives determining the best solution from a set of solutions is not a simple task. In the NSTLBO algorithm, the task of finding the best solution is accomplished by comparing the rank assigned to the solutions based on the non-dominance concept and the crowding distance value.

At the beginning an initial population is randomly generated with P number of solutions (learners). This initial population is then sorted and ranked based on the non-dominance concept. The learner with the highest rank (rank = 1) is selected as the teacher of the class. In case, there exists more than one learner with the same rank then the learner with the highest value of crowding distance is selected as the teacher of the class. Once the teacher is selected the mean of the learners is calculated and the learners are updated based on the teacher phase of the TLBO algorithm, i.e., according to Eqs. (2.1)–(2.3).

After the teacher phase the updated learners (new learners) are recombined with the initial population to obtain a set of $2P$ solutions (learners). These learners are again sorted and ranked based on the non-dominance concept and the crowding distance value for each learner is computed. Based on the new ranking and crowding distance value, P number of best learners are selected. These learners are further updated according to the learner phase of the TLBO algorithm.

In the learner phase, a learner interacts with another randomly chosen learner to enhance his or her knowledge in the subject. The learner with a higher rank is regarded as superior among the two learners. If both the learners hold the same rank then the learner with greater crowding distance value is regarded as superior to the other. Then learners are updated based on Eqs. (2.4) and (2.5) for minimization problems or Eqs. (2.6) and (2.7) for maximization problems.

After the end of learner phase, the new learners are combined with the old learners and again sorted and ranked. Based on the new ranking and crowding distance value, P number of best learners are selected and these learners are directly updated based on the teacher phase in the next iteration.

2.5.1 Non-dominated Sorting of the Population

In this approach the population is sorted into several ranks based on the dominance concept as follows: a solution $x^{(1)}$ is said to dominate other solution $x^{(2)}$ if and only if solution $x^{(1)}$ is no worse than solution $x^{(2)}$ with respect to all the objectives or the solution $x^{(1)}$ is strictly better than solution $x^{(2)}$ in at least one objective. If any of the two conditions are violated then solution $x^{(1)}$ does not dominate solution $x^{(2)}$. Among a set of solutions *P*, the non-dominated set of solutions *n* are those that are not dominated by any member of the set *P* (Deb 2001).

To determine the non-dominated set of solutions from the given population set *P*, each solution *i* in *P* is compared with every other solution in *P*. If solution *i* is not dominated by any other solution in *P* then it is a member of the non-dominated set *n*. The non-dominated solutions identified after the first sorting are ranked as 1 and are deleted from the set *P*. The remaining members of set *P* are sorted in the similar manner and the non-dominated set of solutions identified after second sorting are ranked as 2 and are deleted from the set *P*. This procedure is repeated until entire population is sorted or the set *P* becomes an empty set.

2.5.2 Crowding Distance Computation

The crowding distance computation requires sorting the population according to each objective function value in ascending order of magnitude. Thereafter, for each objective function, the boundary solutions (solutions with smallest and largest function values) are assigned an infinite distance value. All other intermediate solutions are assigned a distance value equal to the absolute normalized difference in the function values of two adjacent solutions. This calculation is continued with the other objective functions. The overall crowding distance value is calculated as the sum of individual distance values corresponding to each objective. Each objective function is normalized before calculating the crowding distance.

2.5.3 Constraint Handling

In order to effectively handle the constraints a constrained dominance concept (Deb 2001) is introduced in the proposed approach. In the presence of constraints, a solution *i* is said to dominate solution *j* if any of the following conditions is true.

- Solution *i* is feasible but solution *j* is not.
- Solution *i* and *j* both are infeasible, but overall constraint violation of solution *i* is less than overall constraint violation of solution.
- Solution *i* and *j* both are feasible but solution *i* dominates solution *j*.

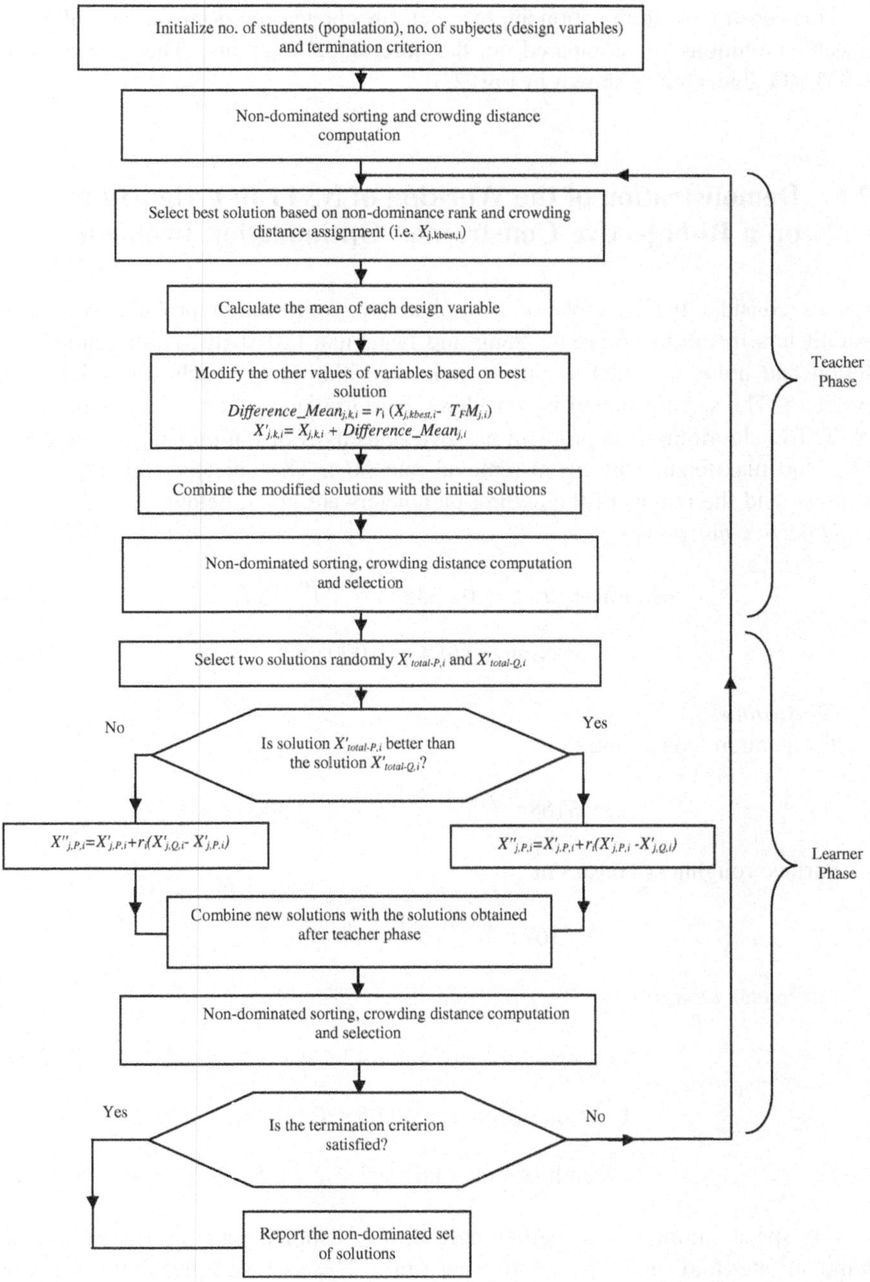

Fig. 2.3 Flowchart of NSTLBO algorithm

This constrained domination approach ensures better non-domination rank to the feasible solutions as compared to the infeasible solutions. The flowchart of NSTLBO algorithm is shown in Fig. 2.3.

2.6 Demonstration of the Working of NSTLBO Algorithm on a Bi-objective Constrained Optimization Problem

Let us consider the example of a bi-objective optimization problem of cutting parameters in turning process. Yang and Natarajan (2010) used differential evolution and non-dominated sorting genetic algorithm-II approachesfor solving the problem. The same problem is considered here to demonstrate the working of the NSTLBO algorithm. The problem has two objectives of minimizing the tool wear (T_w) and maximizing the metal removal rate (M_r). The objective functions, constraints, and the ranges of the cutting parameters are given below.

Objective functions:

$$\text{Minimize }(T_w) = 0.33349\, v^{0.1480} f^{0.4912} d^{0.2898} \tag{2.14}$$

$$\text{Maximize }(M_r) = 1000\, v f d \tag{2.15}$$

Constraints:

Temperature constraint:

$$88.5168\, v^{0.3156} f^{0.2856} d^{0.2250} \le 500 \tag{2.16}$$

Surface roughness constraint:

$$18.5167\, v^{-0.0757} f^{0.7593} d^{0.1912} \le 2 \tag{2.17}$$

Parameter bounds:

$$\text{Cutting speed (m/min)}: 42 \le v \le 201 \tag{2.18}$$

$$\text{Feed rate (mm/rev)}: 0.05 \le f \le 0.33 \tag{2.19}$$

$$\text{Depth of cut (mm)}: 0.5 \le d \le 2.5 \tag{2.20}$$

v = speed (m/min); f = feed (mm/rev); d = depth of cut (mm); M_r = metal removal rate (mm^3/min); T_w = tool wear (mm); T = tool-workpiece interface temperature ($^\circ$C) and R_a = surface roughness (μm).

Now to demonstrate the NSTLBO algorithm, let us assume a population size of 5 (i.e. number of learners), three design variables v, f, and d (i.e., number of subjects) and one iteration as the termination criterion. The initial population is randomly generated within the ranges of the variables and the corresponding values of the objective functions are shown in Table 2.23. The mean values of v, f, and d are also shown.

$$(Z_T)_{max} = 0.000;\ (Z_{Ra})_{max} = 4.597$$

Z' = overall constraint violation and it is given as (Yang and Natarajan 2010),

$$Z' = \frac{Z_T}{(Z_T)_{max}} + \frac{Z_{Ra}}{(Z_{Ra})_{max}} \tag{2.21}$$

In Table 2.23, the values under Z_T and Z_{Ra} represent the values by which these constraints are violated by the candidate solution and $(Z_T)_{max}$ and $(Z_{Ra})_{max}$ represent the maximum values of violations of the constraints of tool-workpiece interface temperature and surface roughness so far in the entire iteration. For example, $(Z_{Ra})_{max} = 6.597 - 2 = 4.597$. The crowding distance CD is 0. Now the teacher tries to improve the mean result of the class. Assuming random numbers $r_1 = 0.91$ for v and $r_2 = 0.67$ for f and $r_3 = 0.25$ for d and $T_f = 1$, the

$$\text{difference\_mean}_v = 0.91 * (171.541 - 149.90) = 19.693$$

$$\text{difference\_mean}_f = 0.67 * (0.0941 - 0.2387) = -0.09688$$

$$\text{difference\_mean}_d = 0.25 * (1.811 - 1.761) = 0.0125$$

The value of difference_mean is added to the corresponding columns of the variables in Tables 2.23 and 2.24 shows the new values of v, f, and d and the corresponding values of the objective functions and the values of constraints.

Now, the initial solutions of Table 2.23 are combined with the solutions obtained in Table 2.24. The Table 2.25 shows the combined population. The ranks are assigned based on the procedure described in Sects. 2.5.1–2.5.3.

$$(Z_T)_{max} = 0.000;\ (Z_{Ra})_{max} = 4.597$$

Now, the selection is done based on the non-dominance rank and the crowding distance. Table 2.26 shows the selected candidate solutions based on the non-dominance rank and the crowding distance. It may be noted that the value of the crowding distance is 0 in Table 2.25.

$$(Z_T)_{max} = 0.000;\ (Z_{Ra})_{max} = 4.597$$

Now, the learner phase starts and any student can interact with any other student for knowledge transfer. This interaction can be done in a random manner. In this example, interactions between learners 1 and 5, 2 and 4, 3 and 1, 4 and 5, and 5 and

Table 2.23 Initial population

Sr. no.	v	f	d	M_r	T_w	T	R_a	Z_T	Z_{Ra}	Z'	Rank	CD
1	171.541	0.0941	1.811	29233.18	0.2657	261.265	2.335	0	0.335	0.073	1	–
2	186.021	0.3217	0.571	34170.33	0.3519	293.677	4.734	0	2.734	0.595	3	–
3	62.1909	0.318	2.198	43469.2	0.4398	280.527	6.597	0	4.597	1	5	–
4	187.226	0.1859	2.367	82384.18	0.4063	346.486	4.095	0	2.095	0.456	2	–
5	142.545	0.274	1.857	72529.46	0.4401	336.292	5.358	0	3.358	0.730	4	–
Mean	149.90	0.2387	1.761									

Table 2.24 New values of the variables, objective functions, constraints, and violations (teacher phase)

Sr. no.	v	f	d	M_r	T_w	T	R_a	Z_T	Z_{Ra}	Z'
1	191.234	0.05[a]	1.8235	17435.76	0.1983	226.057	1.435	0	0	0
2	201[a]	0.2248	0.5835	26365.33	0.3004	272.99	3.6	0	1.6	0.348
3	81.884	0.2211	2.2105	40020.11	0.3838	276.157	4.908	0	2.908	0.633
4	201[a]	0.08902	2.3795	42576.44	0.2865	287.473	2.331	0	0.331	0.072
5	162.238	0.1771	1.8695	53715.13	0.3628	309.727	3.814	0	1.814	0.395

[a]This value has crossed the given range of the variable and hence it is assigned the bound value

2 are considered. Table 2.27 shows the new values of v, f, and d for the learners after the interactions and considering random numbers $r_1 = 0.81$ for v and $r_2 = 0.79$ for f and $r_3 = 0.56$ for d.

$$(Z_T)_{max} = 0.000;\ (Z_{Ra})_{max} = 4.597$$

Now the solutions obtained in the teacher phase (i.e. Table 2.26) are combined with the solutions obtained in the learner phase (i.e. Table 2.27) and are shown in Table 2.28.

The crowding distance values are calculated as described in Sect. 2.5.2. However, for demonstration purpose, sample steps of calculations of the crowding distance are given below.

Step 1: Sort and rank the population of Table 2.28 based on the non-dominated sorting concept.

Step 2: Collect all rank 1 solutions.

Sr. no.	Objective functions		Rank
	M_r	T_W	
1	17435.76	0.1983	1
6	18069.9	0.1989	1
7	25125	0.2189	1
8	20200.25	0.2147	1

Step 3: Determine the minimum and maximum values of both the objective functions for the entire population from Table 2.28 and these are, $(M_r)_{min} = 17435.76$; $(M_r)_{max} = 53715.13$; $(T_w)_{min} = 0.1983$; and $(T_w)_{max} = 0.3628$.

Step 4: Consider only the first objective and sort all the values of the first objective function in the ascending order irrespective of the values of the second objective function.

Table 2.25 Combined population (teacher phase)

Sr. no.	v	f	d	M_r	T_w	T	R_a	Z_T	Z_{Ra}	Z'	Rank	CD
1	171.541	0.0941	1.811	29233.18	0.2657	261.265	2.335	0	0.335	0.073	3	–
2	186.021	0.3217	0.571	34170.33	0.3519	293.677	4.734	0	2.734	0.595	7	–
3	62.1909	0.318	2.198	43469.2	0.4398	280.527	6.597	0	4.597	1	10	–
4	187.226	0.1859	2.367	82384.18	0.4063	346.486	4.095	0	2.095	0.456	6	–
5	142.545	0.274	1.857	72529.46	0.4401	336.292	5.358	0	3.358	0.730	9	–
6	191.234	0.05[a]	1.8235	17435.76	0.1983	226.057	1.435	0	0	0	1	–
7	201[a]	0.2248	0.5835	26365.33	0.3004	272.99	3.6	0	1.6	0.348	4	–
8	81.884	0.2211	2.2105	40020.11	0.3838	276.157	4.908	0	2.908	0.633	8	–
9	201[a]	0.08902	2.3795	42576.44	0.2865	287.473	2.331	0	0.331	0.072	2	–
10	162.238	0.1771	1.8695	53715.13	0.3628	309.727	3.814	0	1.814	0.395	5	–

[a]This value has crossed the given range of the variable and hence it is assigned the bound value

Table 2.26 Candidate solutions based on non-dominance rank and crowding distance (teacher phase)

Sr. no.	v	f	d	M_r	T_w	T	R_a	Z_T	Z_{Ra}	Z'	Rank	CD
1	191.234	0.05[a]	1.8235	17435.76	0.1983	226.057	1.435	0	0	0	1	–
2	201[a]	0.08902	2.3795	42576.44	0.2865	287.473	2.331	0	0.331	0.072	2	–
3	171.541	0.0941	1.811	29233.18	0.2657	261.265	2.335	0	0.335	0.073	3	–
4	201[a]	0.2248	0.5835	26365.33	0.3004	272.99	3.6	0	1.6	0.348	4	–
5	162.238	0.1771	1.8695	53715.13	0.3628	309.727	3.814	0	1.814	0.395	5	–

[a]This value has crossed the given range of the variable and hence it is assigned the bound value

Table 2.27 New values of the variables, objective functions, constraints, and violations (learner phase)

Sr. no.	v	f	d	M_r	T_w	T	R_a	Z_T	Z_{Ra}	Z'	Interaction
1	201[a]	0.05[a]	1.798	18069.9	0.1989	228.912	1.426	0	0	0	1 and 5
2	201[a]	0.05[a]	2.5[a]	25125	0.2189	246.534	1.519	0	0	0	2 and 4
3	187.5	0.05926	1.818	20200.25	0.2147	235.665	1.634	0	0	0	3 and 1
4	201[a]	0.2625	0.5[a]	26381.25	0.31	275.604	3.932	0	1.932	0.4203	4 and 5
5	193.635	0.1075	2.155	44857.97	0.3037	293.22	2.647	0	0.647	0.1407	5 and 2

[a]This value has crossed the given range of the variable and hence it is assigned the bound value

Table 2.28 Combined population (learner phase)

Sr. no.	v	f	d	M_r	T_w	T	R_a	Z_T	Z_{Ra}	Z'	Rank	CD
1	191.234	0.05[a]	1.8235	17435.76	0.1983	226.057	1.435	0	0	0	1	∞
2	201[a]	0.08902	2.3795	42576.44	0.2865	287.473	2.331	0	0.331	0.072	2	–
3	171.541	0.0941	1.811	29233.18	0.2657	261.265	2.335	0	0.335	0.073	3	–
4	201[a]	0.2248	0.5835	26365.33	0.3004	272.99	3.6	0	1.6	0.348	5	–
5	162.238	0.1771	1.8695	53715.13	0.3628	309.727	3.814	0	1.814	0.395	6	–
6	201	0.05	1.798	18069.9	0.1989	228.912	1.426	0	0	0	1	0.1759
7	201	0.05	2.5	25125	0.2189	246.534	1.519	0	0	0	1	∞
8	187.5	0.05926	1.818	20200.25	0.2147	235.665	1.634	0	0	0	1	0.3160
9	201	0.2625	0.5	26381.25	0.31	275.604	3.932	0	1.932	0.4203	7	–
10	193.635	0.1075	2.155	44857.97	0.3037	293.22	2.647	0	0.647	0.1407	4	–

[a]This value has crossed the given range of the variable and hence it is assigned the bound value

Sr. no.	Objective functions		Rank
	M_r	T_W	
1	**17435.76**	0.1983	1
6	**18069.9**	0.1989	1
8	**20200.25**	0.2147	1
7	**25125**	0.2189	1

Step 4a: Assign crowding distance as infinity (∞) to the first and last solutions (i.e., the best and the worst solutions).

Sr. no.	Objective functions		Rank	Crowding distance
	M_r	T_W		
1	**17435.76**	0.1983	1	∞
6	**18069.9**	0.1989	1	
8	**20200.25**	0.2147	1	
7	**25125**	0.2189	1	∞

Step 4b: The crowding distances d_6 and d_8 are calculated as follows (please note that 6 is between 1 and 8; and 8 is between 6 and 7).

$$d_6^{(1)} = 0 + \frac{(M_r)_8 - (M_r)_1}{(M_r)_{\max} - (M_r)_{\min}} = 0 + \frac{20200.25 - 17435.76}{53715.13 - 17435.76} = 0.0762$$

$$d_8^{(1)} = 0 + \frac{(M_r)_7 - (M_r)_6}{(M_r)_{\max} - (M_r)_{\min}} = 0 + \frac{25125 - 18069.9}{53715.13 - 17435.76} = 0.19446$$

Step 5: Consider only the second objective and sort all the values of the second objective function in the ascending order irrespective of the values of the first objective function.

Sr. no.	Objective functions		Rank
	M_r	T_W	
1	17435.76	**0.1983**	1
6	18069.9	**0.1989**	1
8	20200.25	**0.2147**	1
7	25125	**0.2189**	1

Step 5a: Assign crowding distance as infinity (∞) to the first and last solutions (i.e. the best and the worst solutions).

Sr. no.	Objective functions		Rank	Crowding distance
	M_r	T_W		
1	17435.76	**0.1983**	1	∞
6	18069.9	**0.1989**	1	
8	20200.25	**0.2147**	1	
7	25125	**0.2189**	1	∞

Step 5b: The crowding distances d_6 and d_8 are calculated as,

$$d_6^{(2)} = d_6^{(1)} + \frac{(T_W)_8 - (T_W)_1}{(T_W)_{\max} - (T_W)_{\min}} = 0.0762 + \frac{0.2147 - 0.1983}{0.3628 - 0.1983} = 0.1759$$

$$d_8^{(2)} = d_8^{(1)} + \frac{(T_W)_7 - (T_W)_6}{(T_W)_{\max} - (T_W)_{\min}} = 0.19446 + \frac{0.2189 - 0.1989}{0.3628 - 0.1983} = 0.3160$$

Sr. no.	Objective functions		Rank	Crowding distance
	M_r	T_W		
1	17435.76	0.1983	1	∞
6	18069.9	0.1989	1	0.1759
8	20200.25	0.2147	1	0.3160
7	25125	0.2189	1	∞

Similar procedure can be repeated for calculating the crowding distances for the solutions with ranks of 2, 3, 4, 5, 6, and 7. However, in this example, there are only single solutions with the ranks of 2, 3, 4, etc. Hence, no crowding distance is assigned to them.

Now Table 2.29 shows the candidate solutions based on the non-dominance ranks and the crowding distances.

This completes the learner phase and an iteration of the NSTLBO algorithm.

The website for the TLBO algorithm is https://sites.google.com/site/tlborao. The readers may refer to the website for updates on TLBO algorithm.

The next chapter demonstrates the working of TLBO algorithm on complex composite benchmark functions.

Table 2.29 Candidate solutions based on non-dominance ranks and crowding distances (learner phase)

Sr. no.	v	f	d	M_r	T_w	T	R_a	Z_T	Z_{Ra}	Z'	Rank	CD
1	191.234	0.05[a]	1.8235	17435.76	0.1983	226.057	1.435	0	0	0	1	∞
6	201[a]	0.05	1.798	18069.9	0.1989	228.912	1.426	0	0	0	1	0.1759
8	187.5	0.05926	1.818	20200.25	0.2147	235.665	1.634	0	0	0	1	0.3160
7	201	0.05	2.5	25125	0.2189	246.534	1.519	0	0	0	1	∞
2	201[a]	0.08902	2.3795	42576.44	0.2865	287.473	2.331	0	0.331	0.072	2	–

[a]This value has crossed the given range of the variable and hence it is assigned the bound value

References

Deb, K., 2001. Multiobjective Optimization using Evolutionary Algorithms. New York: John Wiley.

Rao, R.V., 2007. Decision Making in the Manufacturing Environment using Graph Theory and Multiple Attribute Decision Making Methods. London: Springer-Verlag.

Rao, R.V., Patel, V., 2012. An elitist teaching-learning-based optimization algorithm for solving complex constrained optimization problems. International Journal of Industrial Engineering Computations 3(4), 535–560.

Rao, R.V., Savsani, V.J., 2012. Mechanical Design Optimization using Advanced Optimization Techniques. London: Springer-Verlag.

Rao, R.V., Savsani, V.J., Vakharia, D.P., 2011. Teaching-learning-based optimization: A novel method for constrained mechanical design optimization problems. Computer-Aided Design 43 (3), 303–315.

Rao, R.V., Savsani, V.J., Vakharia, D.P., 2012a. Teaching-learning-based optimization: A novel optimization method for continuous non-linear large scale problems. Information Sciences 183 (1), 1–15.

Rao, R.V., Savsani, V.J., Balic, J., 2012b. Teaching-learning-based optimization algorithm for unconstrained and constrained real parameter optimization problems. Engineering Optimization 44 (12), 1447–1462.

Yang, S.H., Natarajan, U., 2010. Multiobjective optimization of cutting parameters in turning process using differential evolution and non-dominated sorting genetic algorithm-II approaches. International Journal of Advanced Manufacturing Technology 49, 773–784.

Chapter 3
Application of TLBO and ETLBO Algorithms on Complex Composite Test Functions

Abstract This chapter presents the application of the TLBO and ETLBO algorithms on complex composite test functions each of which is formed by composing the basic standard benchmark functions to construct a more challenging function with randomly located global optimum and several randomly located deep local optima. The results of the applications prove the better competitiveness of the TLBO and ETLBO algorithms.

3.1 Composite Test Functions

In the past two decades, different kinds of optimization algorithms have been designed and applied to solve real-parameter function optimization problems. Some of the popular approaches are real-parameter evolutionary algorithms (EA), evolution strategies (ES), evolutionary programming (EP), differential evolution (DE), particle swarm optimization (PSO), shuffled frog leaping (SFL) algorithm, classical methods such as quasi-Newton method (QN), hybrid evolutionary-classical methods, and other non-evolutionary methods such as simulated annealing (SA), tabu search (TS), and others. Under each category, there exist many different methods varying in their operators and working principles. In most such studies, a subset of the standard benchmark problems (Sphere, Schwefel, Rosenbrock, Rastrigins, etc.) is considered. Although some comparisons are made in some research studies, often they are confusing and limited to the benchmark problems used in the study. In some occasions, the benchmark problem and the chosen algorithm are complementary to each other and the same algorithm may not work in other problems that well. There is definitely a need of evaluating these algorithms in a more systematic manner by specifying a common termination criterion, size of problems, initialization scheme, linkages/rotation, etc. There is also a need to perform a scalability study demonstrating how the running time/evaluations increase with an increase in the problem size.

R.V. Rao, *Teaching Learning Based Optimization Algorithm*,
DOI 10.1007/978-3-319-22732-0_3

In real world optimization, many engineering problems can be classified as multimodal problems. The aim is to locate several globally or locally optimal solutions and then to choose the most appropriate solutions considering practical issues. In recent years, evolutionary algorithms (EA) with various niching techniques have been successfully applied to solve multimodal optimization problems (Liang et al. 2005a, b). In order to compare and evaluate different algorithms, various benchmark functions with various properties have been proposed. Many of these popular benchmark functions possess some properties that have been exploited by some algorithms to achieve excellent results. Some of these problems are given below (Liang et al. 2005a, b).

1. **Problem 1** Global optimum having the same parameter values for different variables/dimensions: Most of the popular benchmark functions have the same parameter values for different dimensions at the global optimum because of their symmetry. For example, the global optima for Rastrigin function and Griewank function are [0, 0, 0, …, 0] and the global optima for Rosenbrock function are [1, 1, 1, …, 1]. In this situation, if there exist some operators to copy one dimension's value to the other dimensions, the global optimum may be found rapidly.
2. **Problem 2** Global optimum at the origin: In this case, the global optimum o is equal to [0, 0, 0, …, 0]. Zhong et al. (2004) proposed the following function:

   $$[l \cdot (1 - \text{sRadius}), l \cdot (1 + \text{sRadius})].$$

   where l is the search center and sRadius is the local search radius to perform the local search. It can be observed that the local search range is much smaller when l is near the origin than when l is far from the origin. This operator is not effective if the global optimum is not at the origin. Hence, this operator is specifically designed to exploit this common property of many benchmark functions.
3. **Problem 3** Global optimum lying in the center of the search range: Some algorithms have the potential to converge to the center of the search range. The mean-centric crossover operator is just a good example for this type. When we randomly generate the initial population uniformly, the mean-centric method will have a trend to lead the population to the center of the search range.
4. **Problem 4** Global optimum on the bounds: This situation is encountered in some multiobjective optimization algorithms as some algorithms set the dimensions moving out of the search range to the bounds (Coello et al. 2004). If the global optimum is on the bounds, as in some multiobjective benchmark functions, the global optimum will be easily found. However, if there are some local optima near the bounds, it will be easy to fall into the local optima and fail to find the global optimum.
5. **Problem 5** Local optima lying along the coordinate axes or no linkage among the variables/dimensions: Most of the benchmark functions, especially high dimensional functions, always have their symmetrical grid structure and local optima are always along the coordinate axes. In this case, the information of the local optima could be used to locate the global optimum. Further, for some functions it is possible to locate the global optimum by using just

D one-dimensional searches for a D dimensional problem. Some co-evolutionary algorithms (Bergh and Engelbrecht 2004) and the one-dimensional mutation operator (Leung and Wang 2001) just use these properties to locate the global optimum rapidly.

By analyzing these problems, it is recommended that the researchers should use the following methods to avoid these problems when they use the benchmark functions suffering from these problems, to test a novel algorithm (Liang et al. 2005a, b).

1. Shift the global optimum to a random position to make the global optimum to have different parameter values for different dimensions for benchmark functions suffering from problems 1 to 3:

$$F(x) = f(x - \text{onew} + \text{oold}). \tag{3.1}$$

where $F(x)$ is the new function, $f(x)$ is old function, oold is the old global optimum and onew is the new setting global optimum which has different values for different dimensions and not in the center of the search range.
2. For Problem 4, considering there are real problems having the global optimum on the bounds, it is an acceptable method for the bounds handling to set the population to the near bounds when they are out of the search range. However, it is recommended that using different kinds of benchmark functions to test the algorithms. For example, it can be used for some problems with the global optimum on bounds, not on bounds and some problems with local optima on bounds. It is not recommended to just test one algorithm that uses this bounds handling method on benchmark functions with the global optimum on bounds and conclude that the algorithm to be good.
3. Rotate the functions with problem 5 as given below:

$$F(x) = f(R \times x), \tag{3.2}$$

where R is an orthogonal rotation matrix. In this way, it can avoid local optima lying along the coordinate axes and retain the benchmark functions' properties at the same time.

Liang et al. (2005a, b) proposed a general framework to construct novel and challenging composite benchmark functions possessing many desirable properties. The idea is to compose the standard benchmark functions to construct a more challenging function with a randomly located global optimum and several randomly located deep local optima. The basic functions used to construct the composite functions are given below.

- Sphere Function

$$f(x) = \sum_{i=1}^{D} x_i^2, x \in [-100, 100]^D. \tag{3.3}$$

- Rastrigin Function

$$f(x) = \sum_{i=1}^{D} \left(x_i^2 - 10\cos(2\Pi x_i) + 10\right), x \in [-5, 5]^D \tag{3.4}$$

- Weierstrass Function

$$f(x) = \sum_{i=1}^{D} \left(\sum_{k=0}^{k_{\max}} \left[a^k \cos\left(2\Pi b^k (x_i + 0.5)\right)\right] \right) - D \sum_{k=0}^{k_{\max}} \left[a^k \cos\left(2\Pi b^k \times 0.5\right)\right] \tag{3.5}$$

$$a = 0.5,\ b = 3,\ k_{\max} = 20,\ x \in [-0.5, 0.5]^D$$

- Griewank Function

$$f(x) = \sum_{i=1}^{D} \frac{x_i^2}{4000} - \prod_{i=1}^{D} \cos\left(\frac{x_i}{\sqrt{i}}\right) + 1, x \in [-100, 100]^D. \tag{3.6}$$

- Ackley Function

$$f(x) = -20 \exp\left(-0.2\sqrt{\frac{1}{D}\sum_{i=1}^{D} x_i^2}\right) - \exp\left(\frac{1}{D}\sum_{i=1}^{D} \cos(2\Pi x_i)\right) + 20 + e, x \in [-32, 32]^D. \tag{3.7}$$

Gaussian functions are used to combine these benchmark functions and blur the individual function's structure. The details are described below.

$F(x)$: new composite function.

$fi(x)$: ith basic function used to construct the composite function.

n: number of basic functions. The bigger is the n the more complex is $F(x)$.

D: dimension.

$[X_{\min}, X_{\max}]^D$: $F(x)$'s search range

$[x_{\min}, x_{\max}]^D$: $f(x)$'s search range

M_i: orthogonal rotation matrix for each $fi(x)$

o_i: new shifted optimum position for each $fi(x)$

$o_i$old: old optimum position for each $fi(x)$

$$F(x) = \sum_{i=1}^{n} \left\{ w_i \times \left[f_i'((x - o_i + o_{i\text{old}})/\lambda_i \times M_i) + \text{bias}_i \right] \right\} + f\_\text{bias} \tag{3.8}$$

w_i: weight value for each *fi*(*x*) is calculated as,

$$w_i = \exp\left(-\frac{\sum_{k=1}^{D}(x_k - o_{\text{ik}} + o_{\text{ikold}})^2}{2D\sigma_i^2}\right)$$

$$w_i = \begin{cases} w_i & \text{if } w_i = \max(w_i) \\ w_i \times (1 - \max(w_i).\wedge 10) & \text{if } w_i \neq \max(w_i) \end{cases} \tag{3.9}$$

Then the weight is normalized.

$$w_i = w_i / \sum_{i=1}^{n} w_i \tag{3.10}$$

σ_i is used to control each *fi*(*x*)'s coverage range. Small σ_i value gives a narrow range for *fi*(*x*). λ_i is used to stretch or compress the function (i.e., $\lambda_i > 1$ means stretch, $\lambda_i < 1$ means compress the composite function). Hence different basic functions have different search range, in order to make full use of the basic function.

$$\lambda_i = \sigma_i * (X_{\max} - X_{\min} / x_{\text{maxi}} - x_{\text{mini}}) \tag{3.11}$$

o_i defines the global and local optima's position, bias defines which optimum is global optimum. The smallest bias_i corresponds to the global optimum. Using o_i, bias_i, a global optimum can be placed anywhere.

If *fi*(*x*) are different functions and different functions have different properties and height then to get a better mixture, the biggest function value f_{maxi} is estimated for 10 functions *fi*(*x*) and then each basic function is normalized to similar height as explained below.

$$|f_{\text{maxi}}| = C * fi(x) / |f_{\text{maxi}}|, \tag{3.12}$$

C is a predefined constant.

$$|f_{\text{maxi}}| \text{ is estimated using} |f_{\text{maxi}}| = fi((\text{z}/\lambda_\text{i}) * \text{M}_\text{i}), \tag{3.13}$$

$$z = X_{\max} \tag{3.14}$$

By controlling *fi*, σ_i, λ_i, bias_i, o_i and M_i, different composite functions with different desired properties can be obtained (Liang et al. 2005a, b). Six composite functions formed by the combination of the basic benchmark functions are described below.

3.1.1 Composite Function 1 (CF1)

This function is formed by combining 10 sphere functions.

$f_1, f_2, \ldots, f_{10}$: Sphere function

$$[\sigma_1, \sigma_2, \ldots, \sigma_{10}] = [1, 1, \ldots, 1]$$

$$[\lambda_1, \lambda_2, \ldots, \lambda_{10}] = \left[\frac{5}{100}, \frac{5}{100}, \ldots, \frac{5}{100}\right]$$

3.1.2 Composite Function 2 (CF2)

This function is formed by combining 10 Griewank functions.

$f_1, f_2, \ldots, f_{10}$: Griewank function

$$[\sigma_1, \sigma_2, \ldots, \sigma_{10}] = [1, 1, \ldots, 1]$$

$$[\lambda_1, \lambda_2, \ldots, \lambda_{10}] = \left[\frac{5}{100}, \frac{5}{100}, \ldots, \frac{5}{100}\right]$$

3.1.3 Composite Function 3 (CF3)

This function is formed by combining 10 Griewank functions.

$f_1, f_2, \ldots, f_{10}$: Griewank function

$$[\sigma_1, \sigma_2, \ldots, \sigma_{10}] = [1, 1, \ldots, 1]$$

$$[\lambda_1, \lambda_2, \ldots, \lambda_{10}] = [1, 1, \ldots, 1]$$

3.1.4 Composite Function 4 (CF4)

This function is formed by combining 2 Ackley functions, 2 Rastrigin functions, 2 Weirstrass functions, 2 Griwank functions and 2 Sphere functions.

$f_{1\text{–}2}\ (x)$: Ackley function

$f_{3\text{–}4}\ (x)$: Rastrigin function

$f_{5\text{–}6}\ (x)$: Weierstrass function

$f_{7\text{–}8}\ (x)$: Griewank function

$f_{9\text{–}10}\ (x)$: Sphere function

$$[\sigma_1, \sigma_2, \ldots, \sigma_{10}] = [1, 1, \ldots, 1].$$

$$[\lambda_1, \lambda_2, \ldots, \lambda_{10}] = \left[\frac{5}{32}, \frac{5}{32}, 1, 1, \frac{5}{0.5}, \frac{5}{0.5}, \frac{5}{100}, \frac{5}{100}, \frac{5}{100}, \frac{5}{100}\right]$$

3.1.5 *Composite Function 5 (CF5)*

This function is formed by combining 2 Rastrigin functions, 2 Weirstrass functions, 2 Griewank functions, 2 Ackley functions and 2 Sphere functions.

$f_{1\text{–}2}$ (x): Rastrigin function
$f_{3\text{–}4}$ (x): Weierstrass function
$f_{5\text{–}6}$ (x): Griewank function
$f_{7\text{–}8}$ (x): Ackley function
$f_{9\text{–}10}$ (x): Sphere function

$$[\sigma_1, \sigma_2, \ldots, \sigma_{10}] = [1, 1, \ldots, 1]$$

$$[\lambda_1, \lambda_2, \ldots, \lambda_{10}] = \left[\frac{1}{5}, \frac{1}{5}, \frac{5}{0.5}, \frac{5}{0.5}, \frac{5}{100}, \frac{5}{100}, \frac{5}{32}, \frac{5}{32}, \frac{5}{100}, \frac{5}{100}\right]$$

3.1.6 *Composite Function 6 (CF6)*

This function is formed by combining 2 Rastrigin functions, 2 Weirstrass functions, 2 Griewank functions, 2 Ackley functions and 2 Sphere functions.

$f_{1\text{–}2}$ (x): Rastrigin function
$f_{3\text{–}4}$ (x): Weierstrass function
$f_{5\text{–}6}$ (x): Griewank function
$f_{7\text{–}8}$ (x): Ackley function
$f_{9\text{–}10}$ (x): Sphere function

$$[\sigma_1, \sigma_2, \ldots, \sigma_{10}] = [0.1, 0.2, 0.3, 0.4, 0.5, 0.6, 0.7, 0.8, 0.9, 1]$$

$$[\lambda_1, \lambda_2, \ldots, \lambda_{10}] = \left[0.1 \times \frac{1}{5}, 0.2 \times \frac{1}{5}, 0.3 \times \frac{5}{0.5}, 0.4 \times \frac{5}{0.5}, 0.5 \times \frac{5}{100}, 0.6 \times \frac{5}{100}, \right.$$
$$\left. 0.7 \times \frac{5}{32}, 0.8 \times \frac{5}{32}, 0.9 \times \frac{5}{100}, 1 \times \frac{5}{100}\right]$$

CF5 and CF6 use the same optima's position o and the same orthogonal matrices $M_1, M_2, \ldots, M_n$. The difference between CF5 and CF6 is the values of σ and λ which makes CF6 to have a narrower coverage area for the basic function with the global optimum and a flatter coverage area for the basic function with the local optima. In this way, the complexity of the function is increased.

3.2 Parameter Settings for the Composite Functions

Liang et al. (2005a, b) defined and used six different optimization algorithms on this group of composite functions. The parameter settings for the composite functions are as follows:

- Basic function number n = 10
- Dimensions D = 10
- C = 2000,
- Search range: $[-5, 5]^{D}$
- f_bias = 0
- bias = [0, 100, 200, 300, 400, 500, 600, 700, 800, 900].

Hence the first function $fi(x)$ is always the function with the global optimum as its bias is zero. o_1, o_2, …, o_9 are all generated randomly in the search range, except o_{10} and is set [0, 0, ……….,0] for trapping algorithms which have a potential to converge to the center of the search range. M_1, M_2,…, M_n are $D \times D$ orthogonal rotation matrixes.

3.3 Results of Different Algorithms on Composite Test Functions

Rao and Waghmare (2013) attempted the composite functions using the TLBO algorithm. Table 3.1 shows the results obtained by using eight algorithms including TLBO and ETLBO on six composite functions.

The values shown in bold in Table 3.1 indicate the best values. For each test function each algorithm is run 20 times and the maximum fitness evaluations are 50,000 for all algorithms. The TLBO algorithm is experimented with different elite sizes, viz. 0, 4, 8, 12, and 16. After making few trials, the elite size of 12 is considered for the ETLBO algorithm. The elite sizes of 12 and 16 have given the same global optimum solutions. The mean values of the results are reported in Table 3.1. As can be seen from the results, TLBO and ETLBO algorithms outperform the other algorithms on all benchmark problems except for test function 1. In that problem, CLPSO is better than the TLBO and ETLBO algorithms. However, the performance of the TLBO and ETLBO algorithms is better than the others (except CLPSO) for test function 1. It can be seen from Table 3.1 that ETLBO and TLBO algorithms find the global optimum solution with better mean for all the six

Table 3.1 Results obtained by using the eight algorithms on six composite functions

Composite function	PSO	CPSO	CLPSO	CMA-ES	G3-PCX	DE	TLBO (Rao and Waghmare 2013)	ETLBO
CF1	1.0000e +002	1.5626e +002	**5.7348e-008**	1.0000e +002	6.0000e +001	6.7459e –002	3.1186e –001	2.5689e –002
CF2	1.5591e +002	2.4229e +002	1.9157e +001	1.6199e +002	9.2699e +001	2.8759e +001	**1.702e +001**	**1.694e +001**
CF3	1.7203e +002	3.6264e +002	1.328e +002	2.1406e +002	3.1980e +002	1.4441e +002	**1.2381e +002**	**1.092e +002**
CF4	3.1430e +002	5.2237e +002	3.2232e +002	6.1640e +002	4.9296e +002	3.2486e +002	**2.9439e +002**	**2.7471e +002**
CF5	8.3450e +001	2.5556e +002	5.370e +000	3.5853e +002	2.6021e +001	1.0789e +001	**5.1815e +000**	**4.9375e +000**
CF6	8.6142e +002	8.5314e +002	5.0116e +002	9.0026e +002	7.7208e +002	4.9094e +002	**2.3018e +002**	**2.1282e +002**

PSO Particle Swarm Optimization, *CPSO* Cooperative PSO, *CLPSO* Comprehensive Learning PSO, *CMA-ES* Evolution Strategy with Covariance Matrix Adaptation, *G3-PCX* G3 model with PCX crossover, *DE* Differential Evolution, *TLBO* Teaching-Learning-Based Optimization algorithm, *ETLBO* Elitist Teaching-Learning-Based Optimization algorithm

composite test functions except for the test function 1. The TLBO and ETLBO algorithms have reduced the global optimum mean from 490.94 obtained by DE to 230.18 and 212.82, respectively, for the most complex composite test function 6 thereby giving improvement over 53 and 57 % using TLBO and ETLBO algorithms.

From the results obtained by the application of different algorithms, it can be observed that all the algorithms have generated their best results for composite function CF1 since it is constructed using the unimodal sphere basic function. With 10 sphere functions, CF1 is a multimodal function with one global optimum and nine local optima. Good areas are easier to find and when the global area has been found, the global optimum is not difficult to achieve. The composite functions CF2 and CF3 are constructed with more complex multimodal Griewank functions. Hence they have more local optima thereby increasing their complexity. The global area is not easy to find and even when the global area has been found the real optimum is difficult to reach. Composite functions CF4, CF5 and CF6 are all constructed with different basic functions and may be called as the hybrid composite functions and each function has different properties. Ackley function, which has a narrow optimum basin, occupies the global area of CF4 and hides the global optimum in a bad fitness area; while in CF5 and CF6, the global areas are occupied by Rastrigin functions whose local optima are so many and it is difficult to reach the global optimum. Based on CF5, σ and λ of CF6 are adjusted to obtain a narrower global optimum area and flatter local optimum areas. From the results, it can be observed that CF6 has become more complex than CF5. The TLBO and ETLBO algorithms are compared with the other six algorithms on these multimodal problems. It can be seen from the results that TLBO and ETLBO algorithms outperform the other six algorithms on five test functions. Moreover, it is observed that the concept of elitism enhances the performance of the TLBO algorithm for the considered complex multimodal composite functions.

In addition to these composite functions, many multiobjective unconstrained and constrained test functions and problems have been attempted and the details are given in the next chapter.

References

Bergh, F., Engelbrecht, A.P., 2004. A cooperative approach to particle swarm optimization. IEEE Transactions on Evolutionary Computation. 8(3), 225–239.

Coello, C.A.C., Pulido, G.T., Lechuga, M.S., 2004. Handling multiple objectives with particle swarm optimization. IEEE Transactions on Evolutionary Computation, 8(3) 256–279.

Leung, Y.W., Wang, Y.P., 2001. An orthogonal genetic algorithm with quantization for global numerical optimization. IEEE Trans. on Evolutionary Computation, 5(1), 41–53.

Liang, J.J., Suganthan, P.N., Deb, K., 2005a. Novel composition test functions for numerical global optimization, IEEE Transactions on Evolutionary Computation, 5(1), 1141–1153.

Liang, J.J., Suganthan, P. N., Deb, K., 2005b. Novel composition test functions for numerical global optimization, Proceedings of IEEE Swarm Intelligence Symposium, SIS 2005, 8–10 June, 68-75.

Rao, R.V., Waghmare, G.G., 2013. Solving composite test functions using teaching-learning-based optimization algorithm. Proceedings of the International Conference on Frontiers of Intelligent Computing: Theory and Applications (FICTA), Advances in Intelligent Systems and Computing 199, 395–403, doi:10.1007/978-3-642-35314-7-45.

Zhong, W.C., Liu, J., Xue, M.Z., 2004. A multiagent genetic algorithm for global numerical optimization. IEEE Transactions on Systems, Man and Cybernetics 34, 1128–1141.

Chapter 4
Application of TLBO and ETLBO Algorithms on Multiobjective Unconstrained and Constrained Test Functions

Abstract Multiobjective optimization is the process of simultaneously optimizing two or more conflicting objectives subject to certain constraints. Real-life engineering designs often contain more than one conflicting objective function and require a multiobjective approach. In a single-objective optimization problem, the optimal solution is clearly defined, while a set of tradeoffs that gives rise to numerous solutions exists in multiobjective optimization problems. Each solution represents a particular performance tradeoff between the objectives and can be considered optimal. In this chapter, the performance of TLBO and ETLBO algorithms are evaluated against the other optimization algorithms over a set of multiobjective unconstrained and constrained test functions and the results are compared. The TLBO and ETLBO algorithms are observed to outperform the other optimization algorithms for the multiobjective unconstrained and constrained benchmark functions.

4.1 Multiobjective Unconstrained Test Functions

There are many different test functions for multiobjective optimization, but a subset of the widely used functions has been tested using TLBO and ETLBO algorithms and the results are compared with those given by the other algorithms including vector-evaluated genetic algorithm (VEGA) (Schaffer 1985), NSGA-II (Deb et al. 2002), multiobjective differential evolution (MODE) (Babu and Gujarathi 2007), differential evolution for multiobjective optimization (DEMO) (Robic and Filipic 2005), multiobjective bees algorithms (Bees) (Pham and Ghanbarzadeh 2007), and strength Pareto evolutionary algorithm (SPEA) (Deb et al. 2002).

1. Shaffer's Min-Min (SCH) test function with convex Pareto front:

$$f_1(x) = x^2, \quad f_2(x) = (x-2)^2, \quad -10^3 \le x \le 10^3. \tag{4.1}$$

R.V. Rao, *Teaching Learning Based Optimization Algorithm*,
DOI 10.1007/978-3-319-22732-0_4

2. ZDT1 function with a convex front:

$$
\begin{aligned}
f_1(x) &= x_1, \quad f_2(x) = g(1 - \sqrt{f_1/g}), \\
g &= 1 + \frac{9\sum_{i=2}^{d} xi}{d-1}, \qquad x_i \in [0,1],\ i = 1,\ldots,30,
\end{aligned}
\tag{4.2}
$$

where d is the number of dimensions.

3. ZDT2 function with a non-convex front:

$$f_1(x) = x_1, \quad f_2(x) = g(1 - f_1/g)^2 \tag{4.3}$$

4. ZDT3 function with a discontinuous front:

$$f_1(x) = x_1, \quad f_2(x) = g\left[1 - \sqrt{f_1/g} - f_1/g\sin(10\pi f_1)\right] \tag{4.4}$$

where g in function ZDT2 and ZDT3 is the same as in function ZDT1.

5. LZ function:

$$
\begin{aligned}
f_1 &= x_1 + \frac{2}{|J_1|}\sum_{j\in J_1}\left[x_j - \sin\left(6\pi x_1 + \frac{j\pi}{n}\right)\right]^2, \\
f_2 &= 1 - \sqrt{x} + \frac{2}{|J_1|}\sum_{j\in J_2}\left[x_j - \sin\left(6\pi x_1 + \frac{j\pi}{n}\right)\right]^2
\end{aligned}
\tag{4.5}
$$

where $J_1 = \{j \mid j \text{ is odd and } 2 \le j \le n\}$, $J_2 = \{j \mid j \text{ is even and } 2 \le j \le n\}$.

This function has a Pareto front $f_2 = 1 - \sqrt{f_1}$ with a Pareto set

$$x_j = \sin\left(6\pi x_1 + \frac{j\pi}{d}\right), \quad j = 2, 3, \ldots, d, x_1 \in [0, 1]. \tag{4.6}$$

After generating 200 points by TLBO, these points are compared with the true front $f_2 = 1 - \sqrt{f_1}$ of ZDT1.

The distance or error between the estimated Pareto front PF^e to its corresponding true fronts PF^t is defined as,

$$E_f = ||\mathrm{PF}^e - \mathrm{PF}^t||^2 = \sum_{j=1}^{N}\left(\mathrm{PF}_j^e - \mathrm{PF}_j^t\right)^2 \tag{4.7}$$

where N is the number of points. The generalized distance (Dg) can be given as,

$$\mathrm{Dg} = \frac{1}{N}\sqrt{\sum_{j=1}^{N}\left(\mathrm{PF}_j^e - \mathrm{PF}_j^t\right)^2} \tag{4.8}$$

Performance metric of inverted-generational distance (IGD) is used. Let $P*$ be a set of uniformly distributed points along the PF (in the objective space). Let A be an approximate set to the PF, then the average distance from $P*$ to A is defined using the following equation:

$$\mathrm{IGD}(A, P^{*}) = \frac{\sum_{\vartheta \epsilon p*}(\vartheta, A)}{|P^{*}|} \tag{4.9}$$

where $d(\vartheta, A)$ is the minimum Euclidean distance between v and the points in A. If $|P*|$ is large enough to represent the Pareto front very well, both the diversity and convergence of the approximated set A could be measured using IGD (A, $P*$). An optimization algorithm will have to try to minimize the value of IGD (A, $P*$) measure.

The performance of TLBO and ETLBO algorithms is also evaluated for few more multiobjective unconstrained benchmark functions (UF1–UF10) presented by Zhang et al. (2009) against the other algorithms. The mathematical representation of these test functions is given in Tables 4.1 and 4.2. The functions UF1–UF7 are unconstrained two-objectives test problems and UF8–UF10 are unconstrained three-objectives test functions.

4.1.1 Computational Results of the Multiobjective Unconstrained Functions and Discussion

The computational results obtained by the TLBO algorithm and ETLBO algorithms are given in Table 4.3. The values shown bold in Table 4.3 indicate the best values. It can be seen from Table 4.3 that the ETLBO and TLBO algorithms have obtained best results on the five multiobjective unconstrained functions ZDT1, ZDT2, ZDT3, SCH, and LZ and obtained the first and second ranks, respectively, among the nine algorithms. The estimated Pareto fronts and true fronts of some functions (SCH, ZDT1, ZDT2, and LZ) are shown in Fig. 4.1. It can be seen from Fig. 4.1 that the TLBO and ETLBO algorithm successfully converges to the optimal Pareto front and the approximation has good distribution. Table 4.4 represents the errors for different benchmark functions for 1000 and 2500 iterations.

In these experiments, the number of function evaluations is set at 25,000 for SCH, ZDT1, ZDT2, ZDT3, and LZ and the TLBO algorithm is evaluated independently 30 times for each test problem. The elitist TLBO algorithm is experimented with different elite sizes, viz. 0, 4, 8, 12, and 16 for SCH, ZDT1, ZDT2, ZDT3, and LZ functions and an elite size of 12 is considered after making few trials.

The comparison of results of seven multiobjective unconstrained functions with different algorithms are given in Table 4.5 and the estimated Pareto fronts and true fronts of unconstrained functions are shown in Fig. 4.2. The number of function

Table 4.1 Mathematical representation of the seven 2-objectives unconstrained test functions

Function	Mathematical representation
UF1	$f_1 = x_1 + \frac{2}{\lvert J_1\rvert}\sum_{j\in J1}\left[x_j - \sin\left(6\pi x_1 + \frac{j\pi}{n}\right)\right]^2,$ $f_2 = 1 - \sqrt{x} + \frac{2}{\lvert J_1\rvert}\sum_{j\in J2}\left[x_j - \sin\left(6\pi x_1 + \frac{j\pi}{n}\right)\right]^2$ $J_1 = \{j \mid j \text{ is odd and } 2 \le j \le n\}$, $J_2 = \{j \mid j \text{ is even and } 2 \le j \le n\}$
UF2	$f_1 = x_1 + \frac{2}{\lvert J_1\rvert}\sum_{j\in J1} y_j^2,\ f_2 = 1 - \sqrt{x} + \frac{2}{\lvert J_1\rvert}\sum_{j\in J2} y_j^2$ $J_1 = \{j \mid j \text{ is odd and } 2 \le j \le n\}$, $J_2 = \{j \mid j \text{ is even and } 2 \le j \le n\}$ $y_j = \begin{cases} x_j - \left[0.3x_1^2\cos\left(24\pi x_1 + \frac{4j\pi}{n}\right) + 0.6x_1\right]\cos\left(6\pi x_1 + \frac{j\pi}{n}\right) j \in j_1 \\ x_j - \left[0.3x_1^2\cos\left(24\pi x_1 + \frac{4j\pi}{n}\right) + 0.6x_1\right]\sin\left(6\pi x_1 + \frac{j\pi}{n}\right) j_2 \end{cases}$
UF3	$f_1 = x_1 + \frac{2}{\lvert J_1\rvert}\left(4\sum_{j\in J1} y_j^2 - 2\prod_{j\in J1}\cos\left(\frac{20y_j\pi}{\sqrt{j}}\right) + 2\right),$ $f_2 = 1 - \sqrt{x_2} + \frac{2}{\lvert J_1\rvert}\left(4\sum_{j\in J1} y_j^2 - 2\prod_{j\in J2}\cos\left(\frac{20y_j\pi}{\sqrt{j}}\right) + 2\right)$ J_1 and J_2 are the same as those of UF1, $y_j = x_j - x_1^{0.5\left(1.0 + \frac{3(j-2)}{n-2}\right)}$, $j = 2, \ldots, n.$
UF4	$f_1 = x_1 + \frac{2}{\lvert J_1\rvert}\sum_{j\in J1} h(y_j),\ f_2 = 1 - x_2 + \frac{2}{\lvert J_2\rvert}\sum_{j\in J2} h(y_j)$ J_1 and J_2 are the same as those of UF1, $y_j = x_j - \sin\left(6\pi x_1 + \frac{j\pi}{n}\right)$, $j = 2, \ldots, n.$
UF5	$f_1 = x_1 + \left(\frac{1}{2N} + \in\right)\lvert\sin(2N\pi x_1)\rvert + \frac{2}{\lvert J_1\rvert}\sum_{j\in J1} h(y_j)$ $f_2 = 1 - x_1 + \left(\frac{1}{2N} + \in\right)\lvert\sin(2N\pi x_1)\rvert + \frac{2}{\lvert J_2\rvert}\sum_{j\in J2} h(y_j)$ J_1 and J_2 are the same as those of UF1, $\varepsilon > 0, y_j = x_j - \sin\left(6\pi x_1 + \frac{j\pi}{n}\right)$, $j = 2, \ldots, n.$ $h(t) = 2t^2 - \cos(4\pi t) + 1$
UF6	$f_1 = x_1 + \max\left\{0.2\left(\frac{1}{2N} + \in\right)\sin(2N\pi x_1)\right\}$ $+ \frac{2}{\lvert j_1\rvert}\left(4\sum_{j\in J1} y_j^2 - 2\prod_{j\in J1}\cos\left(\frac{20j_j\pi}{\sqrt{j}}\right) + 2\right)$ $f_1 = 1 - x_1 + \max\left\{0.2\left(\frac{1}{2N} + \in\right)\sin(2N\pi x_1)\right\}$ $+ \frac{2}{\lvert j_2\rvert}\left(4\sum_{j\in J2} y_j^2 - 2\prod_{j\in J2}\cos\left(\frac{20j_j\pi}{\sqrt{j}}\right) + 2\right)$ J_1 and J_2 are the same as those of UF1, $\in > 0, y_j = x_j - \sin\left(6\pi x_1 + \frac{j\pi}{n}\right), j = 2, \ldots, n.$
UF7	$f_1 = \sqrt[5]{x_1} + \frac{2}{\lvert j_1\rvert}\sum_{j\in J1} y_j^2,\ f_2 = 1 - \sqrt[5]{x_1} + \frac{2}{\lvert j_1\rvert}\sum_{j\in J2} y_j^2$ J_1 and J_2 are the same as those of UF1, $\in > 0, y_j = x_j - \sin\left(6\pi x_1 + \frac{j\pi}{n}\right)$, $j = 2, \ldots, n.$

evaluations is set at 300,000 for UF1–UF10 and evaluated independently 30 times for each test problem. The elitist TLBO algorithm is experimented with different elite sizes, viz. 0, 4, 8, 12, and 16 with different strategies for unconstrained test functions UF1–UF10. After making few trials, the elite size of 12 is considered.

Table 4.2 Mathematical representation of the three 3-objectives unconstrained test functions

Function	Mathematical representation
UF8	$f_1 = \cos(0.5x_1\pi)\cos(0.5x_2\pi) + \frac{2}{\lvert J_1\rvert}\sum_{j\epsilon J_1}\left(x_j - 2x_2\sin\left(2\pi x_1 + \frac{j\pi}{n}\right)\right)^2$, $f_2 = \cos(0.5x_1\pi)\sin(0.5x_2\pi) + \frac{2}{\lvert J_2\rvert}\sum_{j\epsilon J_1}\left(x_j - 2x_2\sin\left(2\pi x_1 + \frac{j\pi}{n}\right)\right)^2$, $f_3 = \sin(0.5x_1\pi) + \frac{2}{\lvert J_1\rvert}\sum_{j\epsilon J_1}\left(x_j - 2x_2\sin\left(2\pi x_1 + \frac{j\pi}{n}\right)\right)^2$, $J_1 = \{j \mid \leq j \leq n$, and $j-1$ is a multiplication of $3\}$, $J_2 = \{j \mid \leq j \leq n$, and $j-2$ is a multiplication of $3\}$, $J_3 = \{j \mid \leq j \leq n$, and j is a multiplication of $3\}$,
UF9	$f_1 = 0.5[\max\{0, (1+\varepsilon)(1-4(2x_1-1)^2)\} + 2x_1]x_2 + \frac{2}{\lvert J_1\rvert}\sum_{j\epsilon J1}\left(x_j - 2x_2\sin\left(2\pi x_1 + \frac{j\pi}{n}\right)\right)^2$, $f_2 = 0.5[\max\{0, (1+\varepsilon)(1-4(2x_1-1)^2)\} + 2x_1]x_2 + \frac{2}{\lvert J_2\rvert}\sum_{j\epsilon J2}\left(x_j - 2x_2\sin\left(2\pi x_1 + \frac{j\pi}{n}\right)\right)^2$, $f_3 = 1 - x_2 + \frac{2}{\lvert J_3\rvert}\sum_{j\epsilon J_3}\left(x_j - 2x_2\sin\left(2\pi x_1 + \frac{j\pi}{n}\right)\right)^2$, $J_1 = \{j \mid \leq j \leq n$, and $j-1$ is a multiplication of $3\}$, $J_2 = \{j \mid \leq j \leq n$, and $j-2$ is a multiplication of $3\}$, $J_3 = \{j \mid \leq j \leq n$, and j is a multiplication of $3\}$ and $\varepsilon = 0.1$,
UF10	$f_1 = \cos(0.5x_1\pi)\cos(0.5x_2\pi) + \frac{2}{\lvert J_1\rvert}\sum_{j\epsilon J1}\left[4y_j^2 - \cos(8\pi y_1) + 1\right]$, $f_2 = \cos(0.5x_1\pi)\sin(0.5x_2\pi) + \frac{2}{\lvert J_1\rvert}\sum_{j\epsilon J1}\left[4y_j^2 - \cos(8\pi y_1) + 1\right]$, $f_3 = \sin(0.5x_1\pi) + \frac{2}{\lvert J_1\rvert}\sum_{j\epsilon J1}\left[4y_j^2 - \cos(8\pi y_1) + 1\right]$, $J_1 = \{j \mid \leq j \leq n$, and $j-1$ is a multiplication of $3\}$, $J_2 = \{j \mid \leq j \leq n$, and $j-2$ is a multiplication of $3\}$, $J_3 = \{j \mid \leq j \leq n$, and j is a multiplication of $3\}$,

Table 4.3 Comparison of results of five multiobjective unconstrained functions with different algorithms

Methods	ZDT1	ZDT2	ZDT3	SCH	LZ
VEGA (Schaffer 1985)	3.79E−02	2.37E−03	3.29E−01	6.98E−02	1.47E−03
NSGA-II (Deb et al. 2002)	3.33E−02	7.24E−02	1.14E−01	5.73E−03	2.77E−02
MODE (Babu and Gujarathi 2007)	5.80E−03	5.50E−03	2.15E−02	9.32E−04	3.19E−03
DEMO (Robic and Filipic 2005)	1.08E−03	7.55E−04	1.18E−03	1.79E−04	1.40E−03
Bees (Pham and Ghanbarzadeh 2007)	2.40E−02	1.69E−02	1.91E−01	1.25E−02	1.88E−02
SPEA (Deb et al. 2002)	1.78E−03	1.34E−03	4.75E−02	5.17E−03	1.92E−03
MOFA (Yang 2012)	1.90E−04	1.52E−04	1.97E−04	4.55E−06	8.70E−04
TLBO (Rao and Waghmare 2014)	**1.12E−07**	**1.70E−06**	**1.61E−06**	**9.99E−07**	**1.27E−06**
ETLBO	**1.05 E−07**	**1.56 E−06**	**1.53 E−06**	**9.67 E−07**	**1.19 E−06**

The TLBO and ETLBO algorithms are compared with archive-based micro Genetic algorithm (AMGA) (Tiwari et al. 2009), clustering multiobjective evolutionary algorithm (ClusteringMOEA) (Wang et al. 2009), Differential Evolution with self-adaptation, and local search algorithm (DECMOSA-SQP) (Zamuda et al. 2009), an improved version of dynamical multiobjective evolutionary algorithm (DMOEADD) (Liu et al. 2009), generalized differential evolution 3 (GDE3) (Kukkonen and Lampinen 2009), LiuLi Algorithm (Liu and Li 2009), multiobjective evolutionary algorithm based on decomposition (MOEAD) (Zhang et al. 2009), enhancing MOEA/D with guided mutation and priority update (MOEADGM) (Chen et al. 2009), multiobjective evolutionary programming (MOEP) (Qu and Suganthan 2009), multiple trajectory search (MTS) (Tseng and Chen 2009), improved algorithm based on an efficient multiobjective evolutionary algorithm (OMOEAII) (Gao et al. 2009), and multiobjective self-adaptive differential evolution algorithm with objective-wise Learning Strategies (OWMOSaDE) (Huang et al. 2009) methods on 10 unconstrained test functions. The values shown bold in Table 4.5 indicate the best values.

Apart from the quantitative comparison of the investigated algorithms, the graphical representations of the Pareto fronts produced by the TLBO and ETLBO algorithms are given in Fig. 4.2. This figure shows the quality of the Pareto fronts produced by the TLBO and ETLBO algorithms. Figure 4.2a shows that the results produced not only have good convergence but also have appropriate distribution over the Pareto front in the objective space. For the UF1 test problem, the TLBO algorithm obtains eighth rank and ETLBO obtains fifth rank among the 16 algorithms. The ETLBO and TLBO algorithms outperform the other algorithms when optimizing the UF2 test problem.

The ETLBO obtains the first rank and TLBO obtains the second rank among the 16 algorithms on the UF2 test problem. Figure 4.2b shows that the Pareto front

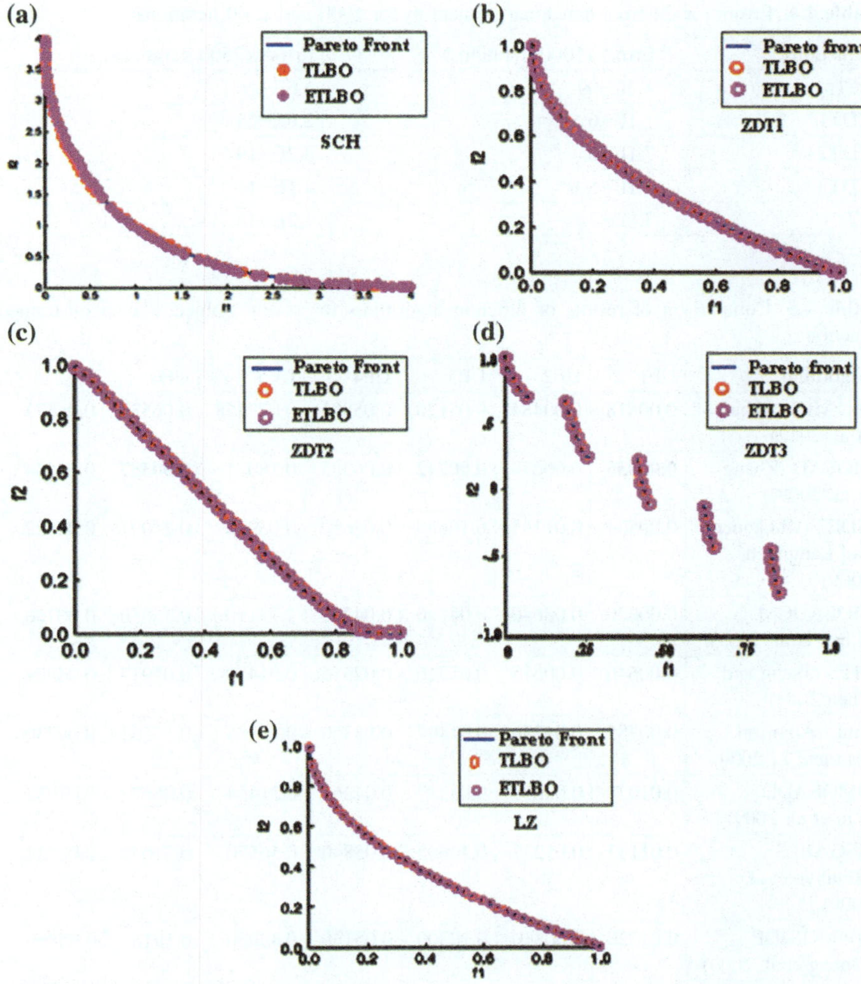

Fig. 4.1 **a–e** The Pareto front obtained by the TLBO and ETLBO algorithms on unconstrained test functions SCH, ZDT1, ZDT2, ZDT3 and LZ

produced has uniform distribution. For the UF3 test problem, the TLBO algorithm obtains eleventh rank and ETLBO obtains third rank among the 16 algorithms. The best convergence is obtained by the MOEAD algorithm. However, the TLBO and ETLBO algorithms have the ability to produce uniformly distributed Pareto fronts as shown in Fig. 4.2c. The ETLBO algorithm obtains the best result on the UF4 test problem and obtains the first rank and the TLBO algorithm obtains the second rank. Figure 4.2d shows the quality of Pareto front for UF4 test problem. It seems that the UF5 presents a hard problem to solve. As can be seen from Fig. 4.2e the TLBO and

Table 4.4 Errors for different benchmark functions for 1000 and 2500 iterations

Functions	Errors (1000 iterations)	Errors (2500 iterations)
SCH	4.3E−09	5.6E−26
ZDT1	1.1E−6	2.6E−23
ZDT2	7.1E−6	3.2E−19
ZDT3	2.1E−5	4.1E−17
LZ	7.8E−7	1.2E−18

Table 4.5 Comparison of results of different algorithms for seven 2-objectives unconstrained functions

Algorithm	UF1	UF2	UF3	UF4	UF5	UF6	UF7
MOABC (Akbari et al. 2012)	0.00618	0.00484	0.05120	0.05801	0.077758	0.06537	0.05573
MOEAD (Zhang et al. 2009)	**0.00435**	0.00679	**0.00742**	0.06385	0.18071	**0.00587**	0.00444
GDE3 (Kukkonen and Lampinen 2009)	0.00534	0.01195	0.10639	0.02650	0.03928	0.25091	0.02522
MOEADGM (Chen et al. 2009)	0.00620	0.00640	0.04290	0.04760	1.79190	0.55630	0.00760
MTS (Tseng and Chen 2009)	0.00646	0.00615	0.05310	0.02356	0.01489	0.05917	0.04079
LiuLi Algoritm (Liu and Li 2009)	0.00785	0.01230	0.01497	0.04350	0.16186	0.17555	0.00730
DMOEADD (Liu et al. 2009)	0.01038	0.00679	0.03337	0.04268	0.31454	0.06673	0.01032
NSGAIILS (Sindhya et al. 2009)	0.01153	0.01237	0.10603	0.05840	0.56570	0.31032	0.02132
OWMOSaDE (Huang et al. 2009)	0.01220	0.00810	0.10300	0.05130	0.43030	0.1918	0.05850
Clustering MOEA (Wang et al. 2009)	0.0299	0.02280	0.05490	0.05850	0.24730	0.08710	0.02230
AMGA (Tiwari et al. 2009)	0.03588	0.01623	0.06998	0.04062	0.09405	0.12942	0.05707
MOEP (Qu and Suganthan 2009)	0.05960	0.01890	0.09900	0.04270	0.22450	0.10310	0.01970
DECMOSA-SQP (Zamunda et al. 2009)	0.07702	0.02834	0.09350	0.03392	0.16713	0.12604	0.02416
DMOEAII (Liu et al. 2009)	0.08564	0.03057	0.27141	0.04624	0.16920	0.07338	0.03354
TLBO (Rao and Waghmare 2014)	0.01021	**0.00478**	0.10049	**0.00546**	0.07651	0.10291	0.010013
ETLBO	0.00634	**0.00399**	0.01684	**0.00528**	**0.01423**	0.01453	**0.00429**

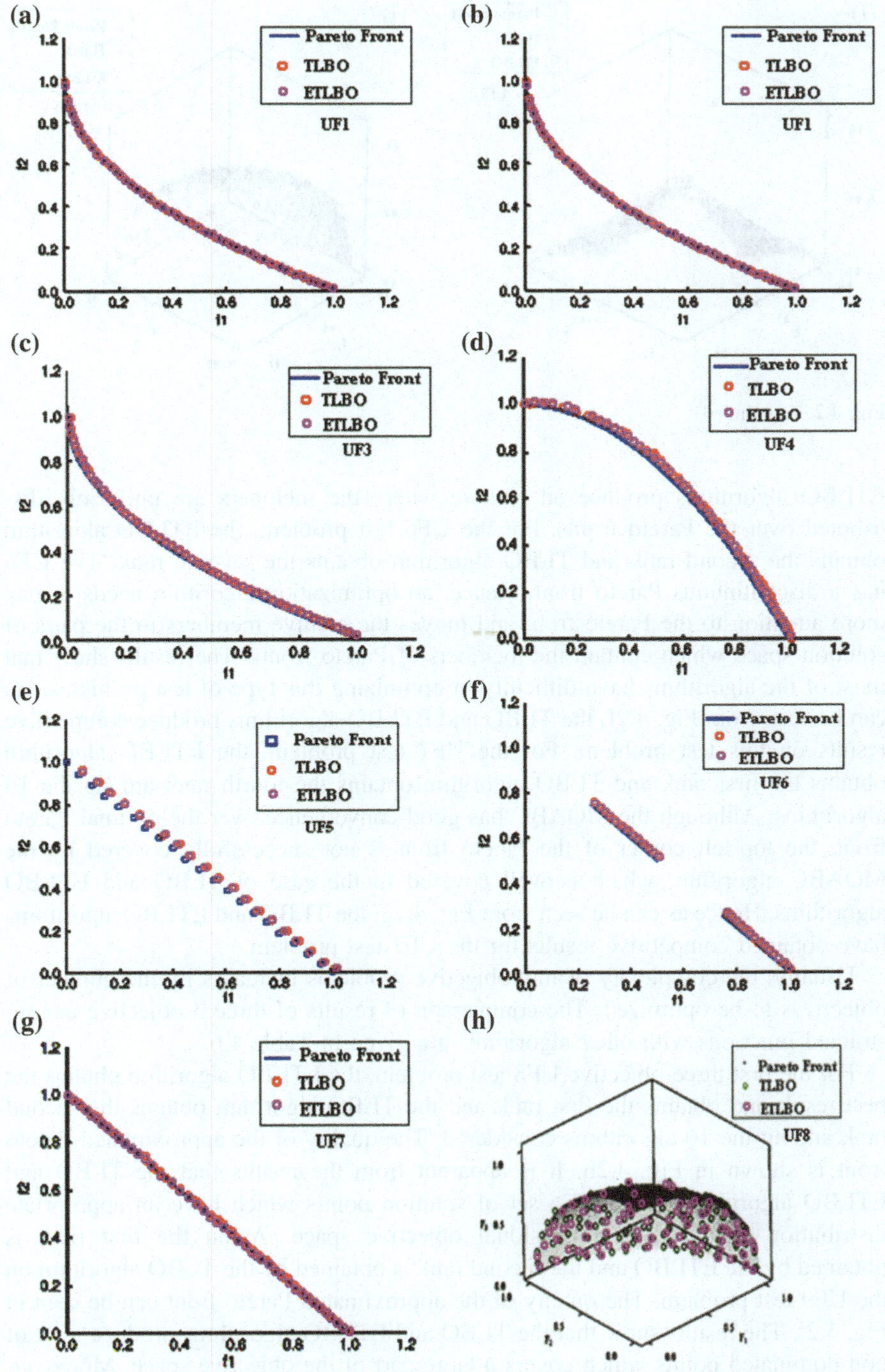

Fig. 4.2 **a–j** The Pareto front obtained by the TLBO and ETLBO algorithms on unconstrained test functions UF1–UF10

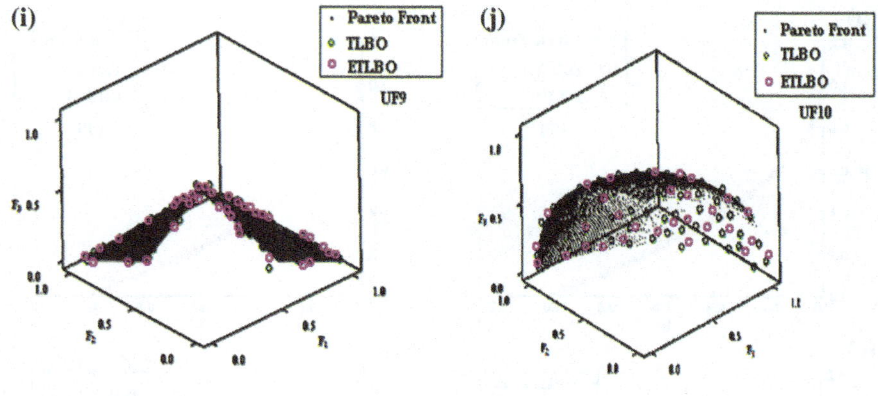

Fig. 4.2 (continued)

ETLBO algorithms produce an archive where the members are uniformly distributed over the Pareto fronts. For the UF6 test problem, the ETLBO algorithm obtains the second rank and TLBO algorithm obtains the seventh rank. The UF6 has a discontinuous Pareto front. Hence, an optimization algorithm needs to pay more attention to the Pareto front and moves the archive members to the parts of solution space which contain the members of Pareto fronts. The results show that most of the algorithms have difficulty in optimizing this type of test problems. As can be seen from Fig. 4.2f, the TLBO and ETLBO algorithms produce competitive results on this test problem. For the UF7 test problem, the ETLBO algorithm obtains the first rank and TLBO algorithm obtains the fourth rank among the 16 algorithms. Although the MOABC has good convergence over the optimal Pareto front, the top-left corner of the Pareto front is not successfully covered by the MOABC algorithm, which is well covered in the case of TLBO and ETLBO algorithms. Hence as can be seen from Fig. 4.2g, the TLBO and ETLBO algorithms have obtained competitive results for the UF7 test problem.

Usually, the complexity of mutiobjective problems increases by the number of objectives to be optimized. The comparison of results of three 3-objective unconstrained functions with other algorithms are given in Table 4.6.

For the first three-objective UF8 test problem, the ETLBO algorithm obtains the best result and obtains the first rank and the TLBO algorithm obtains the second rank among the 16 algorithms considered. The quality of the approximated Pareto front is shown in Fig. 4.2h. It is apparent from the results that the TLBO and ETLBO algorithms produced a set of solution points which have an appropriate distribution in the three-dimensional objective space. Again the first rank is obtained by the ETLBO and the second rank is obtained by the TLBO algorithm on the UF9 test problem. The quality of the approximated Pareto front can be seen in Fig. 5.2i. The results show that the TLBO and ETLBO algorithms produce a set of non-dominated points which covers a large part of the objective space. Moreover, for the UF10 test problem the ETLBO obtains the first rank and obtains the best

Table 4.6 Comparison of results of three 3-objectives unconstrained functions with different algorithms

Algorithm	UF8	UF9	UF10
MOABC (Akbari et al. 2012)	0.06726	0.06150	0.19499
MOEAD (Zhang et al. 2009)	0.05840	0.07896	0.047415
GDE3 (Kukkonen and Lampinen 2009)	0.24855	0.08248	0.43326
MOEADGM (Chen et al. 2009)	0.24460	0.18780	0.5646
MTS (Tseng and Chen 2009)	0.11251	0.11442	0.15306
LiuLi Algoritm (Liu and Li 2009)	0.08235	0.09391	0.44691
DMOEADD (Liu et al. 2009)	0.06841	0.04896	0.32211
NSGAIILS (Sindhya et al. 2009)	0.08630	0.07190	0.84468
OWMOSaDE (Huang et al. 2009)	0.09450	0.09830	0.74300
Clustering MOEA (Wang et al. 2009)	0.23830	0.29340	0.41110
AMGA (Tiwari et al. 2009)	0.17125	0.18861	0.32418
MOEP (Qu and Suganthan 2009)	0.42300	0.34200	0.36210
DECMOSA-SQP (Zamunda et al. 2009)	0.21583	0.14111	0.36985
OMOEAII (Liu et al. 2009)	0.19200	0.23179	0.62754
TLBO (Rao and Waghmare 2014)	**0.004933**	**0.011639**	**0.03823**
ETLBO	**0.004884**	**0.0102746**	**0.03677**

result and TLBO algorithm obtains the second rank. Figure 4.2j shows the quality of the approximated Pareto front by the TLBO and ETLBO algorithms. The results show that the approximated Pareto front covers a large part of the objective space. However, compared to the approximated Pareto fronts of the UF8 and UF9, the TLBO and ETLBO algorithms produced small number of points in the objective space. It can be seen from Fig. 4.2 that the approximated Pareto front for functions UF1–UF10 has good distribution of points and successfully converges to the optimal Pareto front. Table 4.7 presents the IGD statistics over UF1–UF10.

Table 4.7 The IGD statistics over UF1–UF10

Function	Mean (IGD)	Smallest (IGD)	Largest (IGD)	Std. dev. (IGD)
UF1	0.01021	0.01003	0.01214	0.005967
UF2	0.00478	0.00398	0.00507	0.00432
UF3	0.10049	0.09981	0.10118	0.03564
UF4	0.00546	0.00519	0.00598	0.00147
UF5	0.07651	0.07045	0.07855	0.00262
UF6	0.10291	0.10034	0.10987	0.0493
UF7	0.010013	0.010001	0.010181	0.00273
UF8	0.004933	0.004899	0.005348	0.00364
UF9	0.011639	0.01078	0.01461	0.00211
UF10	0.03823	0.03462	0.03976	0.0109

4.2 Multiobjective Constrained Test Functions

The performance of the TLBO and ETLBO algorithms are evaluated against the other algorithms over seven multiobjective constrained test problems CF1–CF7 (Zhang et al. 2009). The mathematical representation of these test functions is given in Table 4.8.

4.2.1 Computational Results of Constrained Multiobjective Functions and Discussion

The number of function evaluations is set at 300,000 for CF1–CF7 and evaluated independently 30 times for each test problem. The TLBO algorithm is experimented with different elite sizes, viz. 0, 4, 8, 12, and 16 for the constrained test functions CF1–CF7 and after making few trials the elite size of 16 is considered. The elite sizes of 12 and 16 have given the same global optimum solutions. The comparison of results of seven multiobjective constrained functions with other algorithms is given in Table 4.9 and the estimated Pareto fronts and true fronts of the constrained functions are shown in Fig. 4.3. The ETLBO algorithm obtains the first rank and the TLBO algorithm obtains the third rank on the CF1 test problem among the 10 algorithms considered. The CF2 test problem is successfully solved by the ETLBO and TLBO algorithms. The ETLBO and TLBO algorithms obtain the first and second rank, respectively, on the CF2 test problem.

The ETLBO algorithm surpasses the other algorithms in solving CF3, CF4, CF5, CF6, and CF7 test functions. The TLBO algorithm obtains the second rank on the CF3, CF4, CF5, CF6, and CF7 test functions. It can be seen from the Fig. 4.3 that the approximated Pareto front for functions CF1–CF7 has good distribution of points and successfully converges to the optimal Pareto front. Table 4.10 presents the IGD statistics over CF1–CF7. The overall performance shows that the TLBO and ETLBO algorithms can be used as effective tools for optimizing the problems with multiple objectives.

4.2.2 Additional Multiobjective Constrained Functions and the Computational Results

Few more computational experiments have been conducted to check the effectiveness of ETLBO algorithm. Different examples are investigated based on the multiobjective benchmark functions taken from the literature. Following multiobjective constrained functions that were presented by Mousa et al. (2012) and Yang and Deb (2012) are attempted now using the ETLBO algorithm.

Table 4.8 Mathematical representation of the seven two-objectives constrained test functions

Function	Mathematical representation
CF1	$f_1(x) = x_1 + \frac{2}{\lvert J_1 \rvert}\sum_{j \epsilon J_1}\left(x_j - x_1^{0.5\left(1.0+\frac{3(j-2)}{n-2}\right)}\right)^2,$ $f_2 = x_1 + \frac{2}{\lvert J_2 \rvert}\sum_{j \epsilon J_2}\left(x_j - x_1^{0.5\left(1.0+\frac{3(j-2)}{n-2}\right)}\right)^2,$ $J_1 = \{j \mid j \text{ is odd and } 2 \le j \le n\}$, $J_2 = \{j \mid j \text{ is even and } 2 \le j \le n\}$ Subject to: $f_1 + f_2 - a\lvert\sin[n\pi(f_1 - f_2 + 1)]\rvert - 1 \ge 0$ where N is an integer and $a \ge \frac{1}{2N}$
CF2	$f_1(x) = x_1 + \frac{2}{\lvert J_1 \rvert}\sum_{j \epsilon J_1}(x_j - \sin(6\pi x_1 + \frac{j\pi}{n}))^2,$ $f_2(x) = 1 - \sqrt{x} + \frac{2}{\lvert J_2 \rvert}\sum_{j \epsilon J_2}(x_j - \cos(6\pi x_1 + \frac{j\pi}{n}))^2$, $J_1 = \{j \mid j \text{ is odd and } 2 \le j \le n\}$, $J_2 = \{j \mid j \text{ is even and } 2 \le j \le n\}$ Subject to: $\frac{t}{1+e^{4\lvert t \rvert}} \ge 0$ where $t = f_2 + \sqrt{f_1} - a\sin[N\pi(\sqrt{f_1} - f_2 - 1)] - 1$
CF3	$f_1(x) = x_1 + \frac{2}{\lvert J_1 \rvert}\left(4\sum_{j \epsilon J_1} y_j^2 - 2\prod_{j \epsilon J_1}\cos\left(\frac{20 y_{j\pi}}{\sqrt{j}}\right) + 2\right),$ $f_2(x) = 1 - x_1^2 + \frac{2}{\lvert J_2 \rvert}\left(4\sum_{j \epsilon J_2} y_j^2 - 2\prod_{j \epsilon J_2}\cos\left(\frac{20 y_{j\pi}}{\sqrt{j}}\right) + 2\right)$, $J_1 = \{j \mid j \text{ is odd and } 2 \le j \le n\}$, $J_2 = \{j \mid j \text{ is even and } 2 \le j \le n\}$, $y_j = x_j - \sin\left(6\pi x_1 + \frac{j\pi}{n}\right)$, $j = 2, \ldots, n$, Subject to: $f_2 + f_1^2 - a\left\lvert\sin\left[n\pi\left(f_1^2 - f_2 + 1\right)\right]\right\rvert - 1 \ge 0$
CF4	$f_1 = x_1 + \sum_{j \epsilon J_1} h_{j(y_j)}$, $f_2 = 1 - x_1 + \sum_{j \epsilon J_2} h_{j(y_j)}$, $J_1 = \{j \mid j \text{ is odd and } 2 \le j \le n\}$, $J_2 = \{j \mid j \text{ is even and } 2 \le j \le n\}$, $h_{2(t)} = \begin{cases} \lvert t \rvert \text{ if } t < \frac{3}{2}\left(1 - \frac{\sqrt{2}}{2}\right) \\ 0.125 + (t-1)^2 \text{otherwise} \end{cases}$ $h_{j(t)=t^2}$ for $j = 3, 4, \ldots, n$ Subject to: $\frac{t}{1+e^{4\lvert t \rvert}} \ge 0$ where $t = x_2 + \sin\left(6\pi x_1 + \frac{2\pi}{n}\right) - 0.5x_1 + 0.25$
CF5	$f_1 = x_1 + \sum_{j \epsilon J_1} h_{j(y_j)}$, $f_2 = 1 - x_1 + \sum_{j \epsilon J_2} h_{j(y_j)}$, $J_1 = \{j \mid j \text{ is odd and } 2 \le j \le n\}$, $J_2 = \{j \mid j \text{ is even and } 2 \le j \le n\}$, $y_j = x_j - \sin\left(6\pi x_1 + \frac{j\pi}{n}\right)$, $j = 2, \ldots, n$, $y_j = \begin{cases} xj - 0.8x_1 cos\left(6\pi x_1 + \frac{j\pi}{n}\right) + 0.6x_1 & if\ j \epsilon J_1 \\ xj - 0.8x_1 sin\left(6\pi x_1 + \frac{j\pi}{n}\right) + 0.6x_1 & if\ j \epsilon J_2 \end{cases}$

(continued)

Table 4.8 (continued)

Function	Mathematical representation
	$h_{2(t)} = \begin{cases} \lvert t \rvert \text{if} t < \frac{3}{2}\left(1 - \frac{\sqrt{2}}{2}\right) \\ 0.125 + (t-1)^2 \text{otherwise} \end{cases} \quad h_{j(t)=t^2} \text{ for } j = 3, 4, \ldots n$ Subject to: $x_2 - 0.8x_1 \sin\left(6\pi x_1 + \frac{2\pi}{n}\right) - 0.5x_1 + 0.25 \geq 0$
CF6	$f_1 = x_1 + \sum_{j \epsilon J_1} y_j^2,\ f_2 = \left(1 - x_1\right)^2 + \sum_{j \epsilon J_2} y_j^2,\ J_1 = \{j \mid j \text{ is odd and } 2 \leq j \leq n\},$ $J_2 = \{j \mid j \text{ is even and } 2 \leq j \leq n\},$ $y_j = \begin{cases} xj - 0.8x_1 \cos\left(6\pi x_1 + \frac{j\pi}{n}\right) + 0.6x_1 \ \mathit{if}\ j \epsilon J_1 \\ xj - 0.8x_1 \sin\left(6\pi x_1 + \frac{j\pi}{n}\right) + 0.6x_1 \ \mathit{if}\ j \epsilon J_2 \end{cases}$ Subject to: $x_2 - 0.8x_1 \sin\left(6\pi x_1 + \frac{2\pi}{n}\right) - \text{sign}\left(0.5(1 - x_1) - \left(1 - x_1\right)^2\right)\sqrt{\left\lvert 0.5(1 - x_1) - \left(1 - x_1\right)^2 \right\rvert} \geq 0$ $x_4 - 0.8x_1 \sin\left(6\pi x_1 + \frac{2\pi}{n}\right) - \text{sign}\left(0.25(1 - x_1) - \left(1 - x_1\right)^2\right)\sqrt{0.2\sqrt{1 - x_1} - 0.5(1 - x_1)} \geq 0$
CF7	$f_1 = x_1 + \sum_{j \epsilon J_1} h_{j(y_j)},\ f_2 = (1 - x_1)^2 + \sum_{j \epsilon J_2} h_{j(y_j)},\ J_1 = \{j \mid j \text{ is odd and } 2 \leq j \leq n\},$ $J_2 = \{j \mid j \text{ is even and } 2 \leq j \leq n\},$ $y_j = \begin{cases} xj - \cos\left(6\pi x_1 + \frac{j\pi}{n}\right) + 0.6x_1 \ \mathit{if}\ j \epsilon J_1 \\ xj - \sin\left(6\pi x_1 + \frac{j\pi}{n}\right) + 0.6x_1 \ \mathit{if}\ j \epsilon J_2 \end{cases}$ $h_2(t) = h_4(t) = t^2,\ h_j(t) = 2t^2 - \cos(4\pi t) + 1, \quad \text{for } j = 3, 5, 6, \ldots, n$ Subject to: $x_2 - \sin\left(6\pi x_1 + \frac{2\pi}{n}\right) - \text{sign}\left(0.5(1 - x_1) - \left(1 - x_1\right)^2\right)\sqrt{\left\lvert 0.5(1 - x_1) - \left(1 - x_1\right)^2 \right\rvert} \geq 0$ $x_4 - \sin\left(6\pi x_1 + \frac{2\pi}{n}\right) - \text{sign}\left(0.25(1 - x_1) - \left(1 - x_1\right)^2\right)\sqrt{\left\lvert 0.2\sqrt{1 - x_1} - 0.5(1 - x_1 \right\rvert} \geq 0$

Table 4.9 Comparison of results of seven multiobjective constrained functions with different algorithms

Algorithm	CF1	CF2	CF3	CF4	CF5	CF6	CF7
MOABC (Akbari et al. 2012)	0.00992	0.01027	0.08621	0.00452	0.06781	0.00483	0.01692
GDE3 (Kukkonen and Lampinen 2009)	0.02940	0.01597	0.12750	0.00799	0.06799	0.06199	0.04169
MOEADGM (Chen et al. 2009)	0.01080	0.00800	0.51340	0.07070	0.54460	0.20710	0.53560
MTS (Tseng and Chen 2009)	0.01918	0.02677	0.10446	0.01109	0.02077	0.01616	0.02469
LiuLi Algoritm (Liu and Li 2009)	0.00850	0.00420	0.18290	0.01423	0.10973	0.01394	0.10446
DMOEADD (Liu et al. 2009)	0.01131	0.00210	0.05630	0.00699	0.01577	0.01502	0.01905
NSGAIILS (Sindhya et al. 2009)	0.00692	0.01183	0.23994	0.01576	0.18420	0.02013	023345
DECMOSA-SQP (Zamuda et al. 2009)	0.10773	0.09460	0.10000	0.15265	0.41275	0.14782	0.26049
TLBO (Rao and Waghmare 2014)	0.0088	**0.000140**	**0.002415**	**0.001305**	**0.01236**	**0.001359**	**0.005270**
ETLBO	**0.00674**	**0.000131**	**0.002392**	**0.001289**	**0.01204**	**0.001281**	**0.005236**

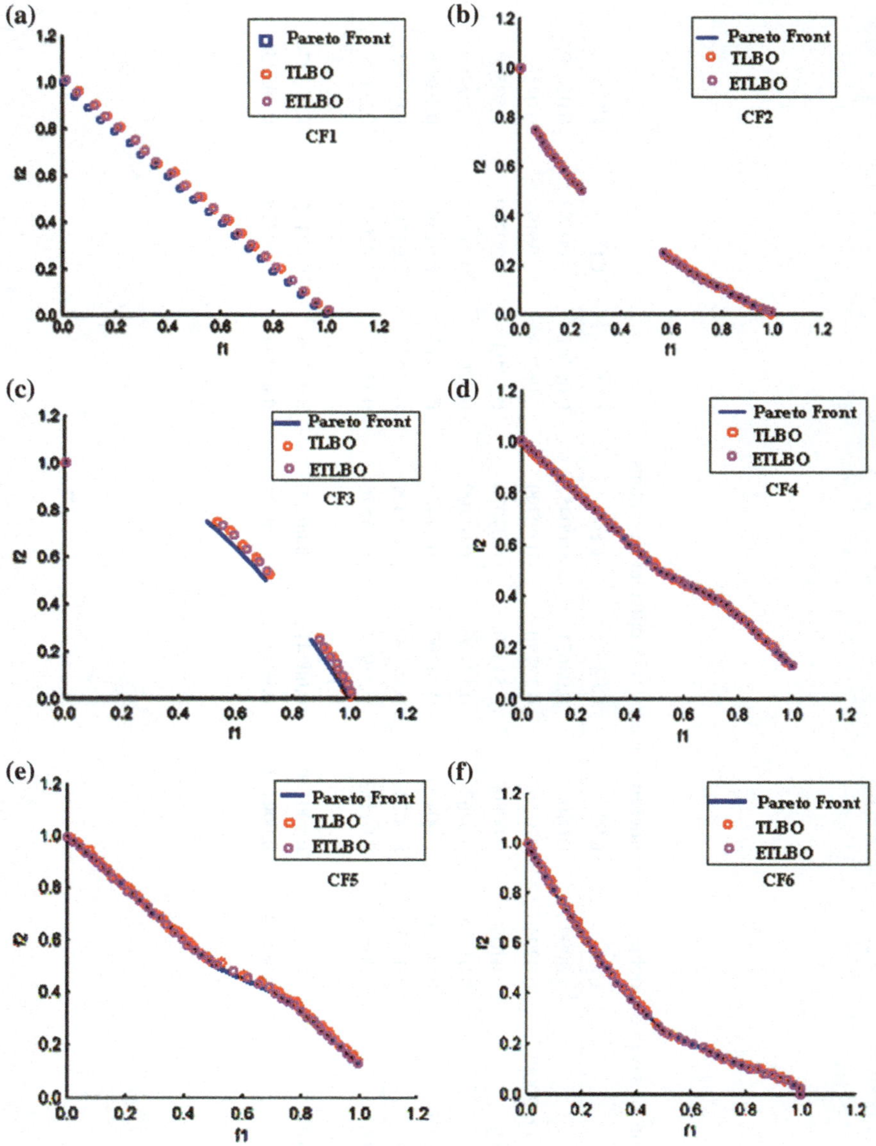

Fig. 4.3 **a–g** The Pareto front obtained by the TLBO and ETLBO algorithms on constrained test functions CF1–CF7

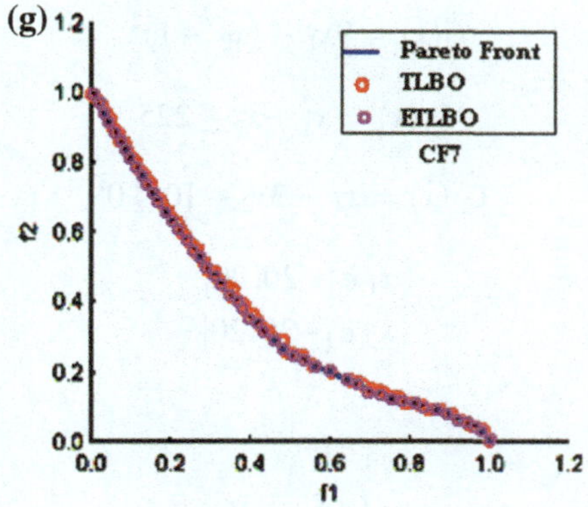

Fig. 4.3 (continued)

Table 4.10 The IGD statistics over CF1–CF7

Problem	Mean (IGD)	Smallest (IGD)	Largest (IGD)	Std. dev.(IGD)
CF1	0.0088	0.0049	0.0098	0.00061
CF2	0.000140	0.000129	0.000187	0.00048
CF3	0.002415	0.002256	0.002578	0.00713
CF4	0.001305	0.001277	0.001401	0.00083
CF5	0.01236	0.01210	0.01293	0.01981
CF6	0.001359	0.001314	0.001388	0.00021
CF7	0.005270	0.005205	0.005314	0.00452

(1) BNH

$$f_1(x) = 4x_1^2 + 4y_1^2 \tag{4.10}$$

$$f_2(x) = (x_1 - 5)^2 + (x_2 - 5)^2 \tag{4.11}$$

$$C_1(x) = (x_1 - 5)^2 + x_2^2 \leq 25 \tag{4.12}$$

$$C_2(x) = (x_1 - 8)^2 + (x_2 - 3)^2 \geq 7.7 \tag{4.13}$$

$$x_1 \in [0, 5]$$
$$x_2 \in [0, 3]$$

(2) SRN

$$f_1(x) = 2 + (x_1 - 2)^2 + (x_2 - 2)^2 \tag{4.14}$$

$$f_2(x) = 9x_1 - (x_2 - 1)^2 \tag{4.15}$$

$$C_1(x) = x_1^2 + x_2^2 \leq 225 \tag{4.16}$$

$$C_2(x) = x_1 - 3x_2 + 10 \leq 0 \tag{4.17}$$

$$x_1 \in [-20, 20]$$
$$x_2 \in [-20, 20]$$

(3) TNK

$$f_1(x) = x_1 \tag{4.18}$$

$$f_2(x) = x_2 \tag{4.19}$$

$$C_1(x) = x_1^2 + x_2^2 - 1 - 0.1\cos\left(16\arctan\left({}^{x_1}/_{x_2}\right)\right) \geq 0 \tag{4.20}$$

$$C_2(x) = (x_1 - 0.5)^2 + (x_2 - 0.5)^2 \leq 0.5 \tag{4.21}$$

$$x_1 \in [0, \pi]$$
$$x_2 \in [0, \pi]$$

(4) OSY

$$f_1(x) = -[25(x_1 - 2)^2 + (x_2 - 2)^2 + (x_3 - 1)^2 + (x_4 - 4)^2 + (x_5 - 1)^2] \tag{4.22}$$

$$f_2(x) = x_1^2 + x_2^2 + x_3^2 + x_4^2 + x_5^2 + x_6^2 \tag{4.23}$$

$$C_1(x) = x_1 + x_2 - 2 \geq 0 \tag{4.24}$$

$$C_2(x) = 6 - x_1 - x_2 \geq 0 \tag{4.25}$$

$$C_3(x) = 2 - x_2 + x_1 \geq 0 \tag{4.26}$$

$$C_4(x) = 2 - x_1 + 3x_2 \geq 0 \tag{4.27}$$

$$C_5(x) = 4 - (x_3 - 3)^2 - x_6 \geq 0 \tag{4.28}$$

$$C_6(x) = (x_5 - 3)^2 + x_6 - 4 \geq 0 \tag{4.29}$$

$$x_1 \in [0, 10]$$
$$x_2 \in [0, 10]$$
$$x_3 \in [1, 5]$$
$$x_4 \in [0, 6]$$
$$x_5 \in [1, 5]$$
$$x_6 \in [0, 10]$$

The estimated Pareto front for BNH function obtained by the ETLBO algorithm with an elite size of 8 is shown in Fig. 4.4. The BNH problem is fairly simple in that the constraints do not introduce additional difficulty in finding the Pareto optimal solution. It is observed that the ETLBO algorithm has performed well and has given a dense sampling of solutions along the true Pareto optimal curve.

The estimated Pareto front for SRN function is shown in Fig. 4.5. The SRN function is simple in nature having continuous convex Pareto front. It can be seen that the ETLBO algorithm has given a good sampling of points along the true curve and gave good distribution of the Pareto optimal solutions.

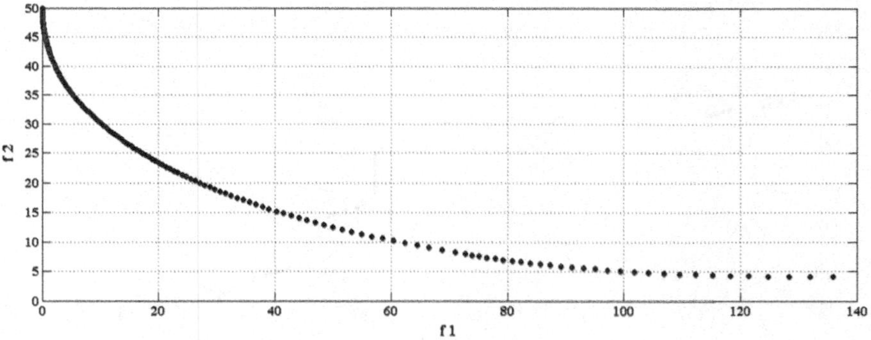

Fig. 4.4 Pareto front obtained by the ETLBO algorithm for the benchmark function BNH

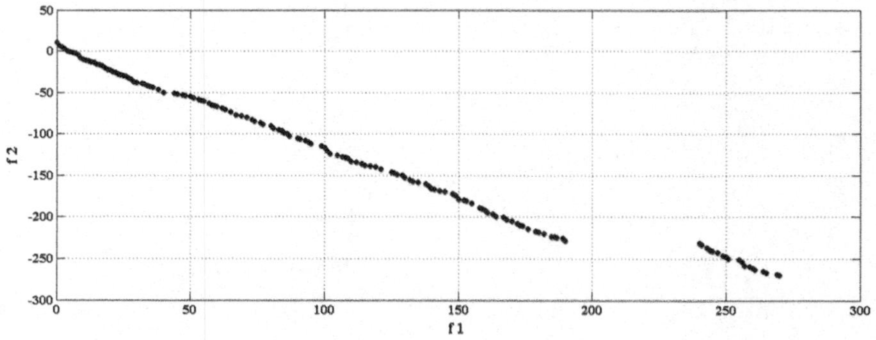

Fig. 4.5 Pareto front obtained by the ETLBO algorithm for the benchmark function SRN

The estimated Pareto front for TNK function is as shown in Fig. 4.6. The TNK function is relatively difficult. The constraints in the TNK problem make the Pareto optimal set discontinuous. As can be seen from the graphs for the TNK problem, the ETLBO algorithm has displayed a better distribution of the Pareto optimal solution but still there are gaps between the non-dominated solutions which make the curve non-smooth.

The estimated Pareto front for OSY function is shown in Fig. 4.7. The OSY function is a relatively difficult function. The constraints in the OSYfunction divide the Pareto optimal set into five regions. For the OSY problem, it can be seen that the ETLBO algorithm has given a good sampling of points at the mid-section of the curve and also found a lot of points at the extreme ends of the curve.

In this chapter, the performance of the TLBO algorithm is verified with the well-known multiobjective optimization methods such as AMGA, clustering MOEA, DECMOSA-SQP, DMOEADD, GDE3, LiuLi Algorithm, MOEAD, MOEADGM, etc. by experimenting with different multiobjective unconstrained and constrained benchmark functions. The experimental results show that the TLBO and ETLBO algorithms perform competitively with the other optimization

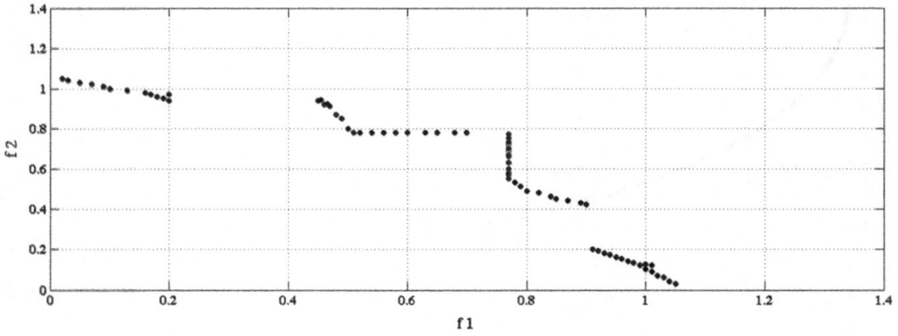

Fig. 4.6 Pareto front obtained by the ETLBO algorithm for the benchmark function TNK

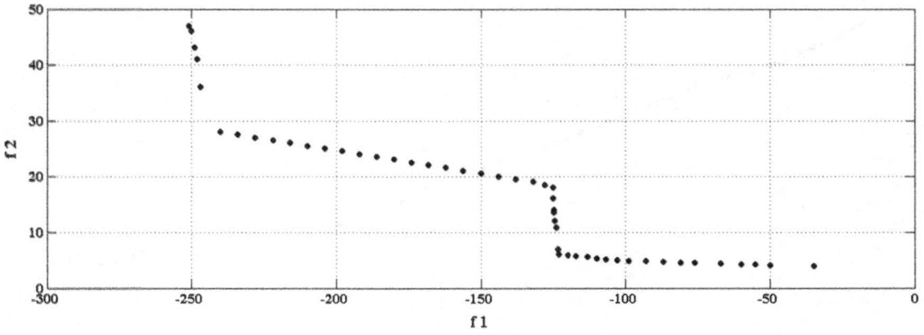

Fig. 4.7 Pareto front obtained by the ETLBO algorithm for the benchmark function OSY

methods reported in the literature. Therefore, the TLBO and ETLBO algorithms are effective and have great potential for solving the multiobjective problems.

The TLBO algorithm is also tested on the constrained design benchmark functions and the details are presented in the next chapter.

References

Akbari, R., Hedayatzadeh, R., Ziarati, K., Hassanizadeh, B., 2012. A multiobjective artificial bee colony algorithm. Swarm and Evolutionary Computation 2, 39–52.

Babu, B.V., Gujarathi, A.M., 2007. Multiobjective differential evolution (MODE) for optimization of supply chain planning and management, in: IEEE Congress on Evolutionary Computation, CEC'07, 2732–2739.

Chen, C.M., Chen, Y., Zhang, Q., 2009. Enhancing MOEA/D with guided mutation and priority update for multiobjective optimization, in: Proceedings of Congress on Evolutionary Computation, CEC'09, 209–216.

Deb, K., Pratap, A., Agarwal, S., Mayarivam, T., 2002. A fast and elitist multiobjective algorithm: NSGA-II. IEEE Trans Evolutionary Computation 6, 182–197.

Gao, S., Zeng, S., Xiao, B., Zhang, L., Shi, Y., Yang, X., Yu. D., Yan, Z., 2009. An orthogonal multiobjective evolutionary algorithm with lower-dimensional crossover, in: Proceeding of Congress on Evolutionary Computation, CEC'09, 1959–1964.

Kukkonen, S., Lampinen, J., 2009. Performance assessment of generalized differential evolution with a given set of constrained multiobjective test problems, in: Proceeding of Congress on Evolutionary Computation, CEC'09, 1943–1950.

Liu, H., Li, X., 2009. The multiobjective evolutionary algorithm based on determined weight and sub-regional search, in: Proceeding of Congress on Evolutionary Computation, CEC'09: 1928–1934.

Liu, M., Zou, X., Chen, Y., Wu, Z., 2009. Performance assessment of DMOEA-DD with CEC 2009 moea competition test instances, in: Proceeding of Congress on Evolutionary Computation, CEC'09, 2913–2918.

Pham, D.T., Ghanbarzadeh, A., 2007. Multiobjective optimization using the bees algorithm. In: 3rd international virtual conference on intelligent production machines and systems (IPROMS 2007) Whittles, Dunnbeath, Scotland.

Qu, B.Y., Suganthan, P.N., 2009. Multiobjective evolutionary programming without non-domination sorting is up to twenty times faster, in: Proceeding of Congress on Evolutionary Computation, CEC'09, 2934–2939.

Rao, R.V., Waghmare, G.G., 2014. A comparative study of a teaching-learning-based optimization algorithm on multiobjective unconstrained and constrained functions. Journal of King Saud University—Computer and Information Sciences 26, 332–346.

Robic, T., Filipic, B., 2005. DEMO: differential evolution for multiobjective optimization. In: Coello Coello CA et al. (eds) EMO, LNCS 2005, 3410, 520–533.

Schaffer, J.D., 1985. Multiple objective optimization with vector evaluated genetic algorithm. in Proceedings of First International Conference. Genetic Algorithms, 93–100.

Sindhya, K., Sinha, K., Deb, K., Miettinen, K., 2009. Local search based evolutionary multiobjective optimization algorithm for constrained and unconstrained problems in: Proceeding of Congress on Evolutionary Computation, CEC'09, 2919–2926.

Tiwari, S., Fadel, G., Koch, P., Deb, K., 2009. Performance assessment of the hybrid archive–based micro genetic algorithm (AMGA) on the CEC09 test problems, in: Proceeding of Congress on Evolutionary Computation, CEC'09, 1935–1942, doi: 10.1109/CEC.2009.4983177.

Tseng, L.Y., Chen, C., 2009. Multiple trajectory search for unconstrained/constrained multiobjective optimization, in: Proceeding of Congress on Evolutionary Computation, CEC'09, 1951–1958.

Wang, Y., Dang, C., Li, H., Han, L., Wei, J., 2009. A clustering multiobjective evolutionary algorithm based on orthogonal and uniform design, in: Proceeding of Congress on Evolutionary Computation, CEC'09, 2927–2933.

Zamuda, A., Brest, J., Boskovic, B., Zumer, V., 2009. Differential evolution with self adaptation and local search for constrained Multiobjective optimization. in: Proceeding of Congress on Evolutionary Computation, CEC'09, 192–202.

Zhang, Q., Zhou, A., Zhao, S., Suganthan, P.N., Liu, W., Tiwari, S., 2009. Multiobjective optimization test instances for the congress on evolutionary computation (CEC'09) 2009 special session and competition, Technical Report, CES-487, School of Computer Science and Electrical Engineering, University of Essex.

Zhang, G., Cheng, J., Gheorghe, M., Meng, Q., 2013. A hybrid approach based on differential evolution and tissue membrane systems for solving constrained manufacturing parameter optimization problems. Applied Soft Computing 13(3), 1528–1542.

Zhong, W.C.,Liu, J., Xue, M.Z., 2004. A multiagent genetic algorithm for global numerical optimization. IEEE Transactions on Systems, Man and Cybernetics 34, 1128–1141.

Chapter 5
Application of TLBO and ETLBO Algorithms on Constrained Benchmark Design Problems

Abstract This chapter presents the performance of the TLBO and the ETLBO algorithms on a class of constrained design optimization problems. The benchmark problems are taken from the research literature related to constrained design optimization. Experimental results show that the TLBO and the ETLBO algorithms are superior or competitive to the other optimization algorithms for the problems considered.

5.1 Constrained Benchmark Design Problems (Zhang et al. 2013; Reprinted with Permission from Elsevier)

5.1.1 Problem 1

$$\min f(x) = 5\sum_{i=1}^{4} x_i - 5\sum_{i=1}^{4} x_i^2 - \sum_{i=5}^{13} x_i \tag{5.1}$$

Subject to the following constraints:

$$g_1(x) = 2x_1 + 2x_2 + x_{10} + x_{11} - 10 \leq 0 \tag{5.2}$$

$$g_2(x) = 2x_1 + 2x_3 + x_{10} + x_{12} - 10 \leq 0 \tag{5.3}$$

$$g_3(x) = 2x_2 + 2x_3 + x_{11} + x_{12} - 10 \leq 0 \tag{5.4}$$

$$g_4(x) = -8x_1 + x_{10} \leq 0 \tag{5.5}$$

$$g_5(x) = -8x_2 + x_{11} \leq 0 \tag{5.6}$$

$$g_6(x) = -8x_3 + x_{12} \leq 0 \tag{5.7}$$

R.V. Rao, *Teaching Learning Based Optimization Algorithm*,
DOI 10.1007/978-3-319-22732-0_5

$$g_7(x) = -2x_4 - x_5 + x_{10} \leq 0 \quad (5.8)$$

$$g_8(x) = -2x_6 - x_7 + x_{11} \leq 0 \quad (5.9)$$

$$g_9(x) = -2x_8 - x_9 + x_{12} \leq 0 \quad (5.10)$$

where, $0 \leq x_i \leq 1$, $i = 1, 2, 3,\ldots, 9$; $0 \leq x_i \leq 100$, $i = 10, 11, 12$; and $0 \leq x_{13} \leq 1$.

The optimal solution is f $(x^*) = -15$ at $x^* = (1, 1, 1, 1, 1, 1, 1, 1, 1, 3, 3, 3, 1)$. The objective of problem 1 is to minimize the objective function value. This problem involves 13 variables and 9 linear inequality constraints.

5.1.2 Problem 2

$$\max f(x) = \left(\sqrt{n}\right)^n \prod_{i=1}^{n} x_i \quad (5.11)$$

Subject to the following constraints:

$$h(x) = \sum_{i=1}^{n} x_i^2 - 1 = 0 \quad (5.12)$$

where, $n = 10$ and $0 \leq x_i \leq 10$, $i = 1, 2, 3,\ldots, n$. The global maximum $f(x^*) = 1$ at $x^* = (1/n^{0.5}, 1/n^{0.5},\ldots)$ Problem 2 is a maximization problem. This problem involves ten variables and a nonlinear constraint.

5.1.3 Problem 3

$$\begin{aligned}\min f(x) = {} & (x_1 - 10)^2 + 5(x_2 - 12)^2 + x_3^4 + 3(x_4 - 11)^2 + 10x_5^6 \\ & + 7x_6^2 + x_7^4 - 4x_6x_7 - 10x_6 - 8x_7\end{aligned} \quad (5.13)$$

Subject to the following constraints:

$$\text{s.t. } g_1(x) = -127 + 2x_1^2 + 3x_2^4 + x_3 + 4x_4^2 + 5x_5 \leq 0 \quad (5.14)$$

$$g_2(x) = -282 + 7x_1 + 3x_2 + 10x_3^2 + x_4 - x_5 \leq 0 \quad (5.15)$$

$$g_3(x) = -196 + 23x_1 + x_2^2 + 6x_6^2 - 8x_7 \leq 0 \quad (5.16)$$

$$g_4(x) = 4x_1^2 + x_2^2 - 3x_1x_2 + 2x_3^2 + 5x_6 - 11x_7 \leq 0 \quad (5.17)$$

where, $-10 \le x_i \le 10$, $i = 1, 2, 3,\ldots,7$. The optimal solution is $f(x^*) = 680.6300573$ at $x^* = (2.330499, 1.951372, -0.4775414, 4.365726, -0.6244870, 1.1038131, 1.594227)$. Problem 3 is a minimization problem. This problem involves seven variables and four nonlinear inequality constraints.

5.1.4 Problem 4

$$\min f(x) = x_1 + x_2 + x_3 \tag{5.18}$$

Subject to the following constraints:

$$g_1(x) = -1 + 0.0025\,(x_4 + x_6) \le 0 \tag{5.19}$$

$$g_2(x) = -1 + 0.0025\,(x_5 + x_7 - x_4) \le 0 \tag{5.20}$$

$$g_3(x) = -1 + 0.01(x_8 - x_5) \le 0 \tag{5.21}$$

$$g_4(x) = -x_1x_6 + 833.3325x_4 + 100x_1 - 83333.333 \le 0 \tag{5.22}$$

$$g_5(x) = -x_2x_7 + 1250x_5 + x_2x_4 - 1250x_4 \le 0 \tag{5.23}$$

$$g_6(x) = -x_3\,x_8 + 1250000 + x_3x_5 - 2500x_5 \le 0 \tag{5.24}$$

where, $100 \le x_1 \le 10{,}000$, $1000 \le x_i \le 10{,}000$, $i = 2, 3$, $100 \le x_i \le 10{,}000$, $i = 4, 5, \ldots, 8$. The optimal solution is $f(x^*) = 7049.248021$ at $x^* = (579.3066, 1359.9709, 5109.9707, 182.0177, 295.601, 217.982, 286.165, 395.6012)$. Problem 4 is a linear minimization problem. This problem involves eight variables and three nonlinear inequality and three linear inequality constraints.

5.1.5 Problem 5

$$\max f(x) = \frac{100 - (x_1 - 5)^2 - (x_2 - 5)^2 - (x_3 - 5)^2}{100} \tag{5.25}$$

Subject to the following constraints:

$$g(x) = (x_1 - p)^2 - (x_2 - q)^2 - (x_3 - r)^2 \le 0 \tag{5.26}$$

where, $0 \le x_i \le 10$, $i = 1, 2, 3$, $p, q, r = 1, 2, 3, \ldots, 9$. The optimal solution is $f(x^*) = 1$ at $x^* = (5, 5, 5)$. Problem 5 is the maximization problem. This problem involves 3 design variables and 729 nonlinear inequality constraints.

5.1.6 Problem 6

This is a welded beam design problem, which is designed for the minimum cost subject to constraints on shear stress (τ), bending stress in the beam (σ), buckling load on the bar (P_c), end deflection of the beam (δ), and the side constraints. There are four design variables as shown in Fig. 5.1, i.e., $h(x_1)$, $L(x_2)$, $t(x_3)$, and $b(x_4)$. This problem can be mathematically formulated as follows:

$$\min f(x) = 1.10471x_1^2x_2 + 0.04811x_3x_4(14.0 + x_2) \tag{5.27}$$

Subject to the following constraints:

$$g_1(x) = \tau(x) - \tau_{max} \le 0 \tag{5.28}$$

$$g_2(x) = \sigma(x) - \sigma_{max} \le 0 \tag{5.29}$$

$$g_3(x) = x_1 - x_4 \le 0 \tag{5.30}$$

$$g_4(x) = 0.10471x_1^2 + 0.04811x_3x_4(14.0 + x_2) - 5.0 \le 0 \tag{5.31}$$

$$g_5(x) = 0.125 - x_1 \le 0 \tag{5.32}$$

$$g_6(x) = \delta(x) - \delta_{max} \le 0 \tag{5.33}$$

$$g_6(x) = P - P_c(x) \le 0 \tag{5.34}$$

where

$$\tau(x) = \sqrt{(\tau')^2 + 2\tau'\tau''\frac{x_2}{2R} + (\tau'')^2} \tag{5.35}$$

$$\tau' = \frac{p}{2^{0.5}x_1x_2} \tag{5.36}$$

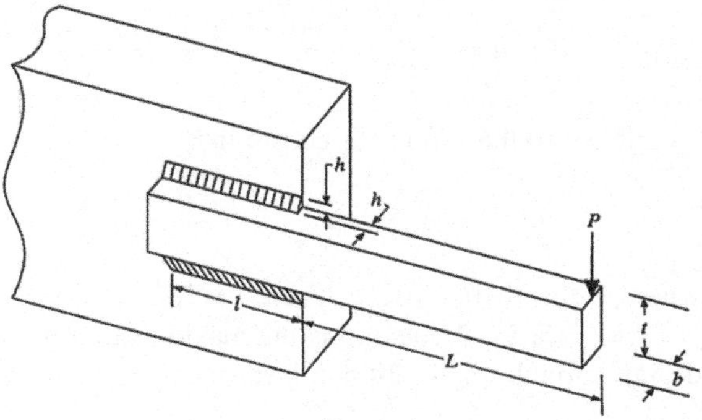

Fig. 5.1 The welded beam design problem (Zhang et al. 2013; Reprinted with permission from Elsevier)

$$\tau'' = \frac{MR}{J} \quad (5.37)$$

$$M = P\left(L + \frac{x_2}{2}\right) \quad (5.38)$$

$$R = \sqrt{\frac{x_2^2}{4} + \left(\frac{x_1 + x_3}{2}\right)^2} \quad (5.39)$$

$$J = 2\left\{2^{0.5} x_1 x_2 \left[\frac{x_2^2}{12} + \left(\frac{x_1 + x_3}{2}\right)\left(\frac{x_1 + x_3}{2}\right)\right]\right\} \quad (5.40)$$

$$\sigma(x) = \frac{6PL}{x_4 x_3^2} \quad (5.41)$$

$$\delta(x) = \frac{4PL^3}{E x_3^3 x_4} \quad (5.42)$$

$$P_c(x) = \frac{4.013E\sqrt{\frac{x_3^2 x_4^6}{36}}}{L^2}\left(1 - \frac{x_3}{2L}\sqrt{\frac{E}{4G}}\right) \quad (5.43)$$

where, $P = 6000$ lb, $L = 14$ in, $E = 30 \times 10^6$ psi, $G = 12 \times 10^6$ psi, $\tau_{max} = 13{,}600$ psi, $\sigma_{max} = 30{,}000$ psi, $\delta_{max} = 0.25$ in, $0.1 \le x_1 \le 2$, $0.1 \le x_2 \le 10$, $0.1 \le x_3 \le 10$ and $0.1 \le x_4 \le 2$. Problem 6 is a welded beam design problem as shown in Fig. 3.12. The objective is to design a welded beam for minimum cost. There are four continuous design variables with two linear and five nonlinear inequality constraints (Zhang et al. 2013).

5.1.7 Problem 7

This is a pressure vessel design problem for minimizing the total cost $f(x)$ of a pressure vessel considering the cost of the material, forming and welding. A cylindrical vessel is capped at both ends by hemispherical heads as shown in Fig. 5.2. There are four design variables: T_s (x_1, thickness of the shell), T_h (x_2, thickness of the head), R (x_3, inner radius), and L (x_4, length of the cylindrical section of the vessel, not including the head). Among the four variables, T_s and T_h which are integer multiples of 0.0625 inches are the available thicknesses of rolled steel plates, and R and L are continuous variables. This problem can be formulated as follows:

$$\min f(x) = 0.6224 x_1 x_3 x_4 + 1.7781 x_2 x_3^2 + 3.1661 x_1^2 x_4 + 19.84 x_1^2 x_3 \quad (5.44)$$

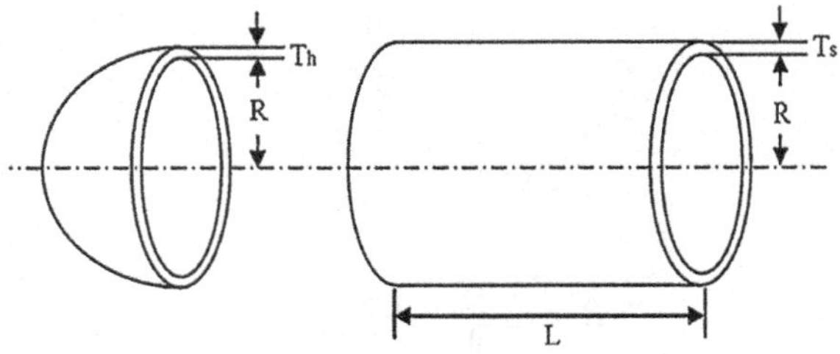

Fig. 5.2 Center and end sections of pressure vessel design problem (Zhang et al. 2013; Reprinted with permission from Elsevier)

Subject to the following constraints:

$$g_1(x) = -x_1 + 0.0193x_3 \leq 0 \quad (5.45)$$

$$g_2(x) = -x_2 + 0.00954x_3 \leq 0 \quad (5.46)$$

$$g_3(x) = -\prod x_3^2 x_4 - \frac{4}{3}\prod x_3^2 + 1296000 \leq 0 \quad (5.47)$$

$$g_4(x) = x_4 - 240 \leq 0 \quad (5.48)$$

where, $1 \leq x_1 \leq 99$, $1 \leq x_2 \leq 99$, $10 \leq x_3 \leq 200$, $10 \leq x_4 \leq 200$.

This problem has a nonlinear objective function with three linear and one nonlinear inequality constraints and two discrete and two continuous design variables.

5.1.8 *Problem 8*

This is a tension/compression spring design problem for minimizing the weight ($f(x)$) of a tension/compression spring (as shown in Fig. 5.3) subject to constraints on minimum deflection, shear stress, surge frequency, and limits on outside diameter and on design variables. The design variables are the wire diameter $d(x_1)$, mean coil diameter $D(x_2)$, and the number of active coils $P(x_3)$. The mathematical formulation of this problem can be described as follows:

$$\min f(x) = (x_3 + 2)x_2 x_1^2 \quad (5.49)$$

Subject to the following constraints:

$$g_1(x) = 1 - \frac{x_2^3 x_3}{71785 x_1^4} \leq 0 \quad (5.50)$$

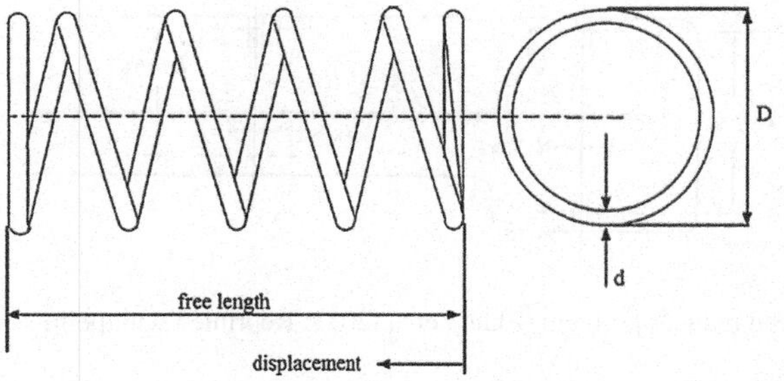

Fig. 5.3 Tension/compression spring design problem (Zhang et al. 2013; Reprinted with permission from Elsevier)

$$g_2(x) = \frac{4x_2^2 - x_1x_2}{12566(x_2x_1^3 - x_1^4)} + \frac{1}{5108x_1^2} - 1 \le 0 \tag{5.51}$$

$$g_3(x) = 1 - \frac{140.45x_1}{x_2^2x_3} \le 0 \tag{5.52}$$

$$g_4(x) = \frac{x_1 + x_2}{1.5} - 1 \le 0 \tag{5.53}$$

where, $0.05 \le x_1 \le 2$, $0.25 \le x_2 \le 1.3$, and $2 \le x_3 \le 15$. The objective is to minimize the weight of the spring subjected to one linear and three nonlinear inequality constraints with three continuous design variables.

5.1.9 Problem 9

This is a speed reducer design problem, shown in Fig. 5.4, for minimizing the weights of the speed reducer subject to constraints on bending stress of the gear teeth, surface stress, and transverse deflections of the shafts and stresses in the shafts. The parameters x_1, x_2,…, x_7 represents the face width (b), module of the teeth (m), number of the teeth in the pinion (z), length of the first shaft between bearings (l_1), length of the second shaft between bearings (l_2), and the diameter of the first shaft (d_1) and the second shaft (d_2), respectively.

$$\begin{aligned} \min f(x) = {} & 0.7854x1\, x_2^2\,(3.3333\,x_3^2 + 14.9334\,x_3 - 43.0934) \\ & - 1.508\,x_1\,(x_6^2 + x_7^2) + 7.4777(x_6^3 + x_7^3) \end{aligned} \tag{5.54}$$

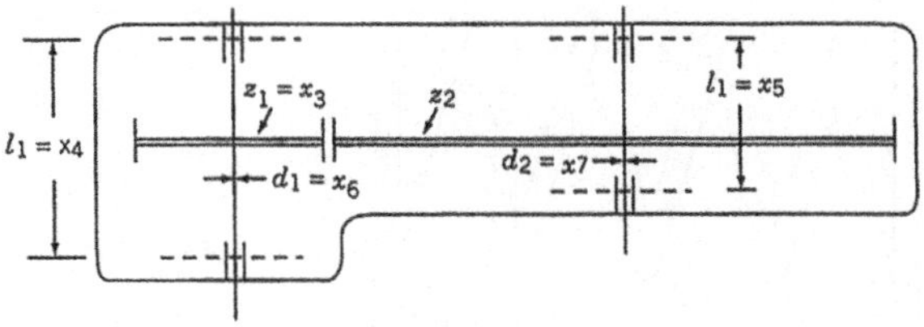

Fig. 5.4 The speed reducer problem (Zhang et al. 2013; Reprinted with permission from Elsevier)

Subject to the following constraints:

$$g_1(x) = \frac{27}{x_1 x_2^2 x_3} - 1 \le 0 \tag{5.55}$$

$$g_2(x) = \frac{397.5}{x_1 x_2^2 x_3^2} - 1 \le 0 \tag{5.56}$$

$$g_3(x) = \frac{1.93\, x_4^3}{x_2 x_3 x_6^4} - 1 \le 0 \tag{5.57}$$

$$g_4(x) = \frac{1.93 x_5^3}{x_2 x_3 x_7^4} - 1 \le 0 \tag{5.58}$$

$$g_5(x) = \sqrt{\frac{\left(\frac{745 x_4}{x_2 x_3}\right)^2 + 16.9 \times 10^6}{110.0 x_6^3}} - 1 \le 0 \tag{5.59}$$

$$g_6(x) = \frac{\sqrt{\left(\frac{745 x_4}{x_2 x_3}\right)^2 + 157.5 \times 10^6}}{85.0 x_6^3} - 1 \le 0 \tag{5.60}$$

$$g_7(x) = \frac{x_2 x_5}{40} - 1 \le 0 \tag{5.61}$$

$$g_8(x) = \frac{5 x_2}{x_1} - 1 \le 0 \tag{5.62}$$

$$g_9(x) = \frac{x_1}{12 x_2} - 1 \le 0 \tag{5.63}$$

$$g_{10}(x) = \frac{1.5 x_6 + 1.9}{x_4} - 1 \le 0 \tag{5.64}$$

$$g_{11}(x) = \frac{1.1x_7 + 1.9}{x_5} - 1 \leq 0 \tag{5.65}$$

where, $2.6 \leq x_1 \leq 3.6$, $0.7 \leq x_2 \leq 0.8$, $17 \leq x_3 \leq 28$, $7.3 \leq x_4 \leq 8.3$, $7.8 \leq x_5 \leq 8.3$, $2.9 \leq x_6 \leq 3.9$, and $5.0 \leq x_7 \leq 5.5$. The objective is to minimize the weight with one discrete and six continuous design variables. There are four linear and seven nonlinear inequality constraints.

5.2 Results of Application of Different Algorithms on the Constrained Benchmark Design Problems

In this section, the ability of TLBO and ETLBO algorithms is demonstrated by implementing the algorithms for the parameter optimization of nine well-defined constrained problems taken from Zhang et al. (2013). These problems were attempted by Rao and Waghmare (2014) using ETLBO algorithm. The objective of problem 1 is to minimize the objective function value. This problem involves 13 variables and 9 linear inequality constraints. Elitist TLBO is compared with other 7 optimization algorithms: differential evolution algorithms and tissue P systems (DETPS) (Zhang et al. 2013), teaching-learning-based optimization (TLBO) (Rao et al. 2011) multimembered evolutionary strategy (M-ES) (Mezura-Montes and Coello 2005), particle evolutionary swarm optimization (PESO) (Zavala et al. 2005), cultural differential evolution (CDE) (Becerra and Coello 2006), co-evolutionary differential evolution (CoDE) (Huang et al. 2007), and artificial bee colony (ABC) (Karaboga and Basturk 2007). For TLBO, only tuning of common control parameters is required as there are no algorithm-specific control parameters. Hence after making a few trials, population size of 75, function evaluations of 14009, and an elite size of 8 are considered (for making fair comparison with the same number of function evaluations) for problem 1. Population sizes of 75 and 100 and elite sizes of 8, 12, and 16 have given the same global optimum solutions. The results of DETPS, TLBO, M-ES, PESO, CDE, CoDE, and ABC are referred from Zhang et al. (2013), Rao et al. (2011), Mezura-Montes and Coello (2005), Zavala et al. (2005), Becerra and Coello (2006), Huang et al. (2007) and Karaboga and Basturk (2007) respectively.

Problem 2 is a maximization problem. This problem involves ten variables and a nonlinear constraint. After making a few trials, a population size of 25, function evaluations of 69996, and an elite size of 0 are considered. Population sizes of 25, 50, 75, and 100 and elite sizes of 0, 4, 8, 12, and 16 have given the same global optimum solutions. Problem 3 is a minimization problem. This problem involves seven variables and four nonlinear inequality constraints. After making a few trials, population size of 50, function evaluations of 30019, and elite size of 8 are considered. Problem 4 is a linear minimization problem. This problem involves eight variables and three nonlinear inequality and three linear inequality constraints. After making a few trials, a population size of 100, function evaluations of 99,987 and an elite size of 4 are considered. Problem 5 is a maximization problem. This problem involves 3 design variables and 729 nonlinear inequality constraints. After making a few trials, a population size of 25, function evaluations of 5011, and an elite size of 0 are considered. Population sizes of 25, 50, 75, and 100 and elite sizes of 0, 4, 8, 12, and 16 have given the same global optimum solutions. Table 5.1 presents the comparison of statistical results of the considered eight algorithms for the test problems 1–5. In Table 5.1, "Best", "Mean," and "Worst" represent best solution, mean best solution, and worst solution, respectively, over 30 independent runs. Table 5.1 presents the comparison of statistical results of eight algorithms for the test problems 1–5. The results shown bold in Table 5.1 indicate the best values.

It can be observed from Table 5.1 that for problem 1 the elitist TLBO finds the same global optimum solution as that given by DETPS, TLBO, M-ES, PESO, CDE, CoDE, and ABC but the elitist TLBO requires approximately 94 %, 94 %, 86 %, 96 %, 44 %, and 33 % fewer function evaluations than ABC, CoDE, CDE, PESO, M-ES, TLBO, and DETPS, respectively. For problem 2, the elitist TLBO finds the better quality of solution than DETPS and CDE. The results of elitist TLBO are same as the results of TLBO, M-ES, and ABC algorithms but it requires approximately 30 %, 71 %, and 71 % fewer function evaluations than TLBO, M-ES, and ABC, respectively. It is, however, surprising to note that the global optimum value obtained by PESO and DETPS is more than 1.000 whereas the global maximum value can be only 1.000 in problem 2. The eitist TLBO requires much less function evaluations than DETPS, TLBO, M-ES, PESO, CDE, and ABC to achieve the global optimum solution. Thus, the elitist TLBO requires less computing time as compared to the other optimization algorithms considered.

For problem 3, solution given by DETPS, PESO, and CDE are better than that given by elitist TLBO, M-ES, CoDE, and ABC algorithms. But DETPS requires much less function evaluations than PESO and CDE to obtain the global optimum solution. More function evaluations means more computing time is required to obtain the global optimum solution. Elitist TLBO finds quality of solution better than M-ES, CoDE, and ABC. The results of elitist TLBO are same as the results of TLBO, PESO, CDE, and DETPS in terms of the best solution but elitist TLBO requires approximately 70 %, 91 %, 70 %, and 8 % fewer function evaluations than TLBO, PESO, CDE, and DETPS, respectively.

For problem 4, elitist TLBO finds the best value of solution better than DETPS, M-ES, PESO, and ABC. Moreover, elitist TLBO requires approximately 58 %, 71 %, and 58 % fewer function evaluations than ABC, PESO, and M-ES, respectively, to obtain the global best solution. However, CDE achieved better quality of solution than the rest seven optimization algorithms. CDE requires same function evaluations as DETPS and TLBO to obtain the better mean and worst solutions. For problem 5, elitist TLBO finds the same global optimum solution as DETPS, TLBO, M-ES, PESO, CDE, CoDE, and ABC but elitist TLBO requires approximately 95 %, 98 %, 95 %, 99 %, 98 %, 90 %, and 24 % fewer function evaluations than ABC, CoDE, CDE, PESO, M-ES, TLBO, and DETPS, respectively.

Table 5.1 Comparison of statistical results of eight algorithms for test problems 1–5

Problems		Elitist TLBO (Rao and Waghmare 2014)	DETPS (Zhang et al. 2013)	TLBO (Rao et al. 2011a)	M-ES Mezura-Montes and Coello 2005)	PESO (Zavala et al. 2005)	CDE (Becerra and Coello 2006)	CoDE (Huang et al. 2007)	ABC (Karaboga and Basturk 2007)
1	Best	–15.0	–15.0	–15.0	–15.0	–15.0	–15.0	–15.0	–15.0
	Mean	–15.0	–15.0	–15.0	–15.0	–15.0	–15.0	–15.0	–15.0
	Worst	–15.0	–15.0	–15.0	–15.0	–15.0	–15.0	–15.0	–15.0
	SD	1.9e-6	–	–	–	–	–	–	–
	FE	**14,009**	20,875	25,000	240,000	350,000	100,100	2,48,000	240,000
2	Best	1.000	1.001	1.000	1.000	**1.005**	0.995	–	1.000
	Mean	1.000	0.992	1.000	1.000	**1.005**	0.789	–	1.000
	Worst	1.000	0.995	1.000	1.000	**1.005**	0.640	–	1.000
	SD	0.000	–	–	–	–	–	–	–
	FE	**69,996**	90,790	100,000	240,000	350,000	100,100	–	240.000
3	Best	**680.630**	**680.630**	**680.630**	680.632	**680.630**	**680.630**	680.771	680.634
	Mean	680.631	**680.630**	680.633	680.643	**680.630**	**680.630**	681.503	680.640
	Worst	680.633	**680.630**	680.638	680.719	**680.630**	**680.630**	685.144	680.653
	SD	3.4e-3	–	–	–	–	–	–	–
	FE	**30,019**	32,586	100,000	240,000	350,000	100,100	240,000	100,000
4	Best	**7049.248**	7049.257	**7049.248**	7051.903	7049.459	**7049.248**	–	7053.904
	Mean	7050.261	7050.834	7083.673	7253.047	7099.101	**7049.248**	–	7224.407
	Worst	7055.481	7063.406	7224.497	7099.101	7251.396	**7049.248**	–	7604.132
	SD	2.8e-2	–	–	–	–	–	–	—
	FE	**99,987**	**100,000**	**100,000**	240,000	350,000	100,100	–	240,000

(continued)

Table 5.1 (continued)

Problems		Elitist TLBO (Rao and Waghmare 2014)	DETPS (Zhang et al. 2013)	TLBO (Rao et al. 2011a)	M-ES Mezura-Montes and Coello 2005)	PESO (Zavala et al. 2005)	CDE (Becerra and Coello 2006)	CoDE (Huang et al. 2007)	ABC (Karaboga and Basturk 2007)
5	Best	1.000	1.000	1.000	1.000	1.000	1.000	1.000	1.000
	Mean	1.000	1.000	1.000	1.000	1.000	1.000	1.000	1.000
	Worst	1.000	1.000	1.000	1.000	1.000	1.000	1.000	1.000
	SD	0.000	–	–	–	–	–	–	–
	FE	**5011**	6,540	50,000	240,000	350,000	100,100	240,000	100,000

Elitist TLBO Elitist Teaching-Learning-Based Optimization; *DETPS* Differential Evolution algorithm and Tissue P Systems; *TLBO* Teaching-Learning-Based Optimization; *M-ES* Multimembered Evolution Strategy; *PESO* Particle Evolutionary Swarm Optimization; *CDE* Cultural Differential Evolution; *CoDE* Co-evolutionary Differential Evolution; *ABC* Artificial Bee Colony; *FE* Function evaluations; result is not available. The results of this table except the results of elitist TLBO are taken from Zhang et al. (2013)

Problem 6 is a welded beam design problem. The objective is to design a welded beam for minimum cost. There are four continuous design variables with two linear and five nonlinear inequality constraints (Zhang et al. 2013; Rao and Waghmare 2014). After making a few trials, a population size of 100, function evaluations of 99991, and an elite size of 4 are considered. Problem 7 is a pressure vessel design problem. The objective is to minimize the total cost of material, forming, and welding. This problem has a nonlinear objective function with three linear and one nonlinear inequality constraints and two discrete and two continuous design variables. After making a few trials, a population size of 75, function evaluations of 4992, and an elite size of 12 are considered.

Problem 8 is a tension/compression spring design problem. The objective is to minimize the weight of the spring subjected to one linear and three nonlinear inequality constraints with three continuous design variables. After making a few trials, population size of 50, function evaluations of 7022 and elite size of 4 are considered. Problem 9 is a speed reducer design problem and the objective is to minimize the weight with one discrete and six continuous design variables. There are four linear and seven nonlinear inequality constraints. After making a few trials, a population size of 100, function evaluations of 9988, and an elite size of 8 are considered. The comparison of statistical results of eight algorithms for the test problems of 6–9 over 30 independent runs are shown in Table 5.2.

The problems described in this section were solved in the past by various researchers using different optimization algorithms. Now the ETLBO algorithm is attempted on the same problems and comparisons are made. It may be mentioned here that the common control parameters of population size, number of generations, and elite sizes for solving different constrained optimization problems presented in this section have been selected after conducting trials for each optimization problem. Like other optimization algorithms (e.g., PSO, ABC, ACO, etc.), TLBO algorithm also has not any special mechanism to handle the constraints. So, for the constrained optimization problems, it is necessary to incorporate any constraint handling techniques with the TLBO algorithm. In the present experiments, Deb's heuristic constrained handling method (Deb 2000) is used to handle the constraints with the TLBO algorithm in which two solutions are selected and compared with each other.

Table 5.2 presents the comparison of statistical results of eight algorithms for test problems 6–9. It can be seen from Table 5.2 that the elitist TLBO is better than the other algorithms for the welded beam problem in terms of quality of solution. The worst solutions obtained by DETPS and MBA algorithms are better than the rest eight optimization algorithms. But DETPS requires less function evaluations than MBA to achieve the better results. MBA is superior to the rest nine optimization algorithms in terms of standard deviation. Hence in terms of robustness MBA is superior to other algorithms considered for this problem. It can also be seen that for the same function evaluations elitist TLBO gives better solution than DETPS and TLBO.

For pressure vessel problem, DETPS is superior to the rest nine algorithms in terms of the quality of solution and uses much smaller number of function

Table 5.2 Comparison of statistical results of eight algorithms for test problems 6–9

Problem		Elitist TLBO (Rao and Waghmare 2014)	DETPS (Zhang et al. 2013)	$(\mu + \lambda)$-ES (Mezura-Montes and Coello 2005)	UPSO (Parsopoulos and Vrahatis 2005)	CPSO (He and Wang 2007)	CoDE (Huang et al. 2007)	PSO-DE (Liu et al. 2010)	ABCA (Akay and Karaboga 2012)	TLBO (Rao et al. 2011)	MBA (Sadollah et al. 2012)
Welded beam	Best	**1.724852**	**1.724852**	**1.724852**	1.92199	1.728024	1.733462	1.724853	**1.724852**	**1.724852**	1.724853
	Mean	**1.724852**	**1.724852**	1.777692	2.83721	1.748831	1.768158	1.724858	1.741913	1.728447	1.724853
	Worst	**1.724853**	**1.724853**	2.074562	4.88360	1.782143	1.824105	1.724881	–	–	**1.724853**
	SD	3.3e-2	2.1e-7	8.8e-2	6.8e-1	1.3e-2	2.2e-2	4.1e-6	3.1e-2	–	**6.9e-19**
	FE	**99,991**	**10,000**	30,000	100,000	200,000	240,000	33,000	30,000	**10,000**	47340
Pressure vessel	Best	**5885.3336**	**5885.3336**	6059.7016	6544.27	6061.0777	6059.7340	6059.7143	6059.7147	6059.7143	5889.3216
	Mean	5887.3338	**5887.3161**	6379.9380	9032.55	6147.1332	6085.2303	6059.7143	6245.3081	6059.7143	6200.64765
	Worst	5956.6921	**5942.3234**	6820.3975	11638.20	6363.8041	6371.0455	6059.7143	–	–	6392.5062
	SD	1.1e + 1	1.0e + 1	2.1e + 2	9.9e + 2	8.6e + 1	4.3e + 1	**1.0e-10**	2.1e + 2	–	160.34
	FE	**4992**	10,000	30,000	100,000	200,000	240,000	42,100	30,000	10,000	70,650
Tension compression spring	Best	**0.012665**	**0.012665**	0.012689	0.013120	0.012675	0.012670	**0.012665**	**0.012665**	**0.012665**	**0.12665**
	Mean	**0.012678**	0.012680	0.013165	0.022948	0.012730	0.012703	0.012665	0.012709	0.012666	0.012713
	Worst	0.012758	0.012769	0.014078	0.050365	0.012924	0.012790	**0.012665**	–	–	0.012900
	SD	4.9e-4	2.7e-5	3.9e-4	7.2e-3	5.2e-5	2.7e-5	**1.2e-8**	1.3e-2	–	6.3e-5
	FE	**7022**	10,000	30,000	100,000	200,000	240,000	24,950	30,000	10,000	7650
Speed reducer	Best	2996.348	2996.348	2996.348	–	–	–	2996.348	2997.058	2996.348	**2994.74421**
	Mean	**2996.348**	**2996.348**	**2996.348**	–	–	–	**2996.348**	2997.058	**2996.348**	2996.769019
	Worst	**2996.348**	**2996.348**	**2996.348**	–	–	–	**2996.348**	–	–	2999.6524
	SD	4.5e-5	5.2e-5	**0.0**	–	–	–	6.4e-6	**0.0**	–	1.56
	FE	9.988	10,000	30,000	–	–	–	54,350	30,000	10,000	**6300**

Elitist TLBO Elitist Teaching-Learning-Based Optimization; *DETPS* Differential Evolution algorithm and Tissue *P* Systems; $(\mu + \lambda)$-ES: $(\mu + \lambda)$-evolutionary strategy; *UPSO* Unified Particle Swarm Optimization; *CPSO* Co-evolutionary Particle Swarm Optimization; *CoDE* Co-evolutionary differential evolution; *PSO-DE* Hybridizing Particle Swarm Optimization with Differential Evolution; *ABCA* Artificial Bee Colony Algorithm; *TLBO* Teaching-Learning-Based Optimization; *MBA* Mine Blast Algorithm; *SD* Standard deviation; *FE* Function evaluations; result is not available. The results of this table except the results of elitist TLBO are taken from Zhang et al. (2013)

evaluations to obtain global optimum solution except elitist TLBO. Elitist TLBO is inferior to DETPS in this problem but it requires much less function evaluations than DETPS and superior to the rest seven optimization algorithms. PSO-DE is superior in terms of standard deviation and shows robustness.

For the tension/compression spring, the elitist TLBO is superior to the rest nine algorithms in terms of quality of solution. Elitist TLBO is superior to UPSO and ABCA and is inferior to DETPS, ($\mu + \lambda$)-ES, CPSO, CoDE, MBA, and PSO-DE in terms of the standard deviations. PSO-DE is the best optimization algorithm in terms of standard deviation for this problem. Elitist TLBO requires less function evaluations than the rest nine algorithms to obtain the best and mean solutions and requires less computational time and effort.

For speed reducer problem, elitist TLBO produces same results as DETPS, TLBO, ($\mu + \lambda$)-ES, and PSO-DE. MBA obtains first rank in terms of best solution among the ten algorithms. Elitist TLBO, DETPS, ($\mu + \lambda$)-ES and PSO-DE provides same better worst value. But DETPS requires less function evaluations than ($\mu + \lambda$)-ES and PSO-DE to achieve the same worst value. ($\mu + \lambda$)-ES and ABCA are the most robust optimization algorithms for the speed reducer problem since zero standard deviation is obtained by these algorithms. Elitist TLBO is inferior to the ($\mu + \lambda$)-ES, PSO-DE, and ABCA in terms of standard deviation in this problem.

From Table 5.3, it may be said that the concept of elitism enhances the performance of the TLBO algorithm for the constrained optimization problems. Similarly, it is observed from the experiments that for majority of the problems the strategy with higher population size produced the better results. Smaller population size required more number of iterations to achieve the global optimum value. For some class of problems the strategy with smaller population size produced the promising results than higher population size. Thus, similar to the other evolutionary- or swarm intelligence-based algorithms, the TLBO algorithm requires proper tuning of the common controlling parameters (i.e. population size, number of generations and elite size) before applying it to any problem. However, unlike

Table 5.3 Population and elite sizes for global solutions with better best, mean, and worst solutions for problems 1–9

Problem number	Population sizes (PS = 25, 50, 75, 100)	Elite sizes (0, 4, 8, 12, 16)
1	75 and 100	8, 12, 16
2	Same value for all population sizes	Same effect for all elite sizes
3	50	8
4	100	4
5	Same value for all population sizes	Same effect for all elite sizes
6	100	4
7	75	12
8	50	4
9	100	8

the other evolutionary or swarm intelligence-based algorithms, TLBO does not require any algorithm-specific control parameters.

The results have shown that the elitist TLBO algorithm is comparatively better or competitive to other optimization algorithms recently reported in the literature. The concept of elitism enhances the performance of the TLBO algorithm for the constrained design optimization problems.

The next chapter presents the design optimization of a gear train using TLBO and ETLBO algorithms.

References

Akay, B., Karaboga, D., 2012. Artificial bee colony algorithm for large-scale problems and engineering design optimization. Journal of Intelligent Manufacturing 23(4), 1001–1014.

Becerra, R.L., Coello, C.A.C., 2006. Cultured differential evolution for constrained optimization. Computer Methods in Applied Mechanics and Engineering 195, 4303–4322.

Deb, K., 2000. An efficient constraint handling method for genetic algorithm, Computer Methods in Applied Mechanics and Engineering 186, 311–338.

He, Q., Wang. L.,2007. An effective co-evolutionary particle swarm optimization for constrained engineering design problems. Engineering Applications of Artificial Intelligence 20, 89–99.

Huang, F.Z., Wang, L., He, Q., 2007. An effective co-evolutionary differential evolution for constrained optimization. Applied Mathematics and Computation 186, 340–356.

Karaboga, D., Basturk, B., 2007. Artificial bee colony (ABC) optimization algorithm for solving constrained optimization problems. in: P. Melin, (Ed.), LNAI (IFSA) 4529, 789–798.

Liu, H., Cai, Z., Wang, Y., 2010. Hybridizing particle swarm optimization with differential evolution for constrained numerical and engineering optimization. Applied Soft Computing 10: 629–640.

Mezura-Montes, E., Coello, C.A.C., 2005. A simple multimembered evolution strategy to solve constrained optimization problems. IEEE Transactions on Evolutionary Computation 9: 1–17.

Parsopoulos, K.E., Vrahatis, M.N., 2005. Unified particle swarm optimization for solving constrained engineering optimization problems. in: L. Wang, (Ed.) LNCS(ICNC) 3612, 582–581.

Rao, R.V., Waghmare, G. G., 2014. Complex constrained design optimization using an elitist teaching-learning-based optimization algorithm. International Journal of Metaheuristics 3(1), 81–102.

Rao, R.V., Savsani, V.J., Vakharia, D.P., 2011. Teaching-learning-based optimization: A novel method for constrained mechanical design optimization problems. Computer-Aided Design 43 (3), 303–315.

Sadollah, A., Bahreininejad, A., Eskandar, H., Hamdi, M., 2013. Mine blast algorithm: A new population based algorithm for solving constrained engineering optimization problems, Applied Soft Computing 13(5), 2592–2612.

Zavala, A.E.M., Aguirre, A.H., Diharce, E.R.V., 2005. Constrained optimization via evolutionary particle swarm optimization algorithm (PESO). in: Proceedings of the 7th Annual conference on Genetic and Evolutionary Computation, June 25–29, Washington DC, 209–216.

Zhang, G., Cheng, J., Gheorghe, M., Meng, Q., 2013. A hybrid approach based on differential evolution and tissue membrane systems for solving constrained manufacturing parameter optimization problems, Applied Soft Computing 13(3), 1528–1542.

Chapter 6
Design Optimization of a Spur Gear Train Using TLBO and ETLBO Algorithms

Abstract Two cases of optimal design of a spur gear train are considered using the TLBO and ETLBO algorithms. The objective considered in both the cases is minimization of weight. The constraints considered in the first case are bending strength of gear, surface durability, torsional strength of the shafts for pinion and gear, and center distance between the pinion and the gear shafts. The second case is the modified form of the first case and it included three more constraints of interference, surface fatigue strength, and width to module ratio. The design variables considered are face width, diameters of pinion and gear shafts, and the number of teeth on pinion and gear module. In addition to these five design variables, the second case considered hardness also as an additional variable. The results of the TLBO and ETLBO algorithms are compared with the GA, SA, and PSO algorithms. The computational results show that the TLBO and ETLBO algorithms have obtained more accurate solutions than those obtained by using the other optimization methods.

6.1 Optimal Weight Design of a Spur Gear Train

Designing a new product consists of several parameters and phases which differ according to the depth of design, input data, design strategy, procedures, and results. Mechanical design includes an optimization process in which designers always consider certain objectives such as strength, deflection, weight, wear, corrosion, etc. depending on the requirements. However, design optimization for a complete mechanical assembly leads to a complicated objective function with a large number of design variables. So it is a good practice to apply optimization techniques for individual components or intermediate assemblies than a complete assembly. For example, in an automobile power transmission system, optimization of a gearbox is computationally and mathematically simpler than the optimization of complete system (Rao and Savsani 2012).

Gears find applications in many mechanical power transmission systems such as automobile, aerospace, machine tools and gear design is still an ongoing activity.

R.V. Rao, *Teaching Learning Based Optimization Algorithm*,
DOI 10.1007/978-3-319-22732-0_6

The complex shape and geometry of gears lead to a large number of design parameters. A traditional gear design involves computations based on tooth bending strength, tooth surface durability, tooth surface fatigue, interference, efficiency, etc. Gear design involves empirical formulas and different graphs and tables which lead to a complicated design. Manual design is very difficult considering these facts and there is a need for the computer-aided design of gears. With the aid of computer, design can be carried out iteratively and the design variables which satisfy the given conditions can be determined. The design so obtained may not be the optimum one, because in the above process the design variables so obtained satisfy only one condition at a time, e.g., if module is calculated based on bending strength, the same module is substituted to calculate the surface durability. It is accepted if it is within the strength limit of surface durability; otherwise, it is changed accordingly. So optimization methods are required to determine the design variables which simultaneously satisfy the given conditions. Moreover, increasing demand for compact, efficient, and reliable gears forces the designer to use optimal design methodology.

There have been a number of studies attempting to optimize gears with the aid of computers. Osyczka (1978) formulated a problem for finding the basic constructional parameters (e.g., modules, numbers of teeth, etc.) of a gearbox to minimize simultaneously four objective functions: volume of elements, peripheral velocity between gears, width of gearbox, and the distance between axes of input and output shafts. The author had used numerical methods considering min–max principle of optimality along with Monte Carlo method and tradeoff studies. A detailed example considering a lathe gearbox optimization problem was presented.

Madhusudan and Vijayasimha (1987) prepared a computer program which was capable of designing a required type of gear under a specific working condition. The authors' work was oriented toward the training and education aspects of the gear design process. Tong and Walton (1987) described an interactive program to design internal gear pairs. The program was having large built-in databases for tooth cutters and materials. A complete design was provided including all the necessary information for manufacture.

The use of computer-aided design gives satisfactory design, but does not guarantee optimum design. Many researches were reported for the optimization of gears which were performed using different optimization techniques. Prayoonrat and Walton (1988) presented the use of direct search method and heuristic iteration approach for minimization of the center distance between input and output shafts. Direct search method was used to obtain primary design and to distinguish unacceptable solutions, while heuristic optimization technique was used to determine final solutions from the range of feasible solutions. Tong and Walton (1987) considered minimization of center distance and gear volume for internal gears as separate objective functions. The complete optimization process was divided into two parts. Belt zone search was used for the first part and half-section algorithm was used for the second part. The number of teeth on pinion, gears, and module were used as the three design variables. The feasible combination of number of teeth on pinion and module were identified using belt zone search and was forwarded to half-section algorithm to find the optimum result.

Jain and Agogino (1990) described a theory of design from an optimization perspective. The theory covers a broad spectrum of the design cycle: from qualitative and innovative design to global numerical parametric design. The emphasis was on a theoretical framework leading to design conceptualization and insight concerning the interrelationships between various parameters of the design, rather than specific numerical solutions. The optimizing design framework was applied for the optimization of gear volume and power for a 18-speed, 5-shafts gear box. Zarefar and Muthukrishnan (1993) used random search algorithm for the weight optimization of helical gear. Four design variables, module, helix angle, number of teeth on pinion, and face width were taken along with constraints on contact stress and bending stress.

Pomrehn and Papalambros (1995a) discussed a method to reduce the solution space by using infeasibility and non-optimality tests in discrete gear optimal design. In another work, Pomrehn and Papalambros (1995b) had presented the discrete optimal design formulations with applications to gear train design. The complex gear train design problem investigated by Pomrehn and Papalambros (1995a) was used for illustrating the strategy. The improvement in the design and the reduction in computational effort gained as a result of designer interaction were demonstrated.

Li et al. (1996) presented a method for the minimization of center distance. They established the upper and lower boundaries for the search interval of center distance by determining the limiting values of American Gear Manufacturers Association (AGMA) geometry factors. Kader et al. (1998) discussed the mode of failure in gears under optimal conditions. Design spaces were formed with module and pinion number of teeth, which were used for the optimal design and to study the mode of failure such as bending, pitting, and scoring. Gear design generally consists of discrete design variables.

Yokota et al. (1998) formulated an optimal weight design problem of a gear for a constrained bending strength of gear, torsional strength of shafts, and each gear dimension as a nonlinear integer programming (NIP) problem and solved the same using an improved genetic algorithm (GA). However, certain constraints were not satisfied and the obtained solution was not optimum. Multiobjective optimization for the gear design was also reported by some researchers. In the present work, an effort is made to verify if any improvement in the solution is possible by employing ETLBO and TLBO algorithms to the same optimization model formulated by Yokota et al. (1998). The gear design problem presented by Yokota et al. (1998) contains five design variables and five constraints. However, in the present work, the design problem is modified according to AGMA equations which contain six design variables and eight constraints. Modified design is optimized using ETLBO and TLBO algorithms for the global optimum.

Khorshid and Seireg (1999) developed a procedure for optimizing discrete nonlinear problems by decomposition of the constraints and the merit function and rearranging the decomposed stages into sequential levels. Forward and backward techniques were used to overcome any infeasibility encountered in the subsystems and to improve the design merit. Thompson et al. (2000) used quasi-Newton minimization method for the optimization of gears considering minimum volume and surface fatigue life as objective functions. Abuid and Ameen (2003) used the combination of min–max method and univariate search method for the optimization of gear volume, center distance, and dynamic factors of shafts and gears.

The same problem was studied by Deb and Jain (2003) by using nondominated sorting genetic algorithm (NSGA-II) in which the problem was experimented by changing design variables and constraints. However, genetic algorithms provide a near optimal solution for a complex problem having large number of variables and constraints. This is mainly due to difficulty in determination of optimum algorithm-specific controlling parameters. Therefore, the efforts are continuing to use more recent optimization algorithms, which are more powerful, robust, and able to provide accurate solution. This chapter is intended to report the application of the TLBO and ETLBO algorithms for solving the optimal spur gear weight design problem.

Huang et al. (2005) developed an interactive physical programming approach to place physical programming into an interactive framework in a natural way. An optimization problem of three-stage spur gear train was presented to illustrate the effectiveness of the above approach. However, the approach is a tedious one. Savsani et al. (2010) applied simulated annealing (SA) and particle swarm optimization (PSO) methods for the optimum design problem of the gear train proposed by Yokata et al. (1998).

In the present work, an effort is made to verfiy if any improvement in the solution is possible by employing the TLBO and ETLBO algorithms to the same optimization model proposed by Yokata et al. (1998), Savsani et al. (2010) and Rao and Savsani (2012). The problem formulation of the gear train is given as.

6.2 Problem Formulation

Figure 6.1 shows the basic geometry of a single stage spur gear train.

The notation used in the design of the gear train is given below.

a	center distance (mm)
b	face width (mm)
b_i	constraint quantities
b_w	thickness of web (mm)
C_l	surface fatigue life factor
C_p	elastic coefficient (MPa)
C_r	surface reliability factor
C_s	surface factor
d_1, d_2	diameter of pinion, gear shaft (mm)
D_i	inside diameter of rim (mm)
d_0	outside diameter of boss (mm)
d_p	drilled hole diameter (mm)
D_r	dedendum circle diameter (mm)
$F(x)$	objective function
F_p	wear load (N)

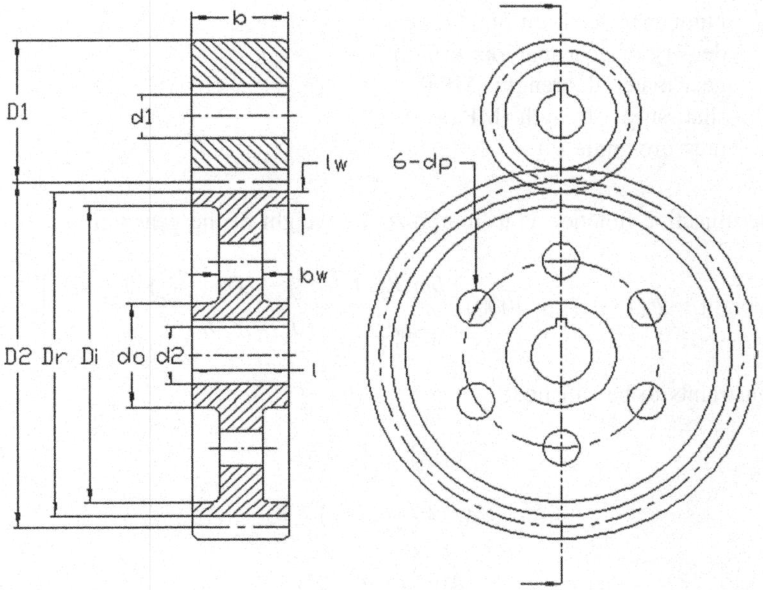

Fig. 6.1 Basic geometry of a single stage spur gear (Yokota et al. 1998; Savsani et al. 2010; Reprinted with permission from Elsevier)

F_s	induced bending load (Lewis formula) (N)
$g_i(x)$	constraints
H	hardness (BHN)
I	geometry factor
J	Lewis geometry factor
K_m	mounting factor
K_{ms}	mean stress factor
K_o	overload factor
K_r	bending reliability factor
K_v	velocity factor
K_w	load factor
l	length of boss (mm)
l_w	thickness of rim (mm)
m	module (mm)
n	number of drilled holes
N_1, N_2	speed of pinion, gear shaft (rpm)
p	power to be transmitted (kW)
S_{fe}	surface fatigue strength (MPa)
S_n	standard moore endurance limit
v	pitch line velocity (m/s)
x	design variable vector
y	Lewis tooth form factor

Z_1, Z_2 number of teeth on pinion, gear
ρ density of gear material (mg/m3)
σ gear material strength (MPa)
τ shaft shear strength (MPa)
φ pressure angle (^{0})

The objective function is to minimize the weight of the gear train.

$$\text{Weight} = F(x) = (\pi\rho/4000)\begin{bmatrix} bm^2Z_1^2(1+a^2) - (D_i^2 - d_o^2)(1-b_w) \\ -nd_p^2b_w - (d_1^2 + d_2^2)b \end{bmatrix} \tag{6.1}$$

Constraints to be satisfied:

$$g_1(x) = F_s \geq b_1 \tag{6.2}$$

$$g_2(x) = (F_s/F_p) \geq b_2 \tag{6.3}$$

$$g_3(x) = d_1^3 \geq b_3 \tag{6.4}$$

$$g_4(x) = d_2^3 \geq b_4 \tag{6.5}$$

$$g_5(x) = (1+a)mZ_1/2 \leq b_5 \tag{6.6}$$

where, $D_r = m(aZ_1 - 2.5)$, $l_w = 2.5m$, $D_i = D_r - 2l_w$, $b_w = 3.5m$, $d_o = d_2 + 25$ $d_p = 0.25(D_i - d_0)$, $D_1 = mZ_1$, $D_2 = mZ_1$, $N_2 = N_1/a$, $Z_2 = Z_1D_2/D_1$, $v = \pi D_1N_1/60{,}000$, $b_1 = 1000\,P/v$, $b_3 = 48.68e6\,P/(N_1\,\tau)$, $b_4 = 48.68e6\,P/(N_2\tau)$, $F_s = \pi K_vK_w\sigma bmy$, $F_p = 2K_vK_wD_1bZ_2/(Z_1 + Z_2)$, a = 4, ρ = 8, P = 7.5, $n = 6$, $\sigma = 294.3$, $y = 0.102$, $b_2 = 0.193$, $\tau = 19.62$, $K_w = 0.8$, $K_v = 0.389$, $g_1(x)$ is for bending strength of tooth, $g_2(x)$ is for surface durability, $g_3(x)$, and $g_4(x)$ are for torsional strength of shafts for pinion and gear, respectively, $g_5(x)$ is for center distance.

Design vector: x = (b, d_1, d_2, Z_1, m)
$20 \leq b \leq 32$
$10 \leq d_1 \leq 30$
$30 \leq d_2 \leq 40$
$18 \leq Z_1 \leq 25$
m = (2.75, 3, 3.5, 4)

Yokata et al. (1998) solved this constrained optimization problem using a genetic algorithm (GA). The above design problem is modified using AGMA standard equations which include many detailed design factors. Value of K_v cannot be constant as it depends on pitch line velocity which is again the function of pitch diameters of pinion/gear. Form factor y depends on the number of teeth and cannot be taken as constant. Moreover, in the above design there is no mention of hardness which plays a very crucial role for the surface fatigue strength. So design is modified considering many additional factors which are practically required for the optimal gear design (Savsani et al. 2010).

Refined design includes six design variables including hardness as an additional design variable and eight constraints which are given below:

Design vector: $x = (b, d_1, d_2, Z_1, m, H)$
$200 \leq H \leq 400$

Constraints to be satisfied:

$$g_1(x) = S_n C_s K_r K_{ms} bJm/K_v K_o K_m \geq b_1 \tag{6.7}$$

$$g_2(x) = S_{fe}^2 C_l^2 C_r^2 bD_1 I/(C_p^2 K_v K_o K_m) \geq b_1 \tag{6.8}$$

$$g_3(x) = \sin^2 \phi D_1(2D_2 + D_1)/(4m) - D_2 - 1 \geq 0 \tag{6.9}$$

$$g_4(x) = b/m \geq 8 \tag{6.10}$$

$$g_5(x) = b/m \leq 16 \tag{6.11}$$

$$g_6(x) = d_1^3 \geq b_3 \tag{6.12}$$

$$g_7(x) = d_2^3 \geq b_4 \tag{6.13}$$

$$g_8(x) = (1 + a)mZ_1/2 \leq b_5 \tag{6.14}$$

where values of J and C_s are determined from Figs. 6.2 and 6.3 respectively. $S_n = 1.7236H$, $kv = (78 + \sqrt{196.85v/78})$, $Sfe = 2.8H - 69$, $Kr = 0.814$, $Kms = 1.4$, $Ko = 1$, $km = 1.3$, $C_p = 191$, $C_1 = 1$,$C_r = 1$, $\phi = 25$. $g_1(x)$ is for bending fatigue strength, $g_2(x)$ is for surface fatigue strength, $g_3(x)$ is for the check for the interference, $g_4(x)$ and $g_5(x)$ are to ensure uniform load distribution, $g_6(x)$ and $g_7(x)$ are for torsional strength of shafts for pinion and gear, respectively, and $g_8(x)$ is the center distance.

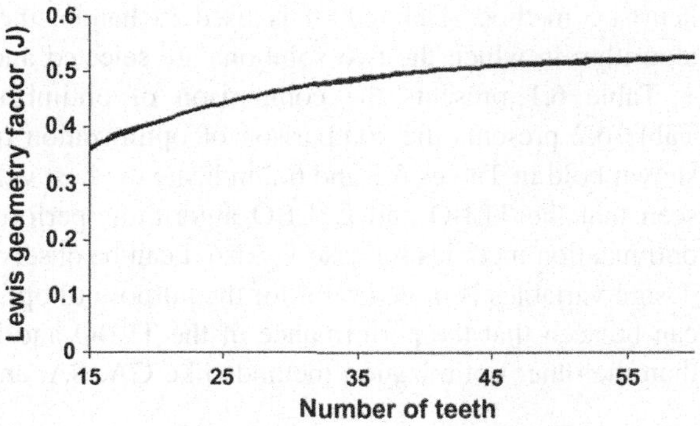

Fig. 6.2 Third order polynomial fit for Lewis geometry factor (Savsani et al. 2010; Reprinted with permission from Elsevier)

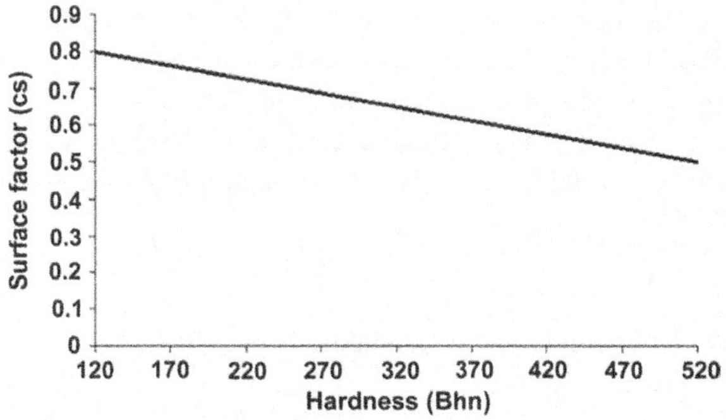

Fig. 6.3 Straight line fit for surface factor (Savsani et al. 2010; Reprinted with permission from Elsevier)

Design vector: $x = (b, d_1, d_2, Z_1, m)$
$10 \leq b \leq 35$
$10 \leq d_1 \leq 30$
$10 \leq d_2 \leq 40$
$18 \leq Z_1 \leq 25$
$m = (1, 1.25, 1.5, 2, 2.75, 3, 3.5, 4)$

6.3 Results and Discussion

To check the effectiveness of the TLBO and ETLBO algorithms, extensive experiments are conducted on gear train from the literature and the results are compared with the other optimization algorithms. Population size of 20 and maximum number of generations of 25 are considered for case 1 while the population size of 20 and maximum number of generations of 150 are considered for case 2. The TLBO algorithm is experimented with different elite sizes, viz. 0, 4, 8, 12, and 16 for the design of gear train. An elite size of 8 is considered after making few trials. Like other optimization algorithms (e.g., PSO, ABC, ACO, etc.), TLBO algorithm also has no special mechanism to handle the constraints. So, for the constrained optimization problems it is necessary to incorporate any constraint handling techniques with the TLBO algorithm. In the present experiments, Deb's heuristic method (Deb 2000) is used to handle the constraints with the TLBO algorithm in which the two solutions are selected and compared with each other.

Table 6.1 presents the comparison of optimization results for case 1 and Table 6.2 presents the comparison of optimization results for case 2. The results shown bold in Tables 6.1 and 6.2 indicate the best values. From Table 6.1 it can be seen that the TLBO and ETLBO algorithms perform well as compared to other optimization methods for case 1. Also it can be observed that the expanded range of design variables is more useful for the purpose of optimal results. From Table 6.2 it can be seen that the performance of the TLBO and ETLBO algorithms are better than the other optimization methods like GA, SA, and PSO algorithms for case 2.

Table 6.1 Comparison of optimization results for case-1

Design variables	GA[a] (Yokata et al. 1998)	SA[a] (Savsani et al. 2010)	PSO[a] (Savsani et al. 2010)	SA[b] (Savsani et al. 2010)	PSO[b] (Savsani et al. 2010)	TLBO[a]	TLBO[b]	ETLBO[a]	ETLBO[b]
Weight (g)	3512.6	3127.71	3127.70	3094.61	3094.60	**3036.45**	**2998.83**	**3029.58**	**2991.41**
b (mm)	24	23.7	23.7	32	32	32	32	32	32
d_1 (mm)	30	30	30	30	36.759	30	30	30	30
d_2 (mm)	30	36.761	36.763	36.756	25	37.290	37.566	37.388	37.748
Z_1	18	18	18	25	2	24.98	25	25	25
m (mm)	2.75	2.75	2.75	2	1.4	3	3	3	3
Active constraints	–	1,4	1,4	1,4	1,4	1,4	1,4	1,4	1,4
Function evaluations	20,000	2200	1000	2200	1000	1000	1000	1000	1000

[a]For the ranges of design variables considered by Yokota et al. (1998)
[b]For the expanded ranges of design variables

Table 6.2 Comparison of optimization results for case 2

Design variables	GA[a] (Yokata et al. 1998)	SA[a] (Savsani et al. 2010)	PSO[a] (Savsani et al. 2010)	GA[b] (Savsani et al. 2010)	SA[b] (Savsani et al. 2010)	PSO[b] (Savsani et al. 2010)	TLBO[a]	TLBO[b]	ETLBO[a]	ETLBO[b]
Weight (g)	2993.70	2993.56	2993.56	1664.30	1661.11	1661.10	**1655.34**	**1649.31**	**1651.57**	**1647.78**
b (mm)	21.999	21.997	21.999	26.87	26.74	26.73	29.67	30.58	30.94	31.47
d_1 (mm)	30	30	30	30	30	30	30	30	30	30
d_2 (mm)	36.751	36.742	36.768	36.75	36.743	36.74	37.42	37.54	38.29	38.18
Z_1	18	18	18	18	18	18	18	18	18	18
m (mm)	2.75	2.75	2.75	2	2	2	2	2	2	2
H (BHN)	341.46	350	338	400	400	400	400	400	400	400
Active constraint	4.7	4.7	4.7	2.7	2.7	2.7	2.7	2.7	2.7	2.7
Function evaluation	6000	3300	3000	6000	3300	3000	2500	2500	2500	2500

[a]For the ranges of design variables considered by Yokota et al. (1998)
[b]For the expanded ranges of design variables

Also it can be observed that the expanded range of design variables is more useful for the purpose of optimal results.

The computational results show that the TLBO and ETLBO algorithms have obtained more accurate solutions than those obtained by using the other optimization methods.

The next chapter presents the optimum design of a plate fin heat sink.

References

Abuid, B.A., Ammen, Y.M., 2003. Procedure for optimum design of a two stage spur gear system, JSME International Journal 46 (4), 1582–1590.

Deb, K., 2000. An efficient constraint handling method for genetic algorithm, Computer Methods in Applied Mechanics and Engineering 186, 311–338.

Deb, K., Jain, S., 2003. Multi-speed gearbox design using multi-objective evolutionary algorithms, ASME Journal of Mechanical Design 125, 609–619.

Huang, H.Z., Tian, Z.G., Zuo, M.J., 2005. Multiobjective optimization of three-stage spur gear trains using interactive physical programming, Journal of Mechanical Science and Technology 19(5) 1080–1086.

Jain, P., Agogino, A.M., 1990. Theory of design: an optimization perspective, Journal of Mechanism and Machine Theory 25 (3), 287–303.

Kader, M.M.A., Nigam, S.P., Grover, G.K., 1998. A study on mode of failures in spur gears under optimized conditions. Mechanism and Machine Theory 33(6), 839–850.

Khorshid, E., Seireg, A., 1999. Discrete nonlinear optimisation by constraint decomposition and designer interaction, International Journal of Computer Applications in Technology 12, 76–82.

Li, X., Symmons, G.R., Cockerhan, G., 1996. Optimal design of involute profile helical gears. Mechanism and Machine Theory 31(6), 717–728.

Madhusudan, G., Vijayasimha, C.R., 1987. Approach to spur gear design, Computer-Aided Design 19 (10), 555–559.

Osyczka, A., 1978. An approach to multicriterion optimization problems for engineering design, Computer Methods in Applied Mechanics and Engineering 15, 309–333.

Pomrehn, L.P., Papalambros, P.Y., 1995a. Infeasibility and non-optimality tests for solution space reduction in discrete optimal design, ASME Journal of Mechanical Design 117, 425–432.

Pomrehn, L.P., Papalambros, P.Y., 1995b. Discrete optimal design formulations with application to gear train design, ASME Journal of Mechanical Design 117, 419–424.

Prayoonrat, S., Walton, D., 1988. Practical approach to optimum gear train design, Computer-Aided Design 20 (2), 83–92.

Rao, R.V. and Savsani, V.J., 2012. Mechanical design optimization using advanced optimization techniques. London:Springer-Verlag.

Savsani, V.J., Rao, R.V., Vakharia, D.P., 2010. Optimal weight design of a gear train using particle swarm optimization and simulated annealing algorithms. Mechanism and Machine Theory 45, 531–541.

Thompson, D.F., Gupta, S., Shukla, A., 2000. Tradeoff analysis in minimum volume design of multi-stage spur gear trains, Mechanism and Machine Theory 35, 609–627.

Tong, B.S., Walton, D., 1987. The optimization of internal gears. International Journal of Machine Tools and Manufacture 27(4), 491–504.

Yokota, T., Taguchi, T., Gen, M., 1998. A solution method for optimal weight design problem of the gear using genetic algorithms, Computers & Industrial Engineering 35, 523–526.

Zarefar, H., Muthukrishnan, S.N., 1993. Computer-aided optimal design via modified adaptive random-search algorithm, Computer-Aided Design 40, 240–248.

Chapter 7
Design Optimization of a Plate Fin Heat Sink Using TLBO and ETLBO Algorithms

Abstract This chapter presents the multiobjective design optimization of a plate fin heat sink equipped with flow-through and impingement-flow air cooling systems using TLBO and ETLBO algorithms. Two objective functions known as entropy generation rate and material cost with five constraints have been taken to measure the performance of the heat sink. Number of fins, height of fins, spacing between the two fins, and oncoming air velocity are considered as the design variables. The results show the better or competitive performance of the TLBO and ETLBO algorithms over the other optimization algorithms considered.

7.1 Design Optimization of Plate Fin Heat Sink

Over the past few decades, the interest of researchers is growing in the field of thermal system optimization in order to improve the thermal performance using advanced optimization techniques. In the present work, thermal optimization of plate fin heat sink is considered. A heat sink is a passive heat exchanger component that cools a device by dissipating heat into the surrounding air. The main objective of designing of heat sink is to increase the surface area in contact with the cooling medium surrounding it. Thus heat sink is a protective device that absorbs and dissipates the excess heat generated by a system. The thermal performance of the heat sink is affected by factors such as approach air velocity, fin design, choice of material, surface treatment, etc. To determine the thermal performance of a heat sink, theoretical, experimental, and numerical methods can be used. In heat sink design, majorly two heat dissipation factors such as heat transfer rate and air resistance are considered. The minimization of entropy generation rate considers the above factors and maximizes the thermal efficiency, surface area, convective coefficients, and able to find optimal flow velocity and viscous dissipation. However, it leads to minimum total heat resistance and larger heat transfer area and requires more fin numbers and larger size of heat sink. Requirement of larger size heat sink leads to more space requirement and wastage of material. Hence,

R.V. Rao, *Teaching Learning Based Optimization Algorithm*,
DOI 10.1007/978-3-319-22732-0_7

in general, simultaneous minimization of the entropy generation rate and material cost of the heat sink is considered as a multiobjective optimization of design of plate fin heat sink (Chen and Chen 2013).

Rao and Waghmare (2015) presented the results of optimization the design of a plate fin heat sink using TLBO algorithm. In this chapter, the results of TLBO (Rao and Waghmar 2015) and the ETLBO algorithm are presented for the design optimization of plate fin heat sink. The nomenclature of the plate fin heat sink is given below.

A	Total surface area of heat sink including the fins and exposed other surface, m^2
A_c	Cross-sectional area for heat conduction of each fin, m^2
b	Spacing between two fins, m
C_p	Heat capacity, J/(Kg K)
D_h	Hydraulic diameter of channel
F_d	Drag force, N
$f\mathrm{Re}_{\mathrm{Dh}}$	Reynolds number group
f_{aap}	The apparent friction factor of a hydrodynamically developing flow
H	Height of the fins, m
h_{eff}	Heat transfer coefficient, (W/m^{2}K)
K	Temperature unit in Kelvin scale
K_{sc}	Coefficient of sudden contraction
K_{se}	Coefficient of sudden expansion
$\underline{k}$	Coefficient of heat conduction, W/(mk)
L	Length of fins, m
N_{ub}	Nusselt number
N	Number of fins
n	Normal vector of a surface
P	Surface area per unit length of fins, m
P_r	Prandtl number
Q	Volume heat source, (W/m^3)
$\underline{Q}$	Heat generation rate, W
R	Thermal resistance, K/W
Re_b	Reynolds number
R_{fin}	Thermal resistance of each fin, K/W
R_{sink}	Overall heat sink thermal resistance, K/W
$\dot{S}_{\mathrm{gen}}$	Entropy generation rate, W/K
T_{amb}	Surrounding temperature, °C
t	Time, s
t_b	Base length of fin, m
t_w	Width of each fin, m
V_{ch}	Channel air velocity, m/s
V_f	Oncoming air velocity, m/s

W	Watt, J/s
w	Width of plate fin, m
η_{fin}	Fin efficiency

In this section, the ability of TLBO and ETLBO algorithms is demonstrated by implementing it on the heat sink optimization problems equipped with flow-through and impingement-flow air cooling system from the literature. Figure 7.1 presents a plate fin heat sink and Table 7.1 represents the operating conditions and physical parameters of the heat sink (Chen and Chen 2013). Assumptions considered while operating the heat sink with impingement-flow and flow-through air cooling systems are: no spreading or constriction resistance; no contact resistance between the mounted heat sink and the base device; uniform and constant heat convection coefficients; no bypassing air flow; adiabatic condition on the fin tip; uniform

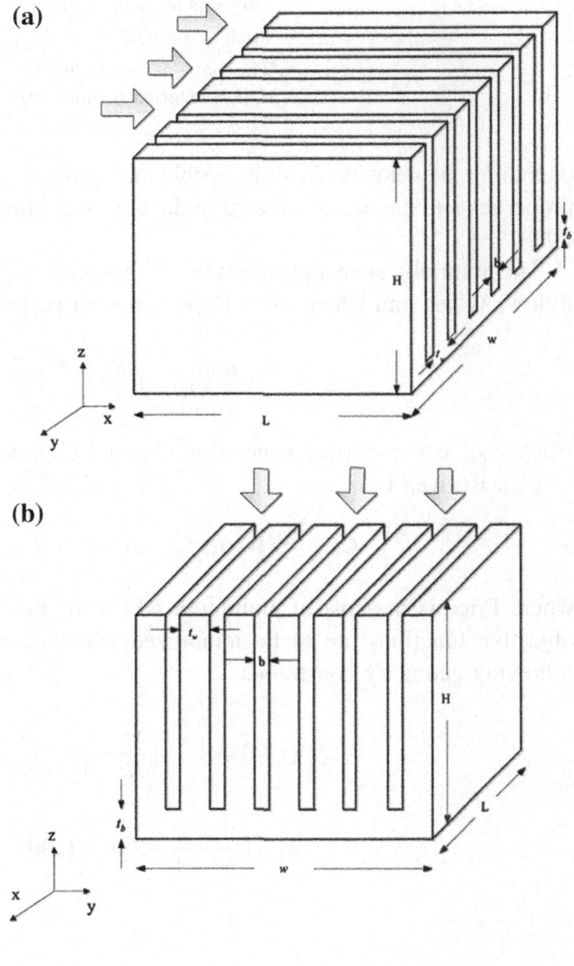

Fig. 7.1 **a** A plate fin heat sink with flow-through air cooling system, **b** A plate fin heat sink with impingement-flow air cooling system (Chen and Chen 2013; Reprinted with permission from Elsevier)

Table 7.1 Operating conditions and physical parameters of the heat sinks (Chen and Chen 2013; Reprinted with permission from Elsevier)

Parameters or conditions	Unit	Value
Base plate area	mm^2	50 × 50
Base plate thickness	Mm	2
Thermal conductivity of solid	W/(mK)	200
Thermal conductivity of device	W/(mK)	163
Thermal conductivity of air	W/(mK)	0.0267
Density of solid (fin)	Kg/m^3	2707
Density of device (heat source)	Kg/m^3	2330
Density of air	Kg/m^3	1.177
Heat capacity of solid	J/(kg K)	905
Heat capacity of device	J/(kg K)	705.5
Heat capacity of air	J/(kg K)	1007
Kinematic viscosity	m^2/s	1.6×10^{-5}
Prandtl number	–	0.703
Heat load	W	30
Surrounding temperature	K	298
Price of aluminum[a]	NTD/kg	65

[a]The price is according to the London Metal Trading Stock Market reported in June 2010

oncoming air velocity; incompressible and laminar air flow and constant physical properties and the heat generated in the device is uniformly distributed (Chen et al. 2008).

The multiobjective optimization of the plate fin heat sinks is formulated as follows (Chen and Chen 2013; Reprinted with permission from Elsevier):

$$\min_{N, H, b, V_f} = \left\{\dot{S}_{\text{gen}},\ C_{\text{mat}}\right\} \tag{7.1}$$

where $\dot{S}_{\text{gen}}$ is the entropy generation rate and C_{mat} is the cost of the heat sink.

C_{mat} is given by,

$$C_{\text{mat}} = (W.\ L.\ t_{\text{bp}} + N.\ H.\ b.\ L)\rho_m \cdot \text{Price} \tag{7.2}$$

where Price is the cost of aluminum which the heat sink is made of. These two objective functions are to be minimized simultaneously while subjecting to the following geometry constraints.

$$g_1 : 0.001 - \left(\frac{w-b}{N-1} - b\right) \leq 0, \tag{7.3}$$

$$g_2 : \left(\frac{w-b}{N-1} - b\right) - 0.005 \leq 0, \tag{7.4}$$

$$g_3 : 0.001 - \frac{H}{(((w-b)/(N-1)) - b)} \leq 0, \tag{7.5}$$

$$g_4 : \frac{H}{(((w-b)/(N-1)) - b)} - 19.4 \leq 0, \tag{7.6}$$

$$g_5 : 0.0001 - \sqrt{\frac{(((w-b)/(N-1)) - b) \times V_{ch}}{\nu} \times \frac{(((w-b)/(N-1)) - b)}{L}} \leq 0 \tag{7.7}$$

In the above constraints, g_1 and g_2 indicate that the fin gap should lie in the range between 0.001 and 0.005 m. The constraints g_3 and g_4 are used the design specification for installation. These two constraints indicate that ratio of the height and thickness of the fins should satisfy the specification between 0.01 and 19.4 due to limited space for installation. The constraint g_5 is added to avoid getting a zero Reynolds number. In addition to the above constraints, the design parameters are confined within the following admissible regions:

$$2 \leq N \leq 40, \tag{7.8}$$

$$0.025 \text{ m} \leq H \leq 0.14 \text{ m}, \tag{7.9}$$

$$2 \text{ x } 10^{-4} \text{ m} \leq b \leq 2.5 \text{ x } 10^{-4} \text{ m}, \tag{7.10}$$

$$0.5 \text{ m/s} \leq V_f \leq 2 \text{ m/s}, \tag{7.11}$$

$$N \times b \leq 0.05 \text{ m}. \tag{7.12}$$

To perform the thermal analysis for the plate-fin heat sinks under forced heat convection, the entropy generation rate can be defined as follows (Bejan 1996):

$$\dot{S}_{gen} = \left(\frac{\dot{Q}}{T_{amb}}\right)^2 R_{sink} + \frac{F_d V_f}{T_{amb}} \tag{7.13}$$

where $\dot{Q}$ denotes the heat generation rate, T_{amb} is the surrounding temperature, F_d and V_f are the air resistance between fins and the oncoming air velocity, respectively. The entropy generation rate is a function of the total heat resistance R_{sink} and head loss. The total heat resistance for plate-fin heat sinks equipped with flow-through or impingement-flow air cooling systems is given as (Kay and London 1984),

$$R_{sink} = \begin{cases} \frac{1}{\left(N/R_{fin}\right) + h_{eff}(N-1)bL} + \frac{t_b}{kLw} & \text{for flow—through air inlet} \\ \frac{1}{h_{eff} A \eta_{fin}} & \text{for impingment—flow air inlet} \end{cases} \tag{7.14}$$

where N is the fin number, A denotes the total surface area heat sink including the fins and exposed other surface, η_{fin} represents the total heat dissipation efficiency. η_{fin} can be defined as follows (Bejan and Morega 1993):

$$\eta_{\text{fin}} = \frac{\tanh(\text{mH})}{\text{mH}} \tag{7.15}$$

The heat resistance in associated with flow-through air cooling system is calculated as (Kay and London 1984),

$$R_{\text{fin}} = \frac{1}{\sqrt{h_{\text{eff}} PkA_c}\,\tanh(\text{mH})} \tag{7.16}$$

$$m = \sqrt{h_{\text{eff}} PkA_c} \tag{7.17}$$

where P and A_c be the perimeter and cross-sectional area of each fin, respectively. By considering the force balance on the heat sink, the total drag force between fins can be determined as (Kay and London 1984),

$$\frac{Fd}{(1/2)\rho V_{\text{ch}}^2} = f_{\text{app}} N(2HL + bL) + K_c(Hw) + K_e(Hw) \tag{7.18}$$

where f_{app} is the friction coefficient. The air velocity V_{ch} in the channel for flow-through and impingement-flow air cooling systems can be given as (Muzychka and Yovanovich 1998),

$$V_{\text{ch}} = \begin{cases} V_f\left(1 + \frac{t_w}{b}\right), & \text{for flow—through air inlet} \\ \frac{V_f L}{2H} & \text{for impingment—flow air inlet} \end{cases} \tag{7.19}$$

The relation between the friction coefficient and the Reynolds number of the rectangular channel is formulated as (Muzychka and Yovanovich 1998),

$$f_{\text{app}}\,\text{Re}_{\text{Dh}} = \left[\left(\frac{3.44}{\sqrt{L}}\right)^2 + (f\,\text{Re}_{\text{Dh}})^2\right] \tag{7.20}$$

where $L^* = L/(D_h R_{eDh})$ and D_h is the hydraulic diameter of the channel. The Reynolds number group $f\text{Re}_{\text{Dh}}$ representing the friction factor of fully developed flow is given as (Muzychka and Yovanovich 1998),

$$f\,\text{Re}_{\text{Dh}} = 24 - 32.527\left(\frac{b}{H}\right) + 46.721\left(\frac{b}{H}\right)^2 - 40.829\left(\frac{b}{H}\right)^3 + 22.954\left(\frac{b}{H}\right)^4 - 6.089\left(\frac{b}{H}\right)^5 \tag{7.21}$$

K_c and K_e are the sudden contraction and expansion coefficients, respectively. K_c and K_e can be calculated using following relations (White 1987).

$$K_c = 0.42\,(1 - \sigma^2) \tag{7.22}$$

$$K_e = (1 - \sigma^2) \tag{7.23}$$

where

$$\sigma = 1 - (\eta t_w / W) \tag{7.24}$$

The convective heat transfer coefficient can be computed as (Teertstra et al. 1999),

$$N_{\mathrm{ub}} = \left[\left(\frac{\mathrm{Re}_b^* P_r}{2} \right)^{-3} + \left(0.664 \sqrt{\mathrm{Re}_b^*} P_r^{1/3} \sqrt{1 + \frac{3.65}{\sqrt{\mathrm{Re}_b^*}}} \right)^{-3} \right]^{-1/3} \tag{7.25}$$

where P_r is the Prandtl number.

$$N_{\mathrm{ub}} = {h_{\mathrm{eff}} b}/{k_f} \tag{7.26}$$

$$\mathrm{Re}_b^* = \mathrm{Re}_b \left({b}/{L} \right) \tag{7.27}$$

$$\mathrm{Re}_b = b\, {V_{\mathrm{ch}}}/{\nu} \tag{7.28}$$

The increments of the fin number and height of the fin have the effects of decreasing the entropy generation rate, but at the same time they increase the material cost. Hence the two objectives, entropy generation rate and material cost are in conflict with each other and thus need to be optimized simultaneously.

In this case of multiobjective design optimization of plate fin heat sink using the priori approach, the equal weights have assigned to the objective functions. The combined objective function in case of two objective functions is formulated by giving equal weightages to the objective functions as:

$$\text{Minimize}\, z = -0.5\,(f_1/f_{1\min}) + 0.5(f_2/f_{2\min}) \tag{7.29}$$

7.2 Results and Discussion

To check the effectiveness of the TLBO and ETLBO algorithms, extensive computational experiments are conducted on heat sink problems equipped with flow-through and impingements-flow air cooling system from the literature and the results are compared with those given by the other optimization algorithms.

Population size of 200 and maximum number of generations of 50 are considered. The TLBO algorithm is experimented with different elite sizes, viz. 0, 4, 8, 12, and 16 on heat sink. After making few trials, elite size of 4 is considered. Also, it is observed that elite size of 0, 4, 8, 12, and 16 have given the same global optimum solutions. In the present experiments, Deb's heuristic constrained handling method (Deb 2000) is used to handle the constraints with the TLBO algorithm. A priori approach is used for the multiobjective optimization by forming a combined objective function and giving equal weightages to the objective functions.

Table 7.1 represents operating conditions and physical parameters of the heat sinks. Table 7.2 presents the comparison for optimized heat sinks with flow-through air inlet system. The results shown bold in Table 7.2 indicate the best values. From Table 7.2 it can be seen that the entropy generation rate of 0.00283, 0.00283, 0.00288, 0.00288, 0.00297, and 0.00290 W/K is obtained using ETLBO algorithm, TLBO algorithm, DBGA (Chen and Chen 2013; Chen et al. 2008; Shih and Liu 2004; Culham and Muzychka 2001), respectively. The objective is to minimize the entropy generation rate and hence it can be concluded that ETLBO and TLBO algorithms achieved the optimum entropy generation rate at $n = 17$, $H = 0.125$ m, $b = 0.0017$ m, and $V_f = 0.72$ m/s. Similarly from Table 7.2 it can be seen that the cost obtained is 31.37 NTD, 31.37 NTD, 32.48 NTD, 33.63 NTD, 38.84 NTD, and 33.62 NTD (the price is according to the London Metal Trading Stock Market reported in June 2010, 1 USD (US dollar) = 30 NTD) using ETLBO algorithm, TLBO algorithm, DBGA (Chen and Chen 2013; Chen et al. 2008; Shih and Liu 2004; Culham and Muzychka 2001), respectively. Minimization of the cost is considered as an objective and hence it can be concluded that ETLBO and TLBO algorithms achieved the optimum cost at $n = 17$, $H = 0.125$ m, $b = 0.0017$ m, and $V_f = 0.72$ m/s. It can also be observed that the cost is improved by 3.53, 7.17, 23.81, and 7.20 % using ETLBO and TLBO algorithms as compared to DBGA (Chen and Chen 2013; Chen et al. 2008; Shih and Liu 2004; Culham and Muzychka 2001), respectively.

From Table 7.2 it can be seen that the ETLBO and TLBO algorithms have outperformed the other algorithms in optimizing the heat sink problem with flow-through air inlet system in terms of entropy generation as well as cost of the material. Table 7.3 shows the comparison for optimized heat sinks with

Table 7.2 Comparison of results of different algorithms for optimum heat sinks with flow-through air inlet system

Methods	N	H (m)	b (m)	V_f (m/s)	$\dot{S}_{gen}$ (W/K)	Cost (NTD)
TLBO (Rao and Waghmare 2015)	17	0.125	0.0017	0.72	**0.00283**	**31.37**
ETLBO	17	0.125	0.0017	0.72	**0.00283**	**31.37**
DBGA (Chen and Chen 2013)	19	0.120	0.0016	1.26	0.00288	32.48
Chen et al. (2008)	18	0.123	0.0017	1.21	0.00288	33.63
Shih and Liu (2004)	20	0.134	0.0016	1.05	0.00297	38.84
Culham and Muzychka (2001)	19	0.122	0.0016	1.21	0.00290	33.62

Table 7.3 Comparison of results of different algorithms for optimum heat sinks with impingement-flow air inlet system

Methods	n	H (m)	b (m)	V_f (m/s)	$\dot{S}_{gen}$ (W/K)	Cost (NTD)
TLBO (Rao and Waghmare 2015)	27	0.025	0.0002	0.5	**0.00546**	**1.344**
ETLBO	27	0.025	0.0002	0.5	**0.00546**	**1.344**
DBGA (Chen and Chen 2013)	40	0.025	0.0002	2	0.00547	2.639
Chen et al. (2008)	38	0.025	0.0002	2	0.00548	2.551
Shih and Liu (2004)	27	0.00392	0.0005	2	0.00577	1.345

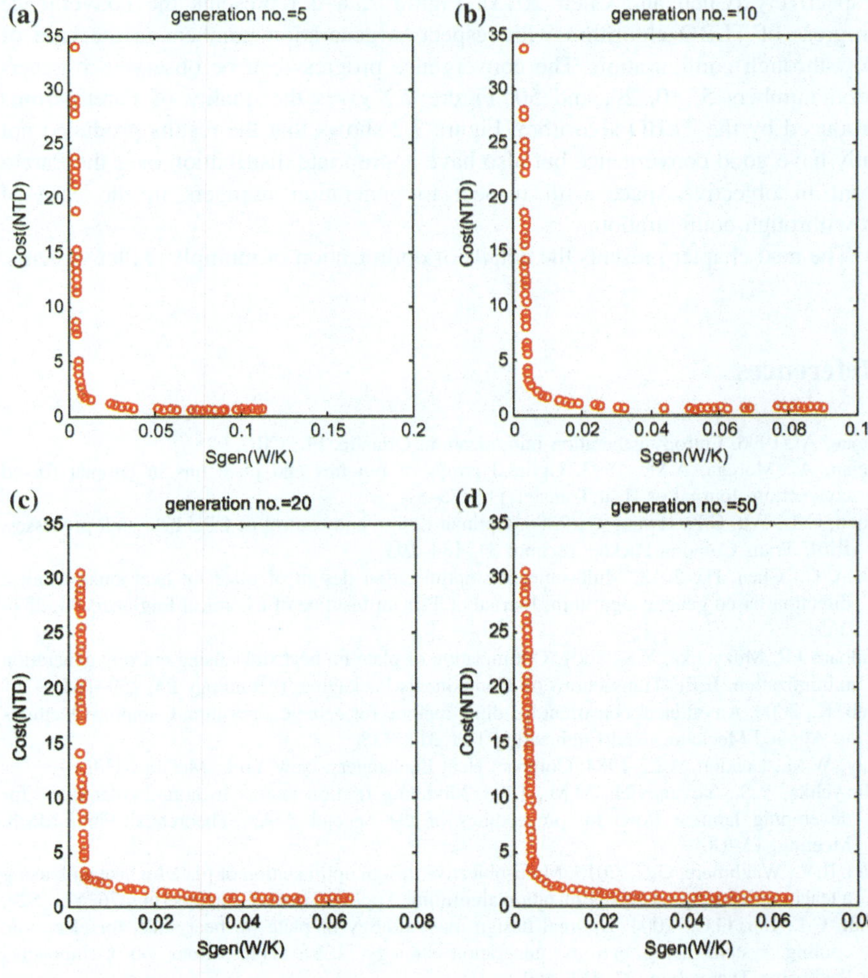

Fig. 7.2 a–d The convergence progress with respect to generation numbers in the case of flow-through configuration (Rao and Waghmare 2015; Reprinted with permission from Elsevier)

impingement-flow air inlet system and it can be seen that the entropy generation rate of 0.00546, 0.00546, 0.00547, 0.00548 and 0.00577 W/K is obtained using ETLBO algorithm, TLBO algorithm, and DBGA (Chen and Chen 2013; Chen et al. 2008; Shih and Liu 2004), respectively. The objective is to minimize the entropy generation rate and hence it can be concluded that ETLBO and TLBO algorithms achieved the optimum entropy generation rate at n = 27, H = 0.025 m, b = 0.0002 m, and V_f = 0.5 m/s. Similarly, it can be seen that the cost obtained is 1.344 NTD, 1.344 NTD, 2.639 NTD, 2.551 NTD, and 1.345 NTD using ETLBO algorithm, TLBO algorithm, and DBGA (Chen and Chen 2013; Chen et al. 2008; Shih and Liu 2004), respectively.

Figures 7.1a and b show a plate fin heat sink with flow-through air cooling system and plate fin heat sink with impingement-flow air cooling system, respectively (Chen and Chen 2013). Figure 7.2a–d represents the convergence progress of TLBO algorithm with respect to generation numbers in the case of flow-through configuration. The convergence progress can be observed at generation numbers 5, 10, 20, and 50. Figure 7.2 gives the quality of Pareto fronts produced by the TLBO algorithm. Figure 7.2 shows that the results produced not only have good convergence but also have appropriate distribution over the Pareto front in objective space with respect to generation numbers in the case of flow-through configuration.

The next chapter presents the details of optimization of multiple chiller systems.

References

Bejan, A., 1996. Entropy generation minimization. Orlando, FL: CRC.

Bejan, A., Morega, A.M., 1993. Optimal arrays of pin fins and plate fins in laminar forced convection. Journal of Heat Transfer 115, 75–81.

Chen, C.T., Wu, C.K, Hwang C., 2008. Optimal design and control of CPU heat sink processes. IEEE Trans Compon Packag Technol 31, 184–195.

Chen, C., Chen, H., 2013. Multi-objective optimization design of plate fin heat sinks using a direction-based genetic algorithm. Journal of Taiwan Institute of Chemical Engineers 44, 257–265.

Culham J.R, Muzychka, Y.S., 2001. Optimization of plate fin heat sinks using entropy generation minimization. IEEE Transactions on Components Packaging Technology 24, 159–165.

Deb, K., 2000. An efficient constraint handling method for genetic algorithm, Computer Methods in Applied Mechanics and Engineering 186, 311–338.

Kay, W.M., London, A.L., 1984. Compact Heat Exchangers. New York: McGraw-Hill.

Muzychka, Y.S., Yovanovich, M.M., 1998. Modeling friction factors in non-circular ducts for developing laminar flow. In: proceedings of the Second AIAA Theoretical Fluid Mech, Meeting, 15–18.

Rao, R.V., Waghmare, G.G., 2015. Multi-objective design optimization of plate fin heat sink using a teaching-learning-based optimization algorithm. Applied Thermal Engineering, 76, 521–529.

Shih, C.J., Liu, G.C., 2004. Optimal design methodology of plate fin heat sinks for electronic cooling system using entropy generation strategy. IEEE Transactions on Components Packaging Technology 27, 551–569.

Teertstra P.M., Yovanovich, M.M., Culham, J.R., Lemczyk, T.F., 1999. Analytical forced convection modeling of plate fin heat sinks. Advances in Electronics Cooling, Journal of Electronics Manufacturing, Vol. 10, No. 4, 2002, pp. 253–261.

White, F.M., 1987. Fluid Mechanics. New York: McGraw-Hill.

Chapter 8
Optimization of Multiple Chiller Systems Using TLBO Algorithm

Abstract Chillers are used in many buildings to provide cooling facility to the building environment. The power consumption is one of the most significant factors that decides the effective maintenance cost of a building and emphasis is given to minimize the power consumption of a multiple chiller system, which is used for cooling the entire building system and simultaneously maintaining the cooling load requirement of the building system. The TLBO algorithm is employed for this study. In order to test the performance of the TLBO algorithm, three case studies are adopted and solved and the results are compared with the results of the previous researchers. The results of the three case studies show the performance supremacy of the TLBO algorithm in terms of power consumption for multiple chiller systems.

8.1 Optimization of Multiple Chiller Systems

A chiller is a machine that removes heat from a liquid via a vapor-compression or absorption refrigeration cycle. This liquid can then be circulated through a heat exchanger to cool air or equipment as required. As a necessary by-product, refrigeration creates waste heat that must be exhausted to ambient or, for greater efficiency, recovered for heating purposes. Many central cooling systems in air-conditioned buildings have multiple chillers operating in parallel to meet various cooling load requirements. The energy performance or coefficient of performance (COP) of chiller systems depends on the heat rejection medium, ambient conditions, compressor efficiency, and perhaps more importantly the load carried by each operating chiller. Considering that the building of cooling load changes hourly from time to time, chillers need to operate at part load with a reduced COP for most of the time. It is worth considering how the building of cooling load should be allocated to individual chillers operating in order to optimize their aggregate COP. Given that most chillers operate with maximum COP at full load, chiller sequencing is a direct approach to enhancing the system performance.

R.V. Rao, *Teaching Learning Based Optimization Algorithm*,
DOI 10.1007/978-3-319-22732-0_8

Under the arrangement of chiller sequencing, all chillers are operating at the same part load ratio, and no additional chillers start to operate until each of the operating chillers is running at full load. In a conventional pumping system, different numbers of chillers and constant-speed pumps operate in pairs to provide aggregate flow of chilled water which satisfies various building cooling loads. Since the flow of chilled water passing through individual chillers is fixed, the load which each of the chillers carries is related directly to the temperature rise of the chilled water. Given the same temperature rise of the chilled water across all the chillers, they have to operate at the same part load ratio, regardless of whether the chillers are of equal size or different sizes. Such an even load sharing strategy has long been used in multiple chiller systems (Chang et al. 2005).

A multiple chiller system consists of two or more chillers connected by parallel or series piping to a distribution system. Using multiple chillers offers some privileges such as operational flexibility, standby capacity, decreasing in-rush current at system start-up, less disruptive maintenance and reduction in power costs at partial load condition. Also, it gives an opportunity to use chillers at their most efficient point. The standby capacity offered by multiple chiller system can be used if repair is needed. During repair or maintenance of one chiller, cooling load can be provided by the remaining units Ardakani et al. (2008).

Water flow should remain constant for stable control. Because the load variations are a function of variations in temperature, it can be sensed by temperature controllers easily. If the water flow changes, the load varies and there is no flow control. This flaw may disturb stable control system and should be eliminated by decoupled system. A decoupled system separates distribution pumping from the production pumping. Although it allows variable flow through the cooling coils, it maintains constant water flow through chillers, allowing good control of multiple chillers. Figure 8.1 shows a typical decoupled system (Ardakani et al. 2008).

Each combination of chiller and pump (as can be seen in Fig. 8.1), operates independently from the other chillers. In this scheme, instead of using temperature as an indicator for variations in demand, relative flow is used. If an increase in demand occurs, greater flow is demanded than that supplied by chiller-pumps. This flow returns water through the bypass to supply header and indicates a need for additional chiller capacity and another chiller pump starts. If water flows on the opposite direction through bypass pipe, it indicates overcapacity and the chiller-pumps turn off. With respect to mentioned privileges for decoupled system, it is widely used in many air-conditioning systems. Chilled water system is used to remove cooling load of an air-conditioned space. The amount of cooling load depends upon temperature of supply chilled water, temperature of returned chilled water, specific heat of chilled water and the flow rate of chilled water. It can be stated as follows:

$$\mathrm{CL} = \dot{m}c(T_{\mathrm{Cr}} - T_{\mathrm{Cs}}) \tag{8.1}$$

where, $\dot{m}$ is the flow rate of chilled water (in kg/s), c is the specific heat of chilled water above freezing (in J/kg K), T_{Cr} is the temperature of returned chilled water (in

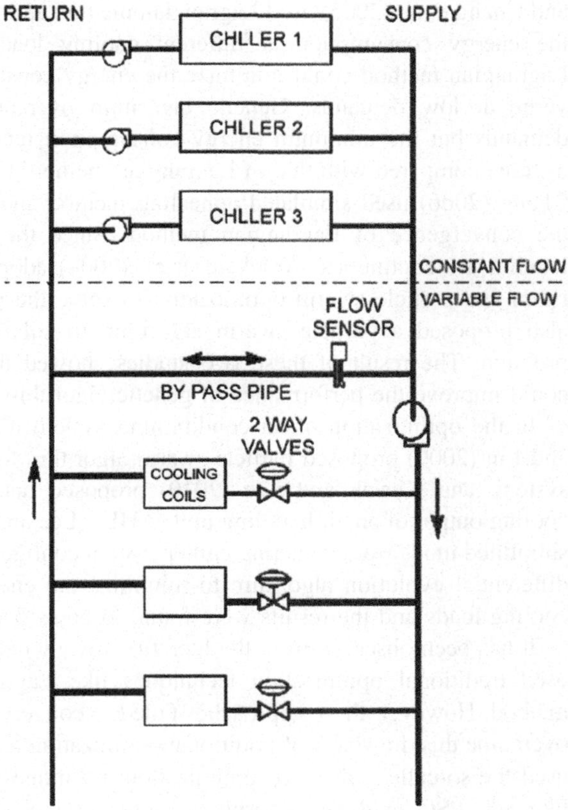

Fig. 8.1 A decoupled system of multi-chiller system (Ardakani et al. 2008; Reprinted with permission from Elsevier)

K), T_{Cs} is the temperature of supplied chilled water (in K), and CL is the cooling load (in kW/s) that must be removed by chilled water system. As mentioned before, in decoupled chiller systems the water flow in a chiller is constant (Ardakani et al. 2008). Instead of operating chillers at the same part load ratio, it is desirable to identify load sharing strategies, properly accounting for their part load performance characteristics which maximize the COP of the entire system. This results in optimization of multiple chiller systems.

In many large buildings, cooling of the building environment is provided with the help of chillers. If the cooling demand is more, multiple chillers are used. The maintenance cost of the building hugely depends on the power consumption of the chiller system. In summer the power consumption is huge, especially in the sub-tropical regions causing peak load in electricity. The multi-chiller system consists of chillers with varying performance characteristics and capacities. If the cooling demand is distributed to the chillers through equal part load ratios, the power consumption will be increased. This leads to selection of optimum chiller loading. Therefore, optimum operation of chiller system is required to minimize the power consumption simultaneously maintaining the desired cooling loads. Chang (2004) and Chang et al. (2005) used Lagrangian method and genetic algorithm to minimize the energy consumption at different cooling loads. The results showed that Lagrangian method could minimize the energy consumption but it could not converge at low demands. Genetic algorithm overcame the convergence at low demands but the minimum energy consumption increased at about 0.4 % on an average compared with that of Lagrangian method. Chang et al. (2006, 2010), and Chang (2006) used simulated annealing method and gradient method to improve the convergence of Lagrangian method. Since the variables of optimal chiller loading are continuous, Ardakani et al. (2008) adopted continuous genetic algorithm and particle swarm optimization to solve the problem. Lee and Lin (2009) also proposed a particle swarm algorithm to solve the optimal chiller loading problem. The result of these two studies showed that particle swarm algorithm could improve the performance of genetic algorithm and Lagrangian method.

In the optimization of air-conditioning system using simulation methods, Lee and Lin (2009) proposed particle swarm algorithm for ice-storage air-conditioning system, and Kusiak and Lee (2010) proposed data mining algorithms for the cooling output of an air handling unit (AHU). Lee and Lee (2007) also developed a simplified model for evaluating chiller system configurations. Lee et al. (2011) used differential evolution algorithm to minimize the energy consumption at different cooling loads and the results were found to be as better as PSO results.

It has been observed from the literature review that some of the researchers had used traditional optimization techniques like Lagrangian method and gradient method. However, these approaches failed to converge at low demands. In order to overcome this drawback of traditional optimization techniques, few researchers had used the so-called advanced optimization techniques such as GA and its versions like SA, PSO, and DE algorithms. Some of the advantages of these advanced optimization techniques are: (i) they are relevant for solving varieties of complex optimization problems and (ii) these are population-based and hence they may give global optimum solution. The case studies presented by Lee et al. (2011) and Ardakani et al. (2008) are considered in this chapter to demonstrate the applicability of the TLBO algorithm for the optimal chiller loading problem to see if any further improvement can be obtained in the results. Lee et al. (2011) considered two case studies where the subjects were a semiconductor plant and a hotel in China and Ardakani et al. (2008) considered a case study where the subject was another semiconductor plant in china.

8.2 Case Studies and Their Results

Three case studies of multiple chiller systems are considered in this work. In case 1, the subject is a semiconductor plant in Hsinchu Science-based Park, Taiwan and the multiple chiller system is composed of three 800RT units (Lee et al. 2011). In case 2, the subject is a hotel in Taipei and the multiple chiller system is composed of two

450RT and two 1000RT units (Lee et al. 2011). In case 3, the subject is another semiconductor plant in china and the multiple chiller system is composed of four 1280RT and two 1250RT units (Ardakani et al. 2008).

8.2.1 Case Study 1

In case study 1, the subject is a semiconductor plant in Hsinchu Science-based Park, Taiwan, and the multiple chiller system is composed of three 800RT units (Lee et al. 2011). Considering a system with all electric cooling, the optimum operation is achieved when the power consumption of the system is minimized and the cooling load is satisfied. Power consumption of the system is given as (Lee et al. 2011)

$$P_i = a_i + b_i * \mathrm{PLR}_i + c_i * \mathrm{PLR}_i^2 + d_i * \mathrm{PLR}_i^3 \tag{8.2}$$

where, a_i, b_i, c_i, d_i are coefficients of interpolation for consumed power versus PLR of *i*th chiller and are predefined. PLR_i (part load ratio) of *i*th chiller is the design variable for that chiller and is defined by the following equation:

$$\mathrm{PLR}_i = \frac{\text{Chiller Load}}{\text{Design capacity}} \tag{8.3}$$

The power every chiller consumes is a function of PLR of the chiller. At low loads, power consumption of centrifugal chillers is high because of motor losses. In high loads, the input power increases due to thermal heat exchange inefficiencies. The objective function is sum of consumed power by each unit, as stated by the following equation:

$$\text{Objective Function} = \text{Minimize} \sum_{i=1}^{\mathrm{NC}} P_i \tag{8.4}$$

where, NC is the number of chillers used and P_i is the consumed power by *i*th chiller. Also, the demanded load must be satisfied by the system. This constraint is stated as follows:

$$\sum_{i=1}^{\mathrm{NC}} (\mathrm{PLR}_i * \overline{\mathrm{RT}_l}) = \mathrm{CL} \tag{8.5}$$

where, $\overline{\mathrm{RT}_l}$ is the capacity of *i*th chiller and CL is demanded cooling load.

Considering the suggestion of the manufacturer, the other constraint is that the partial load of each operating chiller cannot be less than 30 %

Table 8.1 Coefficients of interpolation for consumed power (Lee et al. 2011; Reprinted with permission from Elsevier)

System	Chiller (i)	a_i	b_i	c_i	d_i	Capacity (RT_i)
Case study 1	CH-1	100.95	818.61	–973.43	788.55	800
	CH-2	66.598	606.34	–380.58	275.95	800
	CH-3	130.09	304.50	14.377	99.8	800

$$PLR_i \geq 0.3 \text{ and } PLR_i \leq 1 \tag{8.6}$$

The coefficients of interpolation for the consumed power for case study 1 are given in Table 8.1.

Now, the TLBO algorithm is applied to solve the above optimization problem. The results are compared with the previous results given by Lagrangian method (Chang 2004), genetic algorithm (Chang et al. 2005), particle swarm optimization (Lee and Lin 2009) and differential evolution optimization (Lee et al. 2011).

The population size and the number of generations used for genetic algorithm were 100 and 100, respectively. For the same problem, TLBO is applied with a population size and the number of generations of 50 and 100 so that the number of function evaluations remains the same. It may be noted that as the evaluation is done in both the teacher and the learner phases, the number of function evaluations in TLBO algorithm is calculated as 2 × population size × number of generations. Table 8.2 shows the comparison of TLBO results with genetic algorithm (GA) results and Lagrangian method (LGM). The population size and the number of generations used for particle swarm optimization (PSO) and differential evolution (DE) optimization algorithms were 10 and 100. For the same problem, TLBO is also applied with a population size and the number of generations of 20 and 25 so that the number of function evaluations remains the same. Table 8.3 gives the comparison between TLBO results and the results obtained by the PSO and DE algorithms. The values shown in bold in the table indicate the comparatively better optimum values. The results obtained by TLBO algorithm show that it is achieving better optimal solutions than the genetic algorithm for all cooling loads. The problem of divergence at low loads by LGM has been overcome by TLBO algorithm. It has been observed that the convergence has occurred at 11th iteration at 1583.806 kW.

Table 8.3 gives the comparison between TLBO results and the results obtained by the PSO and DE algorithms. The values shown in bold in the table indicate the comparatively better optimum values. The results obtained show that the TLBO algorithm is giving equally good optimal results as that of the PSO and DE algorithms at all cooling loads. For low demands, one or two chillers are shut down in order to increase the system performance. Here, for loads of 1440 RT, 1200 RT, and 960 RT the first chiller is considered as shut down. It has been observed that the convergence of TLBO algorithm has occurred at 12th iteration at 1583.807 kW. It is observed that the convergence rate is faster in TLBO algorithm than that of DE algorithm.

Table 8.2 Comparison of TLBO results with the results of GA and LGM for case study 1

Desired cooling load	Chiller (*i*)	LGM (Chang 2004)		GA (Chang et al. 2005)		TLBO	
		PLR	Power consumed (kW)	PLR	Power consumed (kW)	PLR	Power consumed (kW)
2160 (90 %)	1	0.73	1583.81	0.81	1590.96	0.725272	**1583.806**
	2	0.97	–	0.93	–	0.974727	–
	3	1.00	–	0.96	–	1.00000	–
1920 (80 %)	1	0.66	1403.20	0.70	1406.02	0.659003	**1403.196**
	2	0.86	–	0.80	–	0.858572	–
	3	0.88	–	0.90	–	0.882423	–
1680 (70 %)	1	0.60	1244.32	0.69	1250.06	0.596074	**1244.314**
	2	0.75	–	0.68	–	0.744966	–
	3	0.76	–	0.73	–	0.758941	–
1440 (60 %)	1	0.53	1102.26	0.52	1107.75	0.530320	**1102.255**
	2	0.62	–	0.74	–	0.615381	–
	3	0.65	–	0.54	–	0.654278	–
1200 (50 %)	1	–	–	0.49	971.21	0.498310	**970.849**
	2	–	–	0.44	–	0.385444	–
	3	–	–	0.57	–	0.616245	–
960 (40 %)	1	–	–	0.31	842.18	0.300387	**841.44**
	2	–	–	0.32	–	0.300200	–
	3	–	–	0.58	–	0.599412	–

8.2.2 *Case Study 2*

In case study 2, the subject is a hotel in Taipei, and the multiple chiller system is composed of two 450 RT and two 1000 RT units and the optimum operation is achieved when the power consumption of the system is minimized and the cooling load is satisfied. Considering a system with all electric cooling, Lee et al. (2011) proposed Eqs. (8.2)–(8.6) for minimizing the power consumption of the system. The coefficients of interpolation for consumed power for case study 2 are given in Table 8.4. The TLBO algorithm is applied on the above problem and the results are compared with the previous results given by Lagrangian method (Chang 2004), Genetic algorithm (Chang et al. 2005), Particle Swarm Optimization (Lee and Lin 2009), and Differential Evolution Optimization (Lee et al. 2011).

The population size and the number of generations used for genetic algorithm were 100 and 100, respectively. For the same problem, TLBO is applied with a population size and the number of generations of 50 and 100 so that the number of function evaluations remains the same. Table 8.5 shows comparison of results of TLBO with GA and LGM. The results obtained by TLBO algorithm show that it has achieved better optimal solutions than genetic algorithm for all loads. The convergence of TLBO algorithm has occurred at 11th iteration at 1857.29 kW.

Table 8.3 Comparison of TLBO results with the results of PSO and DE for case study 1

Desired cooling load	Chiller (i)	PSO (Lee and Lin 2009)		DE (Lee et al. 2011)		TLBO	
		PLR	Power consumed (kW)	PLR	Power consumed (kW)	PLR	Power consumed (kW)
2160 (90 %)	1	0.73	1583.81	0.73	1583.81	0.725540	**1583.807**
	2	0.97	–	0.97	–	0.974511	–
	3	1.00	–	1.00	–	0.999948	–
1920 (80 %)	1	0.66	1403.20	0.66	1403.20	0.659022	**1403.196**
	2	0.86	–	0.86	–	0.859059	–
	3	0.88	–	0.88	–	0.881918	–
1680 (70 %)	1	0.60	1244.32	0.60	1244.32	0.596154	**1244.317**
	2	0.75	–	0.75	–	0.745033	–
	3	0.76	–	0.76	–	0.758791	–
1440 (60 %)	1	0.00	993.60	0.00	993.60	0.00	**993.59**
	2	0.89	–	0.89	–	0.884809	–
	3	0.91	–	0.91	–	0.91517	–
1200 (50 %)	1	0.00	832.33	0.49	832.33	0.00	**832.325**
	2	0.74	–	0.74	–	0.743030	–
	3	0.76	–	0.76	–	0.756969	–
960 (40 %)	1	0.00	692.25	0.00	692.25	0.00	**692.244**
	2	0.57	–	0.57	–	0.569941	–
	3	0.63	–	0.63	–	0.630042	–

Table 8.4 Coefficients of interpolation for consumed power for case study 2 (Lee et al. 2011; Reprinted with permission from Elsevier)

System	Chiller (i)	a_i	b_i	c_i	d_i	Capacity (RT_i)
Case study 2	CH-1	104.09	166.57	–430.13	512.53	450
	CH-2	–67.15	1177.79	–2174.53	1456.53	450
	CH-3	384.71	–779.13	1151.42	–63.2	1000
	CH-4	541.63	413.48	–3626.50	4021.41	1000

The population size and the number of generations used for PSO and DE algorithms were 10 and 100. For the same problem, TLBO is applied with a population size and the number of generations of 20 and 25 so that the number of function evaluations remains the same. Table 8.6 gives the comparison between TLBO results and the results obtained by PSO and DE algorithms. The values shown in bold in the table indicate the comparatively better optimum values. The results obtained show that TLBO algorithm is giving better optimal results than the PSO and DE algorithms at all cooling loads. For low demands, one or two chillers are shut down in order to increase the system performance. Here, for the load of

Table 8.5 Comparison of results of GA and TLBO algorithms for case study 2

Desired cooling load	Chiller (*i*)	LGM (Chang 2004)		GA (Chang et al. 2005)		TLBO	
		PLR	Power consumed (kW)	PLR	Power consumed (kW)	PLR	Power consumed (kW)
2610 (90 %)	1	0.99	1857.30	0.99	1862.18	0.990359	**1857.29**
	2	0.91	–	0.95	–	0.906529	–
	3	1.00	–	1.00	–	1	–
	4	0.76	–	0.74	–	0.756399	–
2320 (80 %)	1	0.83	1455.66	0.86	1457.23	0.829711	**1455.64**
	2	0.81	–	0.81	–	0.805244	–
	3	0.90	–	0.88	–	0.896605	–
	4	0.69	–	0.69	–	0.687646	–
2030 (70 %)	1	0.73	1178.14	0.66	1183.80	0.725695	**1178.12**
	2	0.74	–	0.76	–	0.740364	–
	3	0.72	–	0.76	–	0.721541	–
	4	0.65	–	0.64	–	0.648712	–
1740 (60 %)	1	0.60	998.53	0.60	1001.62	0.604018	**998.52**
	2	0.66	–	0.70	–	0.658195	–
	3	0.56	–	0.57	–	0.564193	–
	4	0.61	–	0.59	–	0.607801	–
1450 (50 %)	1	0.46	904.62	0.60	907.72	0.457024	**904.61**
	2	0.50	–	0.36	–	0.502476	–
	3	0.45	–	0.44	–	0.446798	–
	4	0.57	–	0.58	–	0.571426	–
1160 (40 %)	1	0.30	849.99	0.33	856.30	0.300602	**849.93**
	2	0.30	–	0.32	–	0.300003	–
	3	0.35	–	0.32	–	0.349628	–
	4	0.54	–	0.54	–	0.540098	–

1450 RT the first chiller and for the load of 1160 RT the first and second chillers are considered as shut down. The convergence occurred at about 12th iteration at 1857.298 kW. The convergence occurred for DE after 40 generations, whereas for TLBO the convergence occurred in less than 20 generations for the same function evaluations as of DE.

8.2.3 Case Study 3

In this case, the subject is a semiconductor plant in China. It consists of a total of six chillers which makes the system more challenging for optimization. Four chillers are of 1280 RT and two chillers are of 1250 RT (Ardakani et al. 2008).

Table 8.6 Comparison of TLBO results with the results of PSO and DE for case study 2

Desired cooling load	Chiller (i)	PSO (Lee and Lin 2009)		DE (Lee and Lin 2009)		TLBO	
		PLR	Power consumed (kW)	PLR	Power consumed (kW)	PLR	Power consumed (kW)
2610 (90 %)	1	0.99	1857.30	0.99	1857.30	0.990714	**1857.298**
	2	0.91	–	0.91	–	0.905877	–
	3	1.00	–	1.00	–	1	–
	4	0.76	–	0.76	–	0.756533	–
2320 (80 %)	1	0.83	1455.66	0.83	1455.66	0.827834	**1455.642**
	2	0.81	–	0.81	–	0.805755	–
	3	0.90	–	0.90	–	0.896948	–
	4	0.69	–	0.69	–	0.687917	–
2030 (70 %)	1	0.73	1178.14	0.73	1178.14	0.725695	**1178.13**
	2	0.74	–	0.74	–	0.740364	–
	3	0.72	–	0.72	–	0.721541	–
	4	0.65	–	0.65	–	0.648712	–
1740 (60 %)	1	0.60	998.53	0.60	998.53	0.603547	**998.524**
	2	0.66	–	0.66	–	0.657040	–
	3	0.56	–	0.56	–	0.564573	–
	4	0.61	–	0.61	–	0.608143	–
1450 (50 %)	1	0.61	820.07	0.61	820.07	0.606902	**820.064**
	2	0.00	–	0.00	–	0.00	–
	3	0.57	–	0.57	–	0.568096	–
	4	0.61	–	0.61	–	0.608788	–
1160 (40 %)	1	0.00	651.07	0.00	651.07	0.00	**651.062**
	2	0.00	–	0.00	–	0.00	–
	3	0.56	–	0.56	–	0.555391	–
	4	0.60	–	0.60	–	0.604599	–

Objective function:

$$\sum_{i=1}^{NC} kW_i(\mathrm{PLR}_i) + 10\times\left|\sum_{i=1}^{NC} (\mathrm{PLR}_i * \overline{\mathrm{RT}_i}) - \mathrm{CL}\right| \tag{8.7}$$

The objective is to minimize the above function. CL is demanded cooling load. Expressing it in power consumption form, $kW_i(\mathrm{PLR}_i)$ can be expressed as

$$kW_i(\mathrm{PLR}_i) = P_i = a_i + b_i \times \mathrm{PLR}_i + c_i \times \mathrm{PLR}_i^2 \tag{8.8}$$

Design variables:
Part load ratio (PLR_i) of ith chiller is considered as design variables. For six chillers, a total of six design variables are considered.

$$PLR_i \geq 0.3 \text{ and } PLR_i \leq 1 \tag{8.9}$$

The coefficients of interpolation for consumed power for case study 3 are given in Table 8.7.

The TLBO algorithm is applied to solve the above problem. The results are compared with the previous results of binary Genetic algorithm (B-GA), continuous genetic algorithm (C-GA), and PSO of Ardakani et al. (2008).

The population size and the number of generations used for B-GA and C-GA were 100 and 2000, respectively. For the same problem, TLBO is applied with a population size and the number of generations of 100 and 1000 so that the number of function evaluations remains the same. The comparison of TLBO results with B-GA and C-GA are presented in Table 8.8. The results obtained by TLBO algorithm shows that it is giving good results as compared to those of B-GA for all loads. When compared with C-GA, TLBO produced much better results as load decreases. The convergence of TLBO algorithm has occurred at the 45th iteration at 4738.906 kW.

The population size and the number of generations used for Particle swarm Optimization were 36 and 1000. Hence, for the same problem after a few trials, TLBO is applied with a population size and the number of generations of 50 and 360 so that the number of function evaluations remains same. The comparison of TLBO results along with the results of PSO are presented in Table 8.9. The values shown in bold in the table indicate the comparatively better optimum values. The result obtained by TLBO algorithm show that it is giving good results as compared to those of particle swarm optimization for all loads. As the cooling load decreases, TLBO produced much better results. The convergence of the TLBO algorithm occurred much faster than the PSO algorithm at all loads. The convergence of TLBO algorithm has occurred at 56th iteration at 4739.189 kW.

In this chapter, three case studies of multiple chiller systems are considered for optimization using the TLBO algorithm. The three multi-chiller systems considered have three chillers, four chillers, and six chillers. Among the three case studies, the TLBO results of the first two case studies are compared with those of LGM, GA, PSO, and DE. The results show that the TLBO algorithm has performed better than

Table 8.7 Coefficients of interpolation for consumed power for case study 3 (Ardakani et al. 2008; Reprintesd with permission from Elsevier)

Chiller (i)	a_i	b_i	c_i	Capacity
1	399.345	–122.12	770.46	1280.0
2	287.116	80.04	700.48	1280.0
3	–120.505	1525.99	–502.14	1280.0
4	–19.121	898.76	–98.15	1280.0
5	–95.029	1202.39	–352.16	1250.0
6	191.750	224.86	524.04	1250.0

Table 8.8 Comparison of TLBO results with the results of binary GA and continuous GA for case study 3

Desired cooling load	Chiller (i)	Binary GA (Ardakani et al. 2008)	Power (kW)	Continuous GA (Ardakani et al. 2008)	Power (kW)	TLBO	Power (kW)
		PLR_i	P_i	PLR_i	P_i	PLR_i	P_i
6858 (90 %)	1	0.8473	836.6782	0.8305	829.3918	0.8154931	812.132
	2	0.7266	674.8695	0.7477	738.6028	0.7430411	733.331
	3	0.9996	902.7402	1.0000	903.3450	0.9997120	903.194
	4	0.9989	779.2732	0.9999	781.4819	0.9999120	781.427
	5	0.9992	755.1275	1.0000	755.2010	0.9998750	755.138
	6	0.8287	795.9624	0.8222	730.9420	0.8429710	753.6833
	$\sum$	–	4744.6512	–	4738.9645	6858	**4738.906**
6477 (85 %)	1	0.8261	824.2076	0.8068	802.3774	0.7253411	716.1204
	2	0.5672	557.8702	0.6588	643.9125	0.6587317	643.799
	3	0.9985	902.5429	1.0000	903.3450	0.9999082	903.2970
	4	0.9715	761.4182	1.0000	781.4890	0.9997482	781.3120
	5	0.9972	753.8255	1.0000	755.2010	0.9999162	755.1592
	6	0.7404	645.4848	0.6327	543.8299	0.7167451	622.1279
	$\sum$	–	4445.3493	–	4430.2444	6477	**4421.816**
6096 (80 %)	1	0.7381	728.9208	0.6519	647.1984	0.6470774	642.9222
	2	0.4514	465.9823	0.6147	601.0118	0.5612026	552.6492
	3	0.9856	895.7077	0.9999	903.3449	0.9994113	903.0375
	4	0.9670	758.2284	0.9992	780.9059	0.9999599	781.4602
	5	0.9981	754.2742	0.9999	755.2001	0.9999509	755.1761
	6	0.6612	569.5019	0.5325	460.1566	0.5922143	508.7052
	$\sum$	–	4172.6155	–	4147.8178	6096	**4143.953**
5717 (75 %)	1	0.6139	614.7192	0.5962	600.4008	0.5601624	572.6941
	2	0.4953	498.6126	0.4685	478.3623	0.4689558	478.6999
	3	0.9988	902.7272	1.0000	903.3450	0.9990637	902.8557
	4	0.9413	739.9096	0.9999	781.4888	0.9987152	780.5862
	5	0.9999	755.1976	1.0000	755.2010	0.9995143	754.9589
	6	0.4511	399.8417	0.4353	388.9640	0.4745431	416.4649
	$\sum$	–	3911.0079	–	3907.7607	5717	**3906.261**
5334 (70 %)	1	0.6506	646.0258	0.6237	622.8603	0.6585057	653.0220
	2	0.5206	518.6970	0.4964	499.4574	0.6497915	634.8878
	3	0.9989	902.8016	0.9999	903.3442	0.3013123	293.7053
	4	0.5782	497.7299	0.5733	463.8435	0.9964603	779.0011
	5	0.9873	748.8425	1.0000	755.2010	0.9997973	755.0999
	6	0.4654	409.9351	0.5092	442.1532	0.5987872	514.2856
	$\sum$	–	3694.0319	–	3686.8597	5334	**3630.001**

Table 8.9 Comparison of TLBO results with the results of PSO for case study 3

Desired cooling load	Chiller (i)	PSO (Ardakani et al. 2008)	Power (kW)	TLBO	Power (kW)
		PLR_i	P_i	PLR_i	P_i
6858 (90 %)	1	0.8026	797.6788	0.8183959	815.4339
	2	0.7799	775.6981	0.7633239	756.3555
	3	0.9996	903.1638	0.9998858	903.2850
	4	0.9998	781.3991	0.9998568	781.3878
	5	0.9999	755.1979	0.9997842	755.0934
	6	0.8183	726.6468	0.8191983	727.6304
	$\sum$	–	4739.7845	6858	**4739.189**
6477 (85 %)	1	0.7606	752.2134	0.7534848	744.7491
	2	0.6555	640.5199	0.6670026	652.1406
	3	1.0000	903.3449	0.9999864	903.3377
	4	1.0000	781.4889	0.9998842	781.4075
	5	1.0000	755.2010	0.9998812	755.1417
	6	0.6835	590.2852	0.6792722	586.2565
	$\sum$	–	4423.0534	6477	**4423.0331**
6096 (80 %)	1	0.6591	653.5696	0.6520134	647.2599
	2	0.5798	569.0161	0.5734108	563.3289
	3	0.9991	902.8647	0.9997557	903.2172
	4	0.9979	780.0799	0.9997123	781.2867
	5	0.9921	751.2365	0.9996102	755.0067
	6	0.5710	491.0385	0.5749002	494.2225
	$\sum$	–	4147.8055	6096	**4144.323**
5717 (75 %)	1	0.7713	763.4782	0.7562209	765.6606
	2	0.7177	705.3382	0.70928418	689.8701
	3	0.3000	292.0994	0.3043119	292.4680
	4	0.9991	780.8389	0.9999848	781.4813
	5	1.0000	755.2010	0.9993228	754.8582
	6	0.7187	624.0084	0.7380049	643.3226
	$\sum$	–	3920.9642	5717	**3918.711**
5334 (70 %)	1	0.6418	638.3097	0.6461587	642.1188
	2	0.6621	647.2355	0.6513018	636.3848
	3	0.3301	328.5020	0.3206299	317.1502
	4	0.9906	774.8633	0.9997395	781.3057
	5	0.9990	754.6915	0.9994411	754.9225
	6	0.5806	498.9765	0.5871011	504.3956
	$\sum$	–	3642.5786	5334	**3636.281**

the genetic algorithm at all loads and also has overcome the divergence problem faced by the Lagrangian method at low demands. TLBO gave better optimal results as that of PSO and DE for all demand loads presented and convergence is faster than DE for the same function evaluations.

The TLBO results of the third case study are compared with the results of binary GA, continuous GA, and PSO algorithms. At all demand loads, the TLBO has performed better than the three previously proposed algorithms. The convergence rate is also faster than all other algorithms. Thus, it can be stated that the TLBO algorithm is a competitive optimization algorithm to solve the chiller loading problems.

References

Ardakani, A.J., Ardakani, J.F., Hosseinian, S.H., 2008. A novel approach for optimal chiller loading using particle swarm optimization. Energy and Buildings 40(12), 2177–2187.

Chang, Y.C., 2004. A novel energy conservation method - optimal chiller loading. Electric Power Systems Research 69, 221–226.

Chang, Y.C., Lin, J.K., Chuang, M.H., 2005. Optimal chiller loading by genetic algorithm for reducing energy consumption. Energy and Buildings 37, 147–155.

Chang, Y.C., 2006. An innovative approach for demand side management—optimal chiller loading by simulated annealing. Energy 31, 1883–1896.

Chang, Y.C., Chen, W.C., Lee, C.Y., Huang, C.N., 2006. Simulated annealing based optimal chiller loading for saving energy. Energy Conversion and Management 47, 2044–2058.

Chang, Y.C., Chan, T.S., Lee, W.S., 2010. Economic dispatch of chiller plant by gradient method for saving energy. Applied Energy 87, 1096–1101.

Kusiak, A., Li, M., 2010. Cooling output optimization of an air handling unit. Applied Energy 87, 901–909.

Lee, W.L., Lee, S.H., 2007. Developing a simplified model for evaluating chiller-system configurations. Applied Energy 84, 290–306.

Lee, W.S., Lin, L.C., 2009. Optimal chiller loading by particle swarm algorithm for reducing energy consumption. Applied Thermal Engineering 29, 1730–1734.

Lee, W.S., Chen, Y.T., Wu, T.H., 2009. Optimization for ice-storage air conditioning system using particle swarm algorithm. Applied Energy 86, 1589–1595.

Lee, W.S., Chen, Y.T., Kao, Y., 2011. Optimal chiller loading by differential evolution for reducing energy consumption. Energy and Buildings 43, 599–604.

Chapter 9
Thermoeconomic Optimization of Shell and Tube Condenser Using TLBO and ETLBO Algorithms

Abstract This chapter presents the application of TLBO and ETLBO algorithms for the Thermoeconomic optimization of a shell and tube condenser. The objective function considered is minimization of total cost of the condenser. Five design parameters are considered for the optimization. It is shown that by selecting the optimal design parameters, the total cost of the condenser is reduced using the TLBO and ETLBO algorithms as compared to the design parameters suggested by the GA and PSO algorithms.

9.1 Thermoeconomic Optimization Aspects of Shell and Tube Condenser

Shell and tube condensers are composed of circular pipes and are installed in cylindrical shells. This type of condenser is a well known heat exchanger and is a key component in refrigeration and heat pump systems, thermal system plants, petrochemical plants, refrigeration, and air-conditioning systems. Shell and tube condenser are used to transfer the heat between two or more fluids, between a solid surface and a fluid, or between solid particulates and a fluid at different temperatures and in thermal contact. In heat exchangers, there are usually no external heat and work interactions. There are some effective parameters in shell and tube heat exchanger design such as tube numbers, tube length, tube arrangement, and baffle spacing. Therefore, optimization of this kind of heat exchanger is quite interesting.

In the past few decades, the interest of researchers is growing in the field of Thermoeconomic design optimization of shell and tube condenser. Some of the prominent works include those of Solten et al. (2004), Llopis et al. (2008), Haseli et al. (2008a, b), and Hajabdollahi et al. (2011). Hajabdollahi et al. (2011) presented a Thermoeconomic optimization of a shell and tube condenser using genetic algorithm (GA) and particle swarm optimization (PSO) algorithm. The aim was to find the optimal total cost including the investment and operation costs of the

R.V. Rao, *Teaching Learning Based Optimization Algorithm*,
DOI 10.1007/978-3-319-22732-0_9

condenser. The initial cost included the condenser surface area and operational cost included the pump output power to overcome the pressure loss. The design parameters were the tube number, number of tube passes, inlet and outlet tube diameters, tube pitch ratio, and tube arrangements (30°, 45°, 60°, and 90°). Rao and Waghmare (2014) considered the same problem and applied TLBO algorithm for obtaining the optimal design parameters. Now, this chapter presents the results of application of GA, PSO, TLBO, and ETLBO algorithms for the Thermoeconomic optimization of the shell and tube condenser.

9.1.1 Problem Formulation

The total cost of the shell and tube condenser is considered as an objective function. The total cost includes the cost of heat transfer area as well as the operating cost for the pumping power (Hajabdollahi et al. 2011).

$$C_{\text{total}} = C_{\text{in}} + C_{\text{op}} \tag{9.1}$$

where, C_{in} is the of the investment cost, C_{op} is the annual operating cost, and C_{total} is the total cost in \$.

The investment cost for both shell and tube from stainless steel shell and tubes is (Taal et al. 2003)

$$C_{\text{in}} = 8500 + 409A_{t,0}^{0.85} \tag{9.2}$$

where, A_t is the total tube outside heat transfer area.

The total operating cost related to pumping power to overcome the friction losses is computed from the following equations:

$$C_{\text{op}} = \sum\nolimits_{k=1}^{\text{ny}} \frac{C_o}{(1+i)^k} \tag{9.3}$$

$$C_o = Pk_{el}\tau \tag{9.4}$$

$$P = \frac{1}{\eta}\left(\frac{m_t}{\rho_t}\Delta P_t\right) \tag{9.5}$$

where, ny is the equipment life time, i is the annual discount rate, k_{el}, τ, and η are price of electrical energy, hours of operation per year, and pump efficiency, respectively.

9.1.2 Thermal Modelling

The following assumptions were made (Hajabdollahi et al. 2011; Reprinted with permission from Elsevier):

1. Condensing flow is in the shell direction.
2. Cooling fluid is considered in tube side.
3. The shell pressure loss is negligible.
4. The condenser changes the water state from saturated steam to liquid.

The heat transfer between hot and cold fluid is calculated based on the following relation:

$$Q = \bar{U}mA_{t,0}\Delta T_{\mathrm{lm}} \tag{9.6}$$

where, ΔT_{lm} is the logarithmic mean temperature difference which is defined as

$$\Delta T_{\mathrm{lm}} = \frac{\Delta T_1 - \Delta T_2}{\mathrm{In}\frac{\Delta T_1}{\Delta T_2}} \tag{9.7}$$

Here, $\bar{U}m$ is the mean value of total heat transfer coefficient

$$\Delta T_1 = T_s - T_{t,i} \tag{9.8}$$

$$\Delta T_2 = T_s - T_{t,o} \tag{9.9}$$

Since the total heat transfer area can considerably vary along the heat exchanger, the mean value of this parameter is used as follows:

$$\overline{U}m = \frac{1}{2}\left(\overline{U_1} + \overline{U_2}\right) \tag{9.10}$$

where, $\overline{U_1}$ and $\overline{U_2}$ are the total heat transfer at the inlet and outlet parts of the condenser.

By assuming both $1/\overline{U}$ and ΔT change linearly with $\bar{Q}$, a useful relation for determining the mean value of total heat transfer is obtained as follows:

$$\bar{U}_m = \frac{1}{\bar{\bar{U}}}\frac{\Delta T_{\mathrm{lm}} - \Delta T_2}{\Delta T_1 - \Delta T_2} + \frac{1}{\bar{\bar{U}}}\frac{\Delta T_1 - \Delta T_{\mathrm{lm}}}{\Delta T_1 - \Delta T_2} \tag{9.11}$$

In addition, $\overline{U}$ can be defined as follows:

$$\frac{1}{\bar{\bar{U}}} = R_t + \frac{1}{h_o} \tag{9.12}$$

where, h_o is a convection heat transfer coefficient and R_t is the total inner thermal resistance which are determined as

$$R_t = R_{fo} + \left[\frac{1}{h_i} + R_{fi}\right]\frac{d_o}{d_i} + \frac{t_w}{k_w}\frac{d_o}{D_m} \tag{9.13}$$

$$D_m = \frac{d_o - d_i}{\ln\left[\frac{d_o}{d_i}\right]} \tag{9.14}$$

Here, R_f is the fouling factor and d_o, d_i, k_w, and t_w are the tube outlet diameter, tube inlet diameter, tube conduction conductivity, and tube thickness, respectively. Having known the tube number as well as the mass flow rate in the pipe, the velocity in the pipe is determined as

$$\bar{V} = \frac{4m_t}{\rho\pi d_1^2 N} \rightarrow R_e = \frac{4m_t}{\pi d_1 \mu N} \tag{9.15}$$

where, N is the tube number.

Having known the R_e number, N_u number at the tube side will be defined as a R_e number function (Kakac and Liu 2000):

$$N_u = \frac{0.5f(R_e - 1000)P_r}{1 + 9.7\left(\frac{f}{2}\right)^{0.5}\left(P_r^{2/3} - 1\right)} \text{if } \left(2300 < R_e < 10^4\right) \tag{9.16}$$

$$N_u = \frac{0.5fR_eP_r}{1.07 + 9.7\left(\frac{f}{2}\right)^{0.5}\left(P_r^{2/3} - 1\right)} \text{if } \left(10^4 < R_e < 5 \times 10^6\right) \tag{9.17}$$

$$f = (1.58\ln(R_e) - 3.28)^2 \tag{9.18}$$

Therefore, knowing the N_u number, the convective heat transfer coefficient is determined as

$$h_i = \frac{N_u k_i}{d_i} \tag{9.19}$$

Heat transfer coefficient for the condensate flow (h_o) is obtained based on the following relation (Kakac and Liu 2000)

$$h_o = 0.728\left[\frac{\rho_1 g h_{fg} k_o}{\mu_1 \Delta T_w d_o}\right]^{0.25}\frac{1}{n^{1/6}} \tag{9.20}$$

Here, n is the tube number in a column which may be predicted for each tube arrangement as follows:

For arrangements of 45°, 90°

$$n = \frac{\sqrt{4c_1 p_t^2 N/\pi}}{d_o + p_t} \tag{9.21}$$

For arrangements of 30°, 60°

$$n = \frac{\sqrt{4c_1 p_t^2 N/\pi}}{d_o + \sqrt{3}p_t} \tag{9.22}$$

$$p_t = d_o p_r \tag{9.23}$$

where, p_r is the pitch ratio and c_1 is the tube layout constant that is equal to 1 for 45°, 90°, and 0.87 for 30°, 60° (Dentice and Vanoli 2004).

The shell and tube condenser effectiveness is evaluated as (Kakac and Liu 2000),

$$\epsilon = 1 - e^{-\mathrm{NTU}} \tag{9.24}$$

where, NTU is the number of transfer units that is determined as

$$\mathrm{NTU} = \frac{U_m A_{t,o}}{C_{\min}} \tag{9.25}$$

The tube number, number of tube pass, inlet and outlet diameters, tube pitch ratio, and tube arrangements are considered as design parameters for the optimization process. There are some constraints like shell diameter that should be selected less than 7 m and tube length should be selected less than 15 m. The performance conditions of the shell and tube condenser are: hot fluid temperature (T_h = 125° C); inlet cold fluid temperature (T_{c1} = 12° C); cold fluid mass flow rate ($\dot{m}_c = 400\,\mathrm{kg/s}$), and hot fluid mass flow rate ($\dot{m}_h = 8.7\,\mathrm{kg/s}$).

9.2 Results and Discussion

To check the effectiveness of the TLBO and ETLBO algorithms, extensive computational trials are conducted on shell and tube condenser and the results are compared with those obtained by the other optimization algorithms. The population size of 35 and maximum number of iterations of 200 are considered. The TLBO algorithm is experimented with different elite sizes, viz., 0, 4, 8, 12, and 16 on shell and tube condenser. After making a few trials, elite size of 16 is considered. Computational results show that the TLBO and ETLBO algorithms are better or competitive to the other optimization algorithms considered by the previous researchers.

Table 9.1 gives the range of change in design parameters (Hajabdollahi et al. 2011). Table 9.2 provides the inner and outer diameters of 42 standard tubes (Hajabdollahi et al. 2011). Table 9.3 gives constant parameters for the objective function based on Kakac and Liu (2000).

Table 9.4 represents the optimal design parameters using ETLBO, TLBO, GA, and PSO algorithms. From Table 9.4, it can be seen that the required tube number, inner and outer tube diameters, and tube pitch ratio are reduced by using the TLBO and ETLBO algorithms as compared to the GA and PSO algorithms. Figure 9.1 shows the convergence of the objective function for various numbers of iterations using TLBO and ETLBO algorithms. Table 9.5 represents the optimal value for the condenser total cost using different algorithms. For the optimum set of design parameters, the total condenser cost is 29,068.28 \$/yr using ETLBO algorithm and

Table 9.1 The range of change in design parameters (Hajabdollahi et al. 2011; Reprinted with permission from Elsevier)

Variable	Lower bound	Upper bound
Tube number	100	1000
Number of tube pass	1	3
Inner tube diameter (m)	0.023	0.0254
Outer tube diameter (m)	0.43	0.453
Tube pitch ratio	1.25	1.5
Tube arrangements	30°	90°

Table 9.2 Inner and outer diameters of 42 standard tubes (Hajabdollahi et al. 2011; Reprinted with permission from Elsevier)

Inner diameter (m)	Outer diameter (m)
0.824, 0.742, 1.049, 0.957, 1.380, 1.278, 1.610, 1.500, 2.067, 1.939, 2.469, 2.323, 3.068, 2.900, 3.548, 3.364, 4.026, 3.826, 5.295, 5.047, 4.813, 6.3570, 6.065, 5.761, 8.329, 8.071, 7.625, 10.420, 10.192, 9.750, 12.390, 12.090, 11.750, 13.500, 13.250, 13.000, 15.500, 15.250, 15.000, 17.624, 17.250, 17.000	1.050, 1.050, 1.315, 1.315, 1.660, 1.660, 1.900, 1.900, 2.375, 2.375, 2.875, 2.875, 3.500, 3.500, 4.000, 4.000, 4.500, 4.500, 5.563, 5.563, 5.563, 6.6250, 6.625, 6.6250, 8.625, 8.625, 8.625, 10.750, 10.750, 10.750, 12.750, 12.750, 12.750, 14.000, 14.000, 14.000, 16.000, 16.000, 16.000, 18.000, 18.000, 18.000

Table 9.3 Constant parameters for the objective function based on Kakac and Liu (2000)

	Dimension	Value
Hours of operation	h/year	5000
Price of electrical energy	\$/MWh	20
Pump efficiency	–	0.6
Rate of annual discount	–	0.1
Equipment life	h	5000

Table 9.4 Optimal design parameters using TLBO, ETLBO, GA, and PSO algorithms

Design parameters	Optimal value using TLBO	Optimum value using ETLBO	Optimal value using GA (Hajabdollahi et al. 2011)	Optimal value using PSO (Hajabdollahi et al. 2011)
Tube number	199	198	203	204
Number of tube pass	1	1	1	1
Inner and outer tube diameters (m)	0.02319–0.0432	0.02317–0.0431	0.0406–0.0479	0.0406–0.0479
Tube pitch ratio	1.2516	1.2511	1.2598	1.2527
Tube arrangements	45° or 90°	45° or 90°	45° or 90°	45°–90°

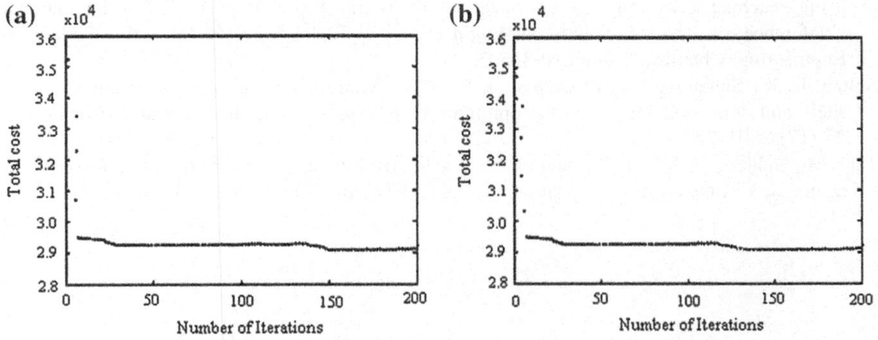

Fig. 9.1 The convergence of the objective function for various number of iterations using **a** TLBO and **b** ETLBO algorithms

Table 9.5 Optimal value for the total cost of the condenser using different algorithms

Algorithm	TLBO	ETLBO	GA (Hajabdollahi et al. 2011)	PSO (Hajabdollahi et al. 2011)
Condenser total cost	29,088.59 \$/yr	29,068.28 \$/yr	29,122.13 \$/yr	29,112.66 \$/yr

29,088.59 \$/yr using TLBO algorithm. From Table 9.5, it can be seen that the total cost of the condenser is reduced by using the ETLBO and TLBO algorithms as compared to the other optimization algorithms considered.

Hence, it can be concluded that the TLBO and ETLBO algorithms can effectively be applied and have a potential to solve the shell and tube condenser design optimization problem. The concept of elitism enhanced the performance of the TLBO algorithm for the optimization of shell and tube condenser.

References

Dentice, D, Vanoli, L., 2004. Thermoeconomic optimization of the condenser in a vapor compression heat pump. International Journal of Refrigeration 27, 433–441.

Hajabdollahi, H., Ahmadi, P., Dincer, I., 2011. Thermoeconomic optimization of a shell and tube condenser using both genetic algorithm and particle swarm. International Journal of Refrigeration 34(4), 1066–1076.

Haseli, Y., Dincer, I., Naterer, G. F., 2008a. Entropy generation of vapor condensation in the presence of a non-condensable gas in a shell and tube condenser. International Journal of Heat and Mass Transfer 51(7–8), 1596–1602.

Haseli, Y., Dincer, I., Naterer, G. F., 2008b. Optimum temperatures in a shell and tube condenser with respect to exergy. International Journal of Heat and Mass Transfer 51 (9–10), 2462–2470.

Kakac, S., Liu, H., 2000. Heat Exchangers Selection Rating and Thermal Design. New York: CRC Press.

Llopis, R., Cabello, R., Torrella, E., 2008. A dynamic model of a shell and tube condenser operating in a vapour compression refrigeration plant. International Journal of Thermal Sciences 47(7), 926–934.

Rao, R.V., Waghmare, G.G., 2014. Thermoeconomic optimization of a shell and tube condenser using teaching-learning-based optimization algorithm. Proceedings of Third International Conference on Recent Trends in Engineering & Technology, K. B. Jain College of Engineering, Chandwad, India, 498–503.

Soltan, B. K., Saffar-Avval, M., Damangir, E., 2004. Minimizing capital and operating costs of shell and tube condensers using optimum baffle spacing. Applied Thermal Engineering 24 (17) 2801–2810.

Taal, M., Bulatov, I., Klemes, J., Stehlik, P., 2003. Cost estimation and energy price forecasts for economic evaluation of retrofit projects. Applied Thermal Engineering 23, 1819–1835.

Chapter 10
Design of a Smooth Flat Plate Solar Air Heater Using TLBO and ETLBO Algorithms

Abstract This chapter presents the application of the TLBO and the ETLBO algorithms for the design optimization of a smooth flat plate solar air heater. The design results obtained by the TLBO and the ETLBO algorithms are found superior or competitive to the results given by GA, SA, and PSO algorithms.

10.1 Design of Smooth Flat Plate Solar Air Heater

Solar air heating is a solar thermal technology in which the energy from the Sun is captured by an absorbing medium and used to heat air. Solar air heating is very extensively used nowadays in commercial and industrial applications. There is an increasing interest among researchers in the design, development, and optimization of a smooth flat plate solar air heater (SFPSAH) over past few decades. In this chapter, the performance of solar air heater is investigated using the TLBO and ETLBO algorithms. The nomenclature used in the problem formulation of the solar air heater is given below.

Nomenclature

A_c Area of absorber plate (m^2)
c_p Specific heat of air (J/kg K)
d Hydraulic diameter of duct (m)
F_o Heat removal factor referred to outlet temperature (dimensionless)
G Mass velocity (kg/s m^2)
h Convective heat transfer coefficient (W/m^2 K)
h_w Wind convection coefficient (W/m^2 K)
S Irradiance (W/m^2)
$\dot{m}$ Mass flow rate of air (kg/s)
N Number of glass covers (dimensionless)
p_r Prandtl number (dimensionless)
R_e Reynolds number (dimensionless)
T Thickness of insulating material (m)

R.V. Rao, *Teaching Learning Based Optimization Algorithm*,
DOI 10.1007/978-3-319-22732-0_10

T_a	Ambient temperature of air (K)
T_i	Inlet temperature of air (K)
T_o	Outlet temperature of air (K)
T_p	Temperature of absorber plate (K)
U_o	Overall loss coefficient (W/m^2 K)
U_t	Top loss coefficient (W/m^2 K)
v	Wind velocity (m/s)
$(\tau\alpha)$	Transmittance–absorptance product (dimensionless)
λ	Thermal conductivity of air (W/m K)
λ_i	Thermal conductivity of insulating material (W/m K)
η_{th}	Thermal efficiency (dimensionless)
ϵ_p	Emissivity of plate (dimensionless)
ϵ_g	Emissivity of glass cover (dimensionless)
β	tilt angle (^{o})

The thermal performance of smooth flat plate solar air heater is investigated using TLBO algorithm by Rao and Waghmare (2015) based on heat transfer phenomena (ASHRAE Standards) and calculation of flat plate collector loss coefficients (Klien 1975). The results of ETLBO algorithm are also included in this chapter along with the results of TLBO algorithm. The thermal performance of a SFPASH can be predicted on the basis of detailed considerations of heat transfer processes and correlations for heat transfer coefficient, heat removal factor, etc. The objective function for thermal performance of SFPSAH is proposed as (Siddhartha et al. 2012; Reprinted with permission from Elsevier)

$$\text{Maximize } \eta_{\text{th}} = F_o\left[\tau\alpha - \left(\frac{T_o - T_i}{S}\right)U_o\right] \tag{10.1}$$

The different relations used for calculating overall loss coefficient (U_o), heat removal factor at outlet (F_o), and temperature rise (T_o–T_i) are computed using the following equations:

$$U_0 = \left[\frac{N}{\left(\frac{C}{t_p}\right)\left[\frac{(t_p - t_a)}{(N + f')}\right]^e} + \frac{1}{h_w}\right]^{-1} + \frac{\sigma(T_p + T_a)\left(T_p^2 + T_a^2\right)}{\left[\varepsilon_p + 0.00591Nh_w\right]^{-1} + \left[\frac{(2N + f' - 1 \pm 0.133\varepsilon_p)}{\varepsilon_p}\right] - N} + \frac{k_i}{t} \tag{10.2}$$

where

$$f' = \left(1 + 0.089h_w - 0.11h_w\varepsilon_p\right)(1 + 0.07866N) \tag{10.3}$$

$$C = 520\left(1 - 0.000051\beta^2\right) \tag{10.4}$$

$$e = 0.43\left(1 - 100T_p^{-1}\right) \tag{10.5}$$

Heat removal factor at outlet (F_o) can be expressed as

$$F_0 = \frac{Gc_p}{U_0}\left[1 - \exp\left(\frac{-U_0 F'}{GC_p}\right)\right] \tag{10.6}$$

where

$$F' = \frac{\left[0.024R_e^{0.8}P_r^{0.4}\frac{\lambda}{d}\right]}{\left[0.024R_e^{0.8}P_r^{0.4}\frac{\lambda}{d} + U_0\right]} \tag{10.7}$$

The temperature rise (T_o–T_i) is computed by following equation:

$$(T_0 - T_i) = \left[\frac{\{(\tau\sigma)S - U_0(T_p - T_a)\}}{\dot{m}c_p}\right]A_c \tag{10.8}$$

The constraints of the problem are:

$$1 \le N \le 3;\ \text{N is varied in steps of 1.} \tag{10.9}$$

$$600 \le S \le 1000;\ \text{S is varied in steps of 200.} \tag{10.10}$$

$$2000 \le \text{Re} \le 20{,}000;\ \text{Re is varied in steps of 2000.} \tag{10.11}$$

The climatic conditions are considered as follows:

$$1 \le v \le 3; \tag{10.12}$$

$$280 \le T_a \le 310; \tag{10.13}$$

The following three different cases are considered (Siddhartha et al. 2012; Rao and Waghmare 2015).

Case 1: Obtain the value of V and T_a through TLBO and ETLBO algorithms and generate ϵ_p (0.85–0.95) and β (0°–70°) randomly.

Case 2: Obtain the value of β and T_a through TLBO and ETLBO algorithms and generate ϵ_p (0.85–0.95) and v (1–3) randomly.

Case 3: Obtain the value of V and β through TLBO and ETLBO algorithms and generate ϵ_p (0.85–0.95) and T_a (280–310 K) randomly for a fixed value of N (1, 2 and 3) and fixed S (600, 800 and 1000 W/m^2) and varying Re ranging from 2000 to 20,000 in an incremental step of 2000.

The next section explains the detained results and discussion.

Table 10.1 Typical values of solar air heater system parameters (Siddhartha et al. 2012; Reprinted with permission from Elsevier)

Collector parameters	Values
Length (*L*) (mm)	1000
Width (*wt*) (mm)	200
Height (*ht*) (mm)	20
Transmittance–absorptance	0.85
Emissivity of glass cover	0.88
Emissivity of glass plate	0.85–0.95
Tilt angle	$0° \leq \beta \leq 70°$

10.2 Results and Discussion

To check the effectiveness of the TLBO and ETLBO algorithms, extensive computational trials are conducted on a flat plate solar air heater and results are compared with those obtained by the other optimization algorithms. Population size of 30 and the maximum number of iterations of 50 are considered. The TLBO algorithm is experimented with different elite sizes, viz. 0, 4, 8, 12, and 16 and after making few trials, an elite size of 12 is considered. In the present experiments, Deb's heuristic constrained handling method (Deb 2000) is used to handle the constraints with the TLBO algorithm.

Table 10.1 represents the typical parameter values of solar air heater system (Siddhartha et al. 2012). Table 10.2 shows the optimum results of thermal performance using ETLBO and TLBO algorithms and comparisons are made with the results of the PSO algorithm at $N = 3$ and $S = 600$ W/m^2. The optimum results of

Table 10.2 Set of optimal results at $N = 3$ and $S = 600$ W/m^2

Cases	Algorithms	v	β	ϵ_p	T_a (K)	Temperature rise (K)	η_{th} (%)
Case 1	PSO (Siddhartha et al. 2012)	1	68.36°	0.89	280.43	10.68	72.42
	TLBO (Rao and Waghmare 2015)	1.23	59.58°	0.8835	293.93	2.1395	**76.6739**
	ETLBO	1.31	63.04°	0.8967	296.39	2.0924	**76.7321**
Case 2	PSO (Siddhartha et al. 2012)	1.02	70.31°	0.94	291.46	10.64	72.19
	TLBO (Rao and Waghmare 2015)	1.84	69.46°	0.92	294.67	10.62	**76.3181**
	ETLBO	1.92	69.67°	0.91	296.02	10.59	**76.4692**
Case 3	PSO (Siddhartha et al. 2012)	1.77	70°	0.90	280.01	10.66	72.31
	TLBO (Rao and Waghmare 2015)	1.98	69.89°	0.91	299.39	10.64	**76.4732**
	ETLBO	1.99	68.45°	0.93	300.12	10.61	**76.7301**

thermal performance are also found at different values of N and S, but for comparison purpose it is mentioned at $N = 3$ and $S = 600$ only since the results for the other settings are not available in Siddhartha et al. (2012). From the table it can be seen that the thermal efficiency is improved by 5.54 % for case 1, 5.39 % for case 2, and 5.44 % for case 3 using TLBO algorithm. The results shown in bold in the Table 10.2 indicate the best values.

The optimal thermal performance corresponding to the optimized set of values of velocity (v), tilt angle (β), emissivity of plate (ϵ_p), and ambient temperature (T_a) is determined using ETLBO and TLBO algorithms and provided in Table 10.2. Table 10.3 presents the range of thermal performance variation for different numbers of glass cover plates. Totally, three sets of glass plates have been considered. Three cases are considered to evaluate thermal performance of solar air heater using ETLBO and TLBO algorithms and the results are compared with those given by the PSO algorithm. From Table 10.3, it can be seen that the thermal efficiency increases as the number of glass cover plate increases. For case 1, the maximum thermal efficiency is obtained at $S = 600$ and Re = 20,000 and is improved by 8.70 %, 7.36 %, and 5.54 % for $N = 1$, $N = 2$, and $N = 3$, respectively, using TLBO algorithm. For case 2, the maximum thermal efficiency is obtained at $S = 600$ and Re = 20,000 and is improved by 8.89 %, 7.35 %, and 5.22 % for $N = 1$, $N = 2$, and $N = 3$, respectively, using ETLBO algorithm. For case 3, the maximum thermal efficiency is obtained at $S = 600$ and Re = 20,000 and is improved by 8.88 %, 7.21 %, and 5.21 % for $N = 1$, $N = 2$, and $N = 3$, respectively, using TLBO algorithm.

Table 10.4 represents a set of optimum results of thermal performance of solar air heater at different Reynolds numbers for $N = 1$ and $S = 600$ W/m^2 using ETLBO and TLBO algorithms and compared with the results obtained by GA, PSO, and SA algorithms. The results shown bold in the table indicate the best values. The thermal performance in terms of thermal efficiency and different operating parameters for different Reynolds numbers varying from 2000 to 20,000 with incremental step of 2000 are estimated and included in Table 10.4. In the table '–' indicates that the results are not available in the cited reference. From Table 10.4 it can be seen that the thermal efficiencies of solar air heater obtained using ETLBO and TLBO algorithms are better than those obtained using other algorithms considered like GA, PSO, and SA. Similarly, Tables 10.5, 10.6, 10.7, 10.8, 10.9, 10.10, 10.11, 10.12, 10.13, 10.14 and 10.15 show a set of optimum results at different Reynolds numbers and the number of glass cover plates (N) is varied in steps of 1 from 1 to 3 and solar radiation intensity (I) is varied in steps of 200 from 600 to 1200 W/m^2. In Tables 10.5, 10.6, 10.7, 10.8, 10.9, 10.10, 10.11, 10.12, 10.13, 10.14 and 10.15, the symbol '–' indicates that the results are not available in the cited reference. From Tables 10.5, 10.6, 10.7, 10.8, 10.9, 10.10, 10.11, 10.12, 10.13, 10.14 and 10.15, it can be seen that ETLBO and TLBO algorithms performed better than the other optimization algorithms considered by the previous researchers.

The thermal efficiency of solar air heater using TLBO and ETLBO algorithms ranges from 31.7385 % to 69.9757 % and 31.8749 % to 69.9992 %, respectively, with increasing Reynolds numbers varying from 2000 to 20,000 with an increasing step of 2000 having single glass cover and irradiance of 600 W/m^2 as shown in

Table 10.3 Range of thermal performance variation for different numbers of glass cover plates

Cases	Algorithms	$N = 1$		$N = 2$		$N = 3$	
		Min. η_{th} (S = 1000, Re = 2000) (%)	Max. η_{th} (S = 600, Re = 20,000) (%)	Min. η_{th} (S = 1000, Re = 2000) (%)	Max. η_{th} (S = 600, Re = 20,000) (%)	Min. η_{th} (S = 1000, Re = 2000) (%)	Max. η_{th} (S = 600, Re = 20,000) (%)
Case 1	PSO (Siddhartha et al. 2012)	17.24	63.88	22.95	69.08	26.36	72.42
	TLBO (Rao and Waghmare 2015)	**29.4882**	**69.9757**	**36.1047**	**74.5783**	**41.3158**	**76.6739**
	ETLBO	**29.6530**	**69.9981**	**36.3895**	**74.6739**	**41.5930**	**76.8940**
Case 2	PSO (Siddhartha et al. 2012)	17.50	63.15	22.54	68.55	27.32	71.63
	TLBO (Rao and Waghmare 2015)	**31.1038**	**69.3214**	**35.8726**	**73.9923**	**42.6736**	**75.5839**
	ETLBO	**31.2309**	**69.5930**	**35.9666**	**74.0583**	**42.8194**	**75.7312**
Case 3	PSO (Siddhartha et al. 2012)	17.65	62.89	22.91	68.88	27.10	72.31
	TLBO (Rao and Waghmare 2015)	**31.7482**	**69.0238**	**36.0827**	**74.2482**	**42.8113**	**76.2941**
	ETLBO	**31.7947**	**69.3937**	**36.4728**	**74.3943**	**42.9487**	**76.3392**

Table 10.4 Set of optimum results at different Reynolds numbers ($N = 1$ and $S = 600$ W/m^2)

S No.	Algorithm	Re	v	T_a	β	ϵ_p	T_o–T_i	η_{th} (%)
1	GA	2000	1.0392	301.6078	41.7255	0.8904	7.7582	29.2294
	PSO	2000	–	–	–	–	–	–
	SA	2000	–	–	–	–	–	19.5737
	TLBO	2000	1.3412	293.3784	36.643	0.8826	6.4286	**31.7385**
	ETLBO	2000	1.4195	294.8680	39.885	0.8921	5.9486	**31.8749**
2	GA	4000	2.9686	295.1765	57.098	0.8806	5.6814	42.1749
	PSO	4000	–	–	–	–	–	–
	SA	4000	–	–	–	–	–	31.9158
	TLBO	4000	2.3485	297.4782	41.3678	0.8698	5.4386	**43.5603**
	ETLBO	4000	2.4593	296.4113	43.5027	0.8738	5.5194	**43.6842**
3	GA	6000	1.6745	299.8824	19.2157	0.8751	4.5364	49.7669
	PSO	6000	–	–	–	–	–	–
	SA	6000	–	–	–	–	–	40.3917
	TLBO	6000	1.8692	303.7547	29.3782	0.9173	4.3298	**50.0247**
	ETLBO	6000	1.8105	301.4932	26.3949	0.9054	4.4168	**50.1159**
4	GA	8000	2.2392	296.3529	25.8039	0.8798	3.7907	55.0673
	PSO	8000	–	–	–	–	–	–
	SA	8000	–	–	–	–	–	46.2107
	TLBO	8000	2.1283	301.8377	45.8828	0.8745	3.7065	**55.9934**
	ETLBO	8000	2.1794	301.9842	39.4890	0.8767	3.6830	**56.0638**
5	GA	10,000	2.7569	302.2353	28	0.8684	3.2359	58.8518
	PSO	10,000	–	–	–	–	–	–
	SA	10,000	–	–	–	–	–	50.9748
	TLBO	10,000	2.2319	299.6739	14.2793	0.8835	3.0531	**60.4672**
	ETLBO	10,000	2.2134	300.5821	12.4890	0.8739	3.1295	**60.5295**
6	GA	12,000	1.1412	290.7059	30.7451	0.8578	2.9532	61.7762
	PSO	12,000	–	–	–	–	–	–
	SA	12,000	–	–	–	–	–	54.8929
	TLBO	12,000	1.7835	299.4882	12.4825	0.8621	2.6854	**62.8964**
	ETLBO	12,000	1.7934	300.2369	14.5894	0.8596	2.7405	**62.9349**
7	GA	14,000	2.8588	307.3333	42.8235	0.8669	2.5477	63.9401
	PSO	14,000	–	–	–	–	–	–
	SA	14,000	–	–	–	–	–	57.6921
	TLBO	14,000	2.5683	302.4248	48.3732	0.8754	2.4579	**64.7136**
	ETLBO	14,000	2.5859	303.6908	47.8991	0.8633	2.4628	**64.8302**
8	GA	16,000	1.2588	302.1569	52.7059	0.8731	2.359	65.9159
	PSO	16,000	–	–	–	–	–	–
	SA	16,000	–	–	–	–	–	60.3319
	TLBO	16,000	1.6735	298.2994	49.3467	0.8634	2.2570	**66.4579**
	ETLBO	16,000	1.7149	300.4880	50.3819	0.8730	2.2945	**66.5796**

(continued)

Table 10.4 (continued)

S No.	Algorithm	Re	v	T_a	β	ϵ_p	$T_o–T_i$	η_{th} (%)
9	GA	18,000	2.8902	308.3529	61.4902	0.8896	2.1111	67.461
	PSO	18,000	–	–	–	–	–	–
	SA	18,000	–	–	–	–	–	62.3207
	TLBO	18,000	1.9827	304.8321	52.3473	0.8739	2.0966	**68.7547**
	ETLBO	18,000	2.4914	305.9487	53.8692	0.8829	2.0588	**68.8399**
10	GA	20,000	1.5725	305.451	39.8039	0.9382	1.9403	68.7416
	PSO	20,000	–	–	–	–	–	63.88
	SA	20,000	–	–	–	–	–	64.0582
	TLBO	20,000	2.1293	302.4858	53.9622	0.8734	1.8965	**69.9757**
	ETLBO	20,000	2.2467	303.8653	49.3890	0.8852	1.8479	**69.9992**

The values corresponding to GA are taken from Varun and Siddhartha (2010). The values corresponding to PSO and SA are taken from Siddhartha et al. (2012). The values corresponding to TLBO are taken from Rao and Waghmare (2015)

Table 10.5 Set of optimum results at different Reynolds numbers (N = 2 and S = 600 W/m^2)

S No.	Algorithm	Re	v	T_a	β	ϵ_p	$T_o–T_i$	η_{th} (%)
1	GA	2000	1.5725	293.3725	54.3529	0.8571	10.2868	36.8908
	PSO	2000	–	–	–	–	–	–
	SA	2000	–	–	–	–	–	25.6859
	TLBO	2000	1.2965	297.9374	37.8745	0.8734	8.9456	**38.1378**
	ETLBO	2000	1.3289	298.8843	41.4783	0.8847	9.8598	**38.2641**
2	GA	4000	1.3765	301.2157	3.5686	0.9461	6.8671	50.324
	PSO	4000	–	–	–	–	–	–
	SA	4000	–	–	–	–	–	39.8566
	TLBO	4000	1.4385	304.7828	18.6564	0.8935	5.7835	**52.6885**
	ETLBO	4000	1.4196	306.3785	15.6933	0.8834	5.8948	**52.7293**
3	GA	6000	2.4118	308.4313	31.5686	0.9492	5.239	57.6223
	PSO	6000	–	–	–	–	–	–
	SA	6000	–	–	–	–	–	48.6497
	TLBO	6000	1.8937	305.8827	49.5543	0.9253	4.9831	**58.8843**
	ETLBO	8000	1.9189	305.9388	45.3998	0.9148	5.1058	**58.9679**
4	GA	8000	1.4	310	56.549	0.8598	4.4204	62.4028
	PSO	8000	–	–	–	–	–	–
	SA	8000	–	–	–	–	–	54.6175
	TLBO	8000	1.5735	309.8372	44.8761	0.9146	3.8736	**63.9765**
	ETLBO	10,000	1.5403	307.3881	48.2870	0.9274	4.2957	**64.0865**
5	GA	10,000	2.098	300.5098	3.2941	0.8755	3.7091	65.5798
	PSO	10,000	–	–	–	–	–	–
	SA	10,000	–	–	–	–	–	58.9721
	TLBO	10,000	1.7732	303.4788	21.3567	0.8921	3.5787	**67.5684**
	ETLBO	12,000	1.7930	304.5883	14.5886	0.9174	3.6771	**67.6882**

(continued)

Table 10.5 (continued)

S No.	Algorithm	Re	v	T_a	β	ϵ_p	T_o–T_i	η_{th} (%)
6	GA	12,000	1.8941	300.6667	9.3333	0.8606	3.2273	67.9671
	PSO	12,000	–	–	–	–	–	–
	SA	12,000	–	–	–	–	–	62.1944
	TLBO	12,000	1.3348	302.1784	16.7833	0.8846	3.0174	**68.8355**
	ETLBO	12,000	1.3689	303.6802	13.5886	0.9037	3.1039	**68.9720**
7	GA	14,000	2.1843	290.3137	40.3529	0.9206	2.8679	69.6984
	PSO	14,000	–	–	–	–	–	–
	SA	14,000	–	–	–	–	–	64.709
	TLBO	14,000	2.4927	296.3229	34.2849	0.8953	2.6583	**70.3568**
	ETLBO	14,000	2.5280	298.4788	37.9773	0.8836	2.6849	**70.4674**
8	GA	16,000	1.7294	298.7843	68.6257	0.8939	2.5842	71.3131
	PSO	16,000	–	–	–	–	–	–
	SA	16,000	–	–	–	–	–	66.7539
	TLBO	16,000	1.8943	301.1673	45.8392	0.8635	2.4846	**72.3462**
	ETLBO	16,000	1.9284	301.6758	51.0482	0.8701	2.5295	**72.4858**
9	GA	18,000	1.0471	302	23.0588	0.8888	2.318	72.4333
	PSO	18,000	–	–	–	–	–	–
	SA	18,000	–	–	–	–	–	68.3109
	TLBO	18,000	1.7139	303.7845	11.2863	0.9265	2.2739	**73.1954**
	ETLBO	18,000	1.7468	302.5903	14.7731	0.9265	2.2923	**73.2499**
10	GA	20,000	2.1608	299.8039	11.5294	0.8939	2.1088	73.4772
	PSO	20,000	–	–	–	–	–	69.08
	SA	20,000	–	–	–	–	–	69.7965
	TLBO	20,000	1.8635	304.6286	6.3283	0.8738	2.0683	**74.5783**
	ETLBO	20,000	1.8395	305.05992	8.2224	0.8836	2.0869	**74.6490**

The values corresponding to GA are taken from Varun and Siddhartha (2010). The values corresponding to PSO and SA are taken from Siddhartha et al. (2012). The values corresponding to TLBO are taken from Rao and Waghmare (2015)

Table 10.4. Similarly, the performance range using TLBO algorithm is from 38.1378 % to 74.5783 % and 43.3532 % to 76.6739 % for the same range of Reynolds number and increasing step size and irradiance having two and three glass covers, respectively. The thermal efficiency ranges from 38.2641 % to 74.6490 % for two glass covers and 43.4167 % to 76.7881 % for three glass covers using ETLBO algorithm for the same range of Reynolds number and irradiance. The maximum value of thermal efficiency is 76.7881 % and is obtained with three glass cover plates and irradiance of 600 W/m^2 at Reynolds number 20,000. The maximum value of efficiency is obtained at V = 1.3194 m/s, tilt angle = 60.4883°, emissivity of plate = 0.8947, ambient temperature = 292.5993 K, and temperature rise of 2.1448 K using ETLBO algorithm. Hence, it can be concluded from Tables 10.5, 10.6, 10.7, 10.8, 10.9, 10.10, 10.11, 10.12, 10.13, 10.14 and 10.15 that the thermal performance of a flat plate solar air heater increases with the increase in Reynolds number.

Table 10.6 Set of optimum results at different Reynolds numbers (N = 3 and S = 600 W/m^2)

S. No.	Algorithm	Re	v	T_a	β	ϵ_p	T_o–T_i	η_{th} (%)
1	GA	2000	1.4	309.2941	41.1765	0.8708	11.35	41.7897
	PSO	2000	1	282.76	14.4	0.92	41.69	28.38
	SA	2000	–	–	–	–	–	19.5737
	TLBO	2000	1.1265	301.3783	22.6394	0.8856	9.2992	**43.3532**
	ETLBO	2000	1.1977	302.5889	27.7832	0.8729	9.3156	**43.4167**
2	GA	4000	2.2627	293.3725	48.8627	0.8908	7.8267	55.1325
	PSO	4000	1	282.38	23.1	0.86	31.89	43.37
	SA	4000	–	–	–	–	–	31.9158
	TLBO	4000	1.4826	299.5738	36.8323	0.8912	7.2369	**56.7543**
	ETLBO	4000	1.5289	300.4992	31.5992	0.8846	7.3146	**56.8976**
3	GA	6000	1.1961	292.7451	64.5098	0.9488	5.9713	61.852
	PSO	6000	1	280.04	53.78	0.9	25.95	52.88
	SA	6000	–	–	–	–	–	40.3917
	TLBO	6000	1.6357	294.4782	45.2376	0.9243	4.9832	**63.4863**
	ETLBO	6000	1.5938	293.4886	49.6770	0.9071	5.0849	**63.5367**
4	GA	8000	1.9725	292.2745	0	0.8633	4.7422	66.1788
	PSO	8000	1.02	280.07	19.46	0.9	21.28	58.25
	SA	8000	–	–	–	–	–	46.2107
	TLBO	8000	1.3789	290.7489	31.5943	0.9074	4.1294	**67.6733**
	ETLBO	8000	1.4277	289.1578	32.5836	0.8846	4.2276	**67.7650**
5	GA	10,000	2.7098	295.5686	2.7451	0.8916	3.9458	68.8946
	PSO	10,000	1	280.11	30.79	0.93	18.23	61.95
	SA	10,000	–	–	–	–	–	50.9748
	TLBO	10,000	1.4937	291.7112	42.6582	0.9247	3.7836	**70.4711**
	ETLBO	10,000	1.5830	292.5988	45.8839	0.9175	3.8682	**70.5652**
6	GA	12,000	1.1961	298.2353	49.9608	0.8516	3.4344	71.0254
	PSO	12,000	1	280.04	62.86	0.94	16.08	65.49
	SA	12,000	–	–	–	–	–	54.8929
	TLBO	12,000	1.7619	300.1388	57.9334	0.9376	3.2694	**72.2885**
	ETLBO	12,000	1.8204	300.6883	61.0693	0.9188	3.2790	**72.3689**
7	GA	14,000	2.0275	292.2745	66.7059	0.9088	3.0221	72.5723
	PSO	14,000	1	280.02	20.45	0.93	14.13	67.72
	SA	14,000	–	–	–	–	–	57.6921
	TLBO	14,000	1.2834	295.7835	39.1178	0.9421	2.8479	**73.6319**
	ETLBO	14,000	1.3384	294.5880	43.5883	0.9375	2.9003	**73.7436**
8	GA	16,000	2.3725	301.2157	65.6078	0.9461	2.6706	73.8123
	PSO	16,000	1.04	280.01	44.17	0.9	12.76	69.28
	SA	16,000	–	–	–	–	–	60.3319
	TLBO	16,000	1.7482	298.5294	61.9326	0.8995	2.5636	**74.3058**
	ETLBO	16,000	1.6927	297.1885	64.2210	0.9048	2.6266	**74.3686**

(continued)

Table 10.6 (continued)

S. No.	Algorithm	Re	v	T_a	β	ϵ_p	T_o–T_i	η_{th} (%)
9	GA	18,000	1.4038	297.6078	20.8627	0.879	2.4098	74.8367
	PSO	18,000	1	280.02	26.52	0.87	11.57	70.84
	SA	18,000	–	–	–	–	–	62.3207
	TLBO	18,000	1.1619	293.8943	21.8311	0.9105	2.3295	**75.9472**
	ETLBO	18,000	1.2184	293.5992	26.5889	0.9274	2.3731	**75.9977**
10	GA	20,000	2.9529	296.1176	65.3333	0.8618	2.2047	75.6454
	PSO	20,000	1	280.43	68.36	0.89	10.68	72.42
	SA	20,000	–	–	–	–	–	64.0582
	TLBO	20,000	1.2729	293.9362	59.5832	0.8835	2.1395	**76.6739**
	ETLBO	20,000	1.3194	292.5993	60.4883	0.8947	2.1448	**76.7881**

The values corresponding to GA are taken from Varun and Siddhartha (2010). The values corresponding to PSO and SA are taken from Siddhartha et al. (2012). The values corresponding to TLBO are taken from Rao and Waghmare (2015)

Table 10.7 Set of optimum results at different Reynolds numbers (N = 1 and S = 800 W/m^2)

S. No.	Algorithm	Re	v	T_a	β	ϵ_p	T_o–T_i	η_{th} (%)
1	GA	2000	2.7725	292.0392	24.4314	0.9229	9.2733	28.2512
	PSO	2000	–	–	–	–	–	–
	SA	2000	–	–	–	–	–	18.7072
	TLBO	2000	2.1139	297.3844	38.3772	0.8895	7.8746	**30.9532**
	ETLBO	2000	2.1743	296.9932	35.8419	0.8736	7.9023	**31.0765**
2	GA	4000	1.2588	296.0392	53.5294	0.9237	7.5112	41.1115
	PSO	4000	–	–	–	–	–	–
	SA	4000	–	–	–	–	–	30.836
	TLBO	4000	1.6726	294.5783	23.4788	0.8631	6.5937	**43.0947**
	ETLBO	4000	1.7289	294.6902	29.4729	0.8573	6.6675	**43.2167**
3	GA	6000	2.098	294.2353	6.8627	0.9135	5.8441	48.9045
	PSO	6000	–	–	–	–	–	–
	SA	6000	–	–	–	–	–	39.2944
	TLBO	6000	1.8371	290.1247	23.9984	0.8953	5.3693	**49.9363**
	ETLBO	6000	1.8749	291.4892	18.6786	0.9041	5.4530	**50.0755**
4	GA	8000	2.7176	290.7059	59.0196	0.9331	4.9948	54.1312
	PSO	8000	–	–	–	–	–	–
	SA	8000	–	–	–	–	–	45.5684
	TLBO	8000	2.2193	298.3667	47.2787	0.9256	4.6932	**55.3843**
	ETLBO	8000	2.1905	299.4886	51.0472	0.9143	4.7858	**55.4976**
5	GA	10,000	1.7294	301.451	57.098	0.881	4.3663	58.2862
	PSO	10,000	–	–	–	–	–	–
	SA	10,000	–	–	–	–	–	50.2908
	TLBO	10,000	1.5632	299.4882	34.5367	0.9173	4.1395	**59.6832**
	ETLBO	10,000	1.5199	300.5119	37.5678	0.9276	4.2227	**59.7754**

(continued)

Table 10.7 (continued)

S. No.	Algorithm	Re	v	T_a	β	ϵ_p	T_o–T_i	η_{th} (%)
6	GA	12,000	1.9176	300.3529	48.0392	0.9069	3.8012	61.0959
	PSO	12,000	–	–	–	–	–	–
	SA	12,000	–	–	–	–	–	54.1293
	TLBO	12,000	1.8936	302.4781	38.4673	0.9268	3.6395	**62.8734**
	ETLBO	12,000	1.8947	302.5882	40.5882	0.9486	3.6629	**62.9852**
7	GA	14,000	1.7765	297.451	57.6471	0.8947	3.4416	63.5274
	PSO	14,000	–	–	–	–	–	–
	SA	14,000	–	–	–	–	–	57.1511
	TLBO	14,000	1.4423	299.4268	27.5764	0.8748	3.1957	**64.2154**
	ETLBO	14,000	1.4793	300.9932	29.5881	0.8663	3.2548	**64.3266**
8	GA	16,000	3.0	291.6471	66.1569	0.9473	3.0741	65.3202
	PSO	16,000	–	–	–	–	–	–
	SA	16,000	–	–	–	–	–	59.6272
	TLBO	16,000	2.7394	295.3278	51.6478	0.9376	2.8643	**66.0115**
	ETLBO	16,000	2.6939	294.8502	55.3995	0.9112	2.8935	**66.1253**
9	GA	18,000	2.5922	291.4118	67.7647	0.9214	2.8268	66.943
	PSO	18,000	–	–	–	–	–	–
	SA	18,000	–	–	–	–	–	61.4528
	TLBO	18,000	2.8846	298.2783	32.9754	0.9475	2.6493	**68.0364**
	ETLBO	18,000	2.8205	298.4729	30.0928	0.9384	2.7754	**68.1352**
10	GA	20,000	2.0353	309.2941	14.2745	0.8606	2.5614	68.2924
	PSO	20,000	–	–	–	–	–	–
	SA	20,000	–	–	–	–	–	63.3172
	TLBO	20,000	2.3732	305.8392	43.4174	0.9146	2.4395	**69.8921**
	ETLBO	20,000	2.4993	304.3955	41.4882	0.9301	2.5038	**69.9675**

The values corresponding to GA are taken from Varun and Siddhartha (2010). The values corresponding to PSO and SA are taken from Siddhartha et al. (2012). The values corresponding to TLBO are taken from Rao and Waghmare (2015)

In this chapter, the results of GA, PSO, SA, TLBO, and ETLBO algorithms are presented for the thermal performance of a smooth flat plate solar air heater (SFPSAH). Maximization of thermal efficiency of SFPSAH is considered as the objective function. The thermal performance is obtained for different Reynolds numbers, irradiance, and number of glass plates. The maximum value of thermal efficiency of 76.7881 % is obtained using ETLBO algorithm with wind velocity of 1.3194 m/s, tilt angle of 60.4883°, plate emissivity of 0.8947, ambient temperature of 292.5993 K, temperature rise of 2.1448 K, irradiance of 600, and Reynolds number of 20,000. The final results obtained by the TLBO and ETLBO algorithms are compared with other optimization algorithms like GA, PSO, and SA and are found to be satisfactory. The results also show that the thermal performance increases with the Reynolds number and the number of glass cover plates but

Table 10.8 Set of optimum results at different Reynolds numbers ($N = 2$ and $S = 800\ \mathrm{W/m^2}$)

S. No.	Algorithm	Re	v	T_a	β	ϵ_p	T_o–T_i	η_{th} (%)
1	GA	2000	1.1098	308.7451	26.0784	0.9186	12.4267	35.6044
	PSO	2000	–	–	–	–	–	–
	SA	2000	–	–	–	–	–	24.5662
	TLBO	2000	1.3847	301.2184	35.7628	0.8845	11.5783	**37.1049**
	ETLBO	2000	1.4285	302.5899	31.5882	0.8954	11.6849	**37.2144**
2	GA	4000	2.2392	294.7843	34.3137	0.9229	9.0186	49.2273
	PSO	4000	–	–	–	–	–	–
	SA	4000	–	–	–	–	–	38.3891
	TLBO	4000	2.8374	291.9345	52.8593	0.8936	8.3491	**50.6738**
	ETLBO	4000	2.8835	291.5993	49.5781	0.8720	8.4395	**50.7362**
3	GA	6000	2.8039	297.7647	26.6275	0.9229	6.9579	56.7569
	PSO	6000	–	–	–	–	–	–
	SA	6000	–	–	–	–	–	47.5704
	TLBO	6000	2.3948	301.8943	41.5675	0.8954	6.1103	**58.2194**
	ETLBO	6000	2.4495	300.4062	38.1048	0.8847	6.3294	**58.3215**
4	GA	8000	2.5529	292.1176	1.098	0.8673	5.7996	61.6262
	PSO	8000	–	–	–	–	–	–
	SA	8000	–	–	–	–	–	53.7162
	TLBO	8000	2.1284	296.3848	18.3573	0.8843	5.2295	**63.3042**
	ETLBO	8000	2.1759	295.3996	15.6994	0.8839	5.2959	**63.3876**
5	GA	10,000	2.6	290.4706	17.5686	0.8884	4.9223	65.0241
	PSO	10,000	–	–	–	–	–	–
	SA	10,000	–	–	–	–	–	58.1612
	TLBO	10,000	2.1839	297.8835	5.6884	0.9256	4.6402	**66.8211**
	ETLBO	10,000	2.2450	295.9950	10.5810	0.9184	4.7732	**66.9485**
6	GA	12,000	1.5255	303.4902	48.3137	0.852	4.2986	67.4832
	PSO	12,000	–	–	–	–	–	–
	SA	12,000	–	–	–	–	–	61.3731
	TLBO	12,000	1.3111	298.7145	12.4678	0.8853	3.9836	**68.7343**
	ETLBO	12,000	1.2857	298.5892	9.4996	0.8951	4.1167	**68.8256**
7	GA	14,000	1.1412	290.0784	11.2549	0.9488	3.7964	69.3503
	PSO	14,000	–	–	–	–	–	–
	SA	14,000	–	–	–	–	–	64.1098
	TLBO	14,000	1.7829	294.7883	21.4573	0.9145	3.5739	**70.2859**
	ETLBO	14,000	1.7693	293.5899	24.5096	09299	3.6680	**70.3965**
8	GA	16,000	1.0314	294.7059	25.8039	0.8869	3.4103	70.8033
	PSO	16,000	–	–	–	–	–	–
	SA	16,000	–	–	–	–	–	66.1263
	TLBO	16,000	1.4388	290.8253	34.8754	0.9354	3.2954	**71.9348**
	ETLBO	16,000	1.4392	291.5900	37.5092	0.9174	3.3045	**71.9998**

(continued)

Table 10.8 (continued)

S. No.	Algorithm	Re	v	T_a	β	ϵ_p	T_o–T_i	η_{th} (%)
9	GA	18,000	1.7686	309.451	5.4902	0.8912	3.0433	72.1471
	PSO	18,000	–	–	–	–	–	–
	SA	18,000	–	–	–	–	–	67.8595
	TLBO	18,000	1.9999	302.7843	21.5784	0.9257	2.9184	**73.7721**
	ETLBO	18,000	2.1048	301.4995	17.5572	0.9184	2.9876	**73.8976**
10	GA	20,000	1.0941	301.2157	21.1373	0.9382	2.8031	73.1107
	PSO	20,000	–	–	–	–	–	–
	SA	20,000	–	–	–	–	–	69.3062
	TLBO	20,000	1.6583	307.2738	7.8345	0.8853	2.6754	**74.4992**
	ETLBO	20,000	1.7294	305.9992	10.4983	0.8923	2.7341	**74.5786**

The values corresponding to GA are taken from Varun and Siddhartha (2010). The values corresponding to PSO and SA are taken from Siddhartha et al. (2012). The values corresponding to TLBO are taken from Rao and Waghmare (2015)

Table 10.9 Set of optimum results at different Reynolds numbers (N = 3 and S = 800 W/m^2)

S. No.	Algorithm	Re	v	T_a	β	ϵ_p	T_o–T_i	η_{th} (%)
1	GA	2000	2.3804	299.3333	12.902	0.9229	14.4294	40.3751
	PSO	2000	–	–	–	–	–	–
	SA	2000	–	–	–	–	–	18.7072
	TLBO	2000	1.8947	293.8122	32.3632	0.8954	12.4839	**41.8493**
	ETLBO	2000	1.9594	294.6839	28..5881	0.8823	12.6943	**41.9643**
2	GA	4000	1.8078	298.7059	49.6863	0.9088	10.1644	54.124
	PSO	4000	–	–	–	–	–	–
	SA	4000	–	–	–	–	–	30.836
	TLBO	4000	2.1378	293.7229	21.4673	0.8842	8.9937	**55.9932**
	ETLBO	4000	2.0948	292.5902	19.4774	0.8740	9.0485	**56.0958**
3	GA	6000	2.7725	299.098	29.6471	0.8524	7.694	61.1521
	PSO	6000	–	–	–	–	–	–
	SA	6000	–	–	–	–	–	39.2944
	TLBO	6000	2.4296	290.1283	36.9761	0.8824	6.8746	**62.4839**
	ETLBO	6000	2.3960	291.5009	33.3252	0.8990	6.9341	**62.5386**
4	GA	8000	2.3412	307.098	70	0.939	6.2278	65.5463
	PSO	8000	–	–	–	–	–	–
	SA	8000	–	–	–	–	–	45.5684
	TLBO	8000	2.6510	304.9392	58.3468	0.8964	5.7385	**67.2395**
	ETLBO	8000	2.7395	303.5681	62.6925	0.8834	5.8852	**67.3664**

(continued)

Table 10.9 (continued)

S. No.	Algorithm	Re	v	T_a	β	ϵ_p	T_o–T_i	η_{th} (%)
5	GA	10,000	1.4863	293.5294	14.8235	0.8751	5.2581	68.485
	PSO	10,000	–	–	–	–	–	–
	SA	10,000	–	–	–	–	–	50.2908
	TLBO	10,000	1.5738	301.3935	34.4674	0.9365	4.8624	**69.8883**
	ETLBO	10,000	1.6905	301.9593	31.4883	0.9267	4.8959	**69.9765**
6	GA	12,000	1.6039	296.4314	45.2941	0.941	4.5261	70.5836
	PSO	12,000	–	–	–	–	–	–
	SA	12,000	–	–	–	–	–	54.1293
	TLBO	12,000	1.2399	300.3782	25.2485	0.9156	4.3638	**72.0012**
	ETLBO	12,000	1.3957	299.4896	28.6773	0.9267	4.4185	**72.1076**
7	GA	14,000	1.7059	309.6863	21.9608	0.8614	3.9446	72.2261
	PSO	14,000	–	–	–	–	–	–
	SA	14,000	–	–	–	–	–	57.1511
	TLBO	14,000	1.3911	304.5692	9.2474	0.8951	3.7953	**73.4924**
	ETLBO	14,000	1.4589	303.5896	11.5224	0.9076	3.8377	**73.5633**
8	GA	16,000	1.4627	305.7647	61.2157	0.9124	3.5459	73.5504
	PSO	16,000	–	–	–	–	–	–
	SA	16,000	–	–	–	–	–	59.6272
	TLBO	16,000	1.6784	299.7832	48.3466	0.8627	3.3681	**74.1775**
	ETLBO	16,000	1.7395	300.5881	44.6767	0.8870	3.4189	**74.2436**
9	GA	18,000	2.4824	296.2745	44.1961	0.9331	3.1948	74.5676
	PSO	18,000	–	–	–	–	–	–
	SA	18,000	–	–	–	–	–	61.4528
	TLBO	18,000	2.0384	291.5622	29.4673	0.8776	3.0193	**75.3221**
	ETLBO	18,000	2.1449	292.5993	31.3775	0.8821	3.1139	**75.4428**
10	GA	20,000	1.2275	309.2157	51.8824	0.9422	2.9059	75.3931
	PSO	20,000	–	–	–	–	–	–
	SA	20,000	–	–	–	–	–	63.3172
	TLBO	20,000	1.6293	302.7223	32.4672	0.9189	2.7646	**76.5913**
	ETLBO	20,000	1.5639	303.5992	30.5883	0.9034	2.8294	**76.6857**

The values corresponding to GA are taken from Varun and Siddhartha (2010). The values corresponding to PSO and SA are taken from Siddhartha et al. (2012). The values corresponding to TLBO are taken from Rao and Waghmare (2015)

slightly decreases with the increase in irradiance. Hence, it can be concluded that TLBO and ETLBO algorithms are effective algorithms and have potential for finding the optimal set of design and operating parameters at which the thermal performance of a smooth flat plate solar air heater is maximum. Again, it is observed that the concept of elitism enhances the performance of the TLBO algorithm for the optimization of smooth flat plate solar air heater.

Table 10.10 Set of optimum results at different Reynolds numbers (N = 1 and S = 1000 W/m^2)

S. No.	Algorithm	Re	v	T_a	β	ϵ_p	T_o–T_i	η_{th} (%)
1	GA	2000	2.2471	298.9412	35.6863	0.8665	11.444	27.4864
	PSO	2000	–	–	–	–	–	17.24
	SA	2000	–	–	–	–	–	17.6966
	TLBO	2000	1.8831	293.2967	47.3782	0.8965	10.3842	**29.4882**
	ETLBO	2000	1.9478	294.6893	41.4885	0.8829	10.5638	**29.5744**
2	GA	4000	2.3176	291.4118	10.4314	0.8539	8.9379	40.1084
	PSO	4000	–	–	–	–	–	–
	SA	4000	–	–	–	–	–	30.0172
	TLBO	4000	2.1293	295.2434	4.8932	0.8834	7.8343	**41.9837**
	ETLBO	4000	2.0845	296.9913	8.4991	0.8654	7.9945	**42.0754**
3	GA	6000	2.6941	298.3137	4.3922	0.85	7.1593	48.1196
	PSO	6000	–	–	–	–	–	–
	SA	6000	–	–	–	–	–	38.1824
	TLBO	6000	2.1934	299.8387	19.3893	0.8924	6.9837	**49.8212**
	ETLBO	6000	2.1048	300.5992	16.4886	0.8894	7.0184	**49.9374**
4	GA	8000	1.2667	303.4902	1.9216	0.932	6.0706	53.5817
	PSO	8000	–	–	–	–	–	–
	SA	8000	–	–	–	–	–	44.6621
	TLBO	8000	1.3847	297.3289	18.4923	0.8999	5.7184	**55.2149**
	ETLBO	8000	1.2759	297.5891	15.9235	0.9043	5.7941	**55.3298**
5	GA	10,000	1.8471	295.2549	26.3529	0.8633	5.3605	57.657
	PSO	10,000	–	–	–	–	–	–
	SA	10,000	–	–	–	–	–	49.3149
	TLBO	10,000	2.2389	299.5263	43.6722	0.9135	4.9285	**59.1038**
	ETLBO	10,000	2.1859	300.5892	38.4624	0.9265	4.9960	**59.1957**
6	GA	12,000	1.3765	304.7451	22.2353	0.9484	4.6377	60.5783
	PSO	12,000	–	–	–	–	–	–
	SA	12,000	–	–	–	–	–	53.4182
	TLBO	12,000	1.2184	300.3784	11.3832	0.9145	4.3795	**62.4895**
	ETLBO	12,000	1.3240	299.5899	13.5822	0.9268	4.5018	**62.5893**
7	GA	14,000	2.7725	309.2941	17.8431	0.9288	4.0572	63.0055
	PSO	14,000	–	–	–	–	–	–
	SA	14,000	–	–	–	–	–	56.5065
	TLBO	14,000	2.3495	303.5638	36.2781	0.8936	3.8947	**64.1783**
	ETLBO	14,000	2.4885	302.5063	32.5992	0.9076	3.9048	**64.2305**
8	GA	16,000	2.1451	293.3725	5.2157	0.9371	3.7709	65.0191
	PSO	16,000	–	–	–	–	–	–
	SA	16,000	–	–	–	–	–	59.0541
	TLBO	16,000	2.4183	297.4567	23.6638	0.8954	3.6397	**65.9987**
	ETLBO	16,000	2.4967	298.3906	21.5993	0.9108	3.7583	**66.0632**

(continued)

Table 10.10 (continued)

S. No.	Algorithm	Re	v	T_a	β	ϵ_p	T_o–T_i	η_{th} (%)
9	GA	18,000	1.7529	291.2549	27.1765	0.9127	3.4971	66.6081
	PSO	18,000	–	–	–	–	–	–
	SA	18,000	–	–	–	–	–	61.1254
	TLBO	18,000	2.0847	290.8743	8.2563	0.8845	3.3865	**67.6341**
	ETLBO	18,000	2.1759	289.5993	10.4934	0.8999	3.4069	**67.7203**
10	GA	20,000	2.4039	292.1961	55.7255	0.9182	3.2241	67.7749
	PSO	20,000	–	–	–	–	–	–
	SA	20,000	–	–	–	–	–	62.9629
	TLBO	20,000	2.1194	296.9633	36.6728	0.9043	3.1975	**69.1102**
	ETLBO	20,000	2.0489	297.4063	38.9593	0.9158	3.2149	**69.2360**

The values corresponding to GA are taken from Varun and Siddhartha (2010). The values corresponding to PSO and SA are taken from Siddhartha et al. (2012). The values corresponding to TLBO are taken from Rao and Waghmare (2015)

Table 10.11 Set of optimum results at different Reynolds numbers ($N = 2$ and $S = 1000$ W/m^2)

S. No.	Algorithm	Re	v	T_a	β	ϵ_p	T_o–T_i	η_{th} (%)
1	GA	2000	2.9294	308.8235	27.1765	0.9249	14.3719	34.5239
	PSO	2000	–	–	–	–	–	22.95
	SA	2000	–	–	–	–	–	23.5642
	TLBO	2000	2.7839	302.5673	39.2781	0.9054	13.7954	**36.1047**
	ETLBO	2000	2.5893	303.4893	31.5883	0.8993	13.8749	**36.1857**
2	GA	4000	2.0196	296.5098	11.5294	0.8696	11.0482	48.311
	PSO	4000	–	–	–	–	–	–
	SA	4000	–	–	–	–	–	37.5379
	TLBO	4000	2.7329	301.3485	29.4882	0.8756	9.9535	**49.8348**
	ETLBO	4000	2.6830	301.4892	30.0353	0.8847	9.9837	**49.9941**
3	GA	6000	2.6706	295.8824	67.2549	0.941	8.7547	56.0124
	PSO	6000	–	–	–	–	–	–
	SA	6000	–	–	–	–	–	46.5347
	TLBO	6000	2.2937	300.3962	48.3257	0.8842	8.1482	**57.9234**
	ETLBO	6000	2.1859	299.3995	45.9583	0.8912	8.2759	**58.1286**
4	GA	8000	1.5333	292.3529	9.6078	0.9245	7.1766	61.0463
	PSO	8000	–	–	–	–	–	–
	SA	8000	–	–	–	–	–	52.6971
	TLBO	8000	1.6621	290.4852	27.6829	0.8944	6.8637	**62.8342**
	ETLBO	8000	1.5824	288.3995	21.9593	0.9048	6.9402	**62.9930**

(continued)

Table 10.11 (continued)

S. No.	Algorithm	Re	v	T_a	β	ϵ_p	T_o–T_i	η_{th} (%)
5	GA	10,000	2.2627	303.4118	52.7059	0.8755	6.0997	64.3693
	PSO	10,000	–	–	–	–	–	–
	SA	10,000	–	–	–	–	–	57.3815
	TLBO	10,000	2.0382	297.3584	32.5710	0.8685	5.6786	**66.1038**
	ETLBO	10,000	1.9451	298.4893	27.5883	0.8805	5.7858	**66.2695**
6	GA	12,000	2.5765	303.1961	67.2549	0.9429	5.2672	67.0317
	PSO	12,000	–	–	–	–	–	–
	SA	12,000	–	–	–	–	–	60.7765
	TLBO	12,000	2.2146	302.2468	53.2892	0.8821	4.8953	**68.8937**
	ETLBO	12,000	2.2054	301.3992	52.5883	0.8989	4.9185	**68.9638**
7	GA	14,000	1.4549	303.6471	5.2157	0.9029	4.6683	69.0025
	PSO	14,000	–	–	–	–	–	–
	SA	14,000	–	–	–	–	–	63.3947
	TLBO	14,000	1.7816	294.6549	32.8267	0.9421	4.4168	**70.0127**
	ETLBO	14,000	1.7938	293.4902	29.0953	0.9375	4.4596	**70.1748**
8	GA	16,000	2.1686	307.8824	56	0.8692	4.2096	70.5238
	PSO	16,000	–	–	–	–	–	–
	SA	16,000	–	–	–	–	–	65.6112
	TLBO	16,000	1.9725	302.3473	34.9392	0.9184	3.9994	**71.6739**
	ETLBO	16,000	1.9982	301.3892	31.4883	0.9274	4.1994	**71.8375**
9	GA	18,000	1.7294	304.1961	65.0588	0.9375	3.8271	71.7789
	PSO	18,000	–	–	–	–	–	–
	SA	18,000	–	–	–	–	–	67.4154
	TLBO	18,000	1.6427	308.3697	47.1003	0.8756	3.6379	**72.5611**
	ETLBO	18,000	1.5888	307.2994	46.3898	0.8868	3.7584	**72.6739**
10	GA	20,000	2.0196	302.4706	4.6667	0.9492	3.4616	72.8854
	PSO	20,000	–	–	–	–	–	–
	SA	20,000	–	–	–	–	–	68.8388
	TLBO	20,000	2.3991	298.7654	18.3877	0.9257	3.3982	**74.2392**
	ETLBO	20,000	2.4859	299.3899	13.5967	0.9177	3.4059	**74.3958**

The values corresponding to GA are taken from Varun and Siddhartha (2010). The values corresponding to PSO and SA are taken from Siddhartha et al. (2012). The values corresponding to TLBO are taken from Rao and Waghmare (2015)

Table 10.12 Set of optimum results at different Reynolds numbers (N = 3 and S = 1000 W/m^2)

S. No.	Algorithm	Re	v	T_a	β	ϵ_p	T_o–T_i	η_{th} (%)
1	GA	2000	2.2314	296.8235	69.451	0.9073	18.1867	39.2811
	PSO	2000	–	–	–	–	–	26.36
	SA	2000	–	–	–	–	–	17.6966
	TLBO	2000	1.8366	299.3458	56.2886	0.9256	16.7855	**41.3158**
	ETLBO	2000	1.7868	297.6775	58.5843	0.9475	16.8942	**41.4567**
2	GA	4000	1.5725	303.098	47.7647	0.8614	12.4564	53.1766
	PSO	4000	–	–	–	–	–	–
	SA	4000	–	–	–	–	–	30.0172
	TLBO	4000	1.3857	300.4832	34.5673	0.8953	11.2398	**55.0012**
	ETLBO	4000	1.2885	301.5782	31.5678	0.9099	11.3752	**55.1849**
3	GA	6000	2.1451	304.5882	29.098	0.888	9.4275	60.5002
	PSO	6000	–	–	–	–	–	–
	SA	6000	–	–	–	–	–	38.1824
	TLBO	6000	1.8453	307.9765	16.4873	0.8745	8.6536	**62.2119**
	ETLBO	6000	1.7499	306.9883	18.5356	0.8748	8.7395	**62.3485**
4	GA	8000	1.9333	301.1373	48.3137	0.8739	7.7438	64.9469
	PSO	8000	–	–	–	–	–	–
	SA	8000	–	–	–	–	–	44.6621
	TLBO	8000	2.1183	296.3696	62.5893	0.8951	7.0964	**66.3957**
	ETLBO	8000	2.2820	299.4883	58.4210	0.9048	7.2395	**66.4796**
5	GA	10,000	2.4431	294.3922	24.9804	0.9175	6.4845	68.0499
	PSO	10,000	–	–	–	–	–	–
	SA	10,000	–	–	–	–	–	49.3149
	TLBO	10,000	2.1827	298.4594	8.3462	0.9054	5.996	**69.7367**
	ETLBO	10,000	2.2591	296.8048	11.4827	0.9012	6.1087	**69.8106**
6	GA	12,000	2.1294	301.7647	33.4902	0.8904	5.5936	70.27
	PSO	12,000	–	–	–	–	–	–
	SA	12,000	–	–	–	–	–	53.4182
	TLBO	12,000	2.6748	297.2470	48.3672	0.8797	5.3478	**71.7732**
	ETLBO	12,000	2.6849	299.1900	43.5729	0.8867	5.4187	**71.9256**
7	GA	14,000	1.0627	303.4118	3.0196	0.9014	4.9224	71.9223
	PSO	14,000	–	–	–	–	–	–
	SA	14,000	–	–	–	–	–	56.5065
	TLBO	14,000	1.3882	297.6738	23.6482	0.9262	4.6855	**73.3937**
	ETLBO	14,000	1.4700	299.3851	21.7573	0.9191	4.7231	**73.5058**
8	GA	16,000	1.8	295.2549	7.1373	0.8782	4.4093	73.2418
	PSO	16,000	–	–	–	–	–	–
	SA	16,000	–	–	–	–	–	59.0541
	TLBO	16,000	1.5638	299.4537	19.3678	0.8963	4.2378	**74.1189**
	ETLBO	18,000	1.5194	300.0278	23.6984	0.8921	4.3956	**74.2484**

(continued)

Table 10.12 (continued)

S. No.	Algorithm	Re	v	T_a	β	ϵ_p	T_o–T_i	η_{th} (%)
9	GA	18,000	1.4941	290.7059	58.1961	0.8594	4.0209	74.2811
	PSO	18,000	–	–	–	–	–	–
	SA	18,000	–	–	–	–	–	61.1254
	TLBO	18,000	1.2811	293.2882	23.7748	0.8848	3.8267	**75.1932**
	ETLBO	18,000	1.3960	295.5831	26.4572	0.8885	3.9284	**75.2958**
10	GA	20,000	1	291.3333	21.6863	0.9433	3.6408	75.2149
	PSO	20,000	–	–	–	–	–	–
	SA	20,000	–	–	–	–	–	62.3454
	TLBO	20,000	1.3294	294.9836	47.2782	0.9145	3.5796	**76.4188**
	ETLBO	20,000	1.2229	295.3782	43.6783	0.9286	3.6239	**76.5739**

The values corresponding to GA are taken from Varun and Siddhartha (2010). The values corresponding to PSO and SA are taken from Siddhartha et al. (2012). The values corresponding to TLBO are taken from Rao and Waghmare (2015)

Table 10.13 Set of optimum results at different Reynolds numbers ($N = 1$ and $S = 1200$ W/m^2)

S. No.	Algorithm	Re	v	T_a	β	ϵ_p	T_o–T_i	η_{th} (%)
1	GA	2000	1.1098	298	12.0784	0.8524	13.7712	26.6961
	PSO	2000	–	–	–	–	–	–
	SA	2000	–	–	–	–	–	17.4604
	TLBO	2000	1.5248	294.7832	23.3772	0.8738	12.7467	**28.3282**
	ETLBO	2000	1.4953	295.6993	19.4775	0.8947	12.8943	**28.4784**
2	GA	4000	2.6784	301.8431	30.7451	0.8669	10.178	39.3842
	PSO	4000	–	–	–	–	–	–
	SA	4000	–	–	–	–	–	29.0595
	TLBO	4000	2.4261	296.3782	49.2628	0.8634	9.4684	**41.6739**
	ETLBO	4000	2.3859	295.7830	48.3779	0.8747	9.5738	**41.8304**
3	GA	6000	2.0431	304.4314	68.3529	0.9022	8.6299	47.2705
	PSO	6000	–	–	–	–	–	–
	SA	6000	–	–	–	–	–	37.6487
	TLBO	6000	1.8131	301.7223	57.2837	0.8854	7.9854	**49.7638**
	ETLBO	6000	1.7499	299.4880	57.3675	0.8720	8.2885	**49.8495**
4	GA	8000	1.7765	294.7843	58.7451	0.8947	7.4196	52.8345
	PSO	8000	–	–	–	–	–	–
	SA	8000	–	–	–	–	–	43.9187
	TLBO	8000	1.3989	297.3629	62.4629	0.9252	6.8594	**54.7834**
	ETLBO	8000	1.4527	296.3712	57.3885	0.9184	6.9937	**54.9121**

(continued)

Table 10.13 (continued)

S. No.	Algorithm	Re	v	T_a	β	ϵ_p	T_o–T_i	η_{th} (%)
5	GA	10,000	1.4941	297.2941	64.7843	0.8947	6.4737	57.0424
	PSO	10,000	–	–	–	–	–	–
	SA	10,000	–	–	–	–	–	48.8642
	TLBO	10,000	1.7712	301.8264	48.2563	0.9065	5.8832	**59.1021**
	ETLBO	10,000	1.8294	300.5684	45.9682	0.9117	5.9258	**59.2498**
6	GA	12,000	1.9333	309.6863	53.2549	0.8653	5.5802	60.1529
	PSO	12,000	–	–	–	–	–	–
	SA	12,000	–	–	–	–	–	52.7259
	TLBO	12,000	2.2316	302.3782	35.2645	0.8812	5.2394	**61.8847**
	ETLBO	12,000	2.3295	300.6739	32.5883	0.8994	5.3675	**61.9649**
7	GA	14,000	2.3333	300.2745	23.6078	0.8739	4.9692	62.4509
	PSO	14,000	–	–	–	–	–	–
	SA	14,000	–	–	–	–	–	55.7717
	TLBO	14,000	2.4712	298.3782	43.2842	0.8848	4.7183	**63.9456**
	ETLBO	14,000	2.5893	300.6884	39.5883	0.8846	4.8948	**64.0849**
8	GA	16,000	1.3608	302.4706	21.9608	0.8598	4.5627	64.4455
	PSO	16,000	–	–	–	–	–	–
	SA	16,000	–	–	–	–	–	58.3572
	TLBO	16,000	1.7629	299.2774	36.6721	0.8932	4.3752	**65.7832**
	ETLBO	16,000	1.6948	300.8833	34.7583	0.8949	4.4896	**65.8993**
9	GA	18,000	2.7255	300.5098	24.1569	0.8951	4.0956	66.2157
	PSO	18,000	–	–	–	–	–	–
	SA	18,000	–	–	–	–	–	60.3263
	TLBO	18,000	2.0527	303.2567	13.8288	0.9382	3.9611	**67.4527**
	ETLBO	18,000	2.1049	299.4883	11.4953	0.9249	4.0185	**67.5352**
10	GA	20,000	1.1961	301.451	55.1765	0.9006	3.8724	67.5588
	PSO	20,000	–	–	–	–	–	–
	SA	20,000	–	–	–	–	–	62.3454
	TLBO	20,000	1.2836	304.8453	29.2752	0.9161	3.7923	**69.0637**
	ETLBO	20,000	1.2903	301.4783	25.6994	0.9225	3.8175	**69.1785**

The values corresponding to GA are taken from Varun and Siddhartha (2010). The values corresponding to PSO and SA are taken from Siddhartha et al. (2012). The values corresponding to TLBO are taken from Rao and Waghmare (2015)

Table 10.14 Set of optimum results at different Reynolds numbers ($N = 2$ and $S = 1200$ W/m^2)

S. No.	Algorithm	Re	v	T_a	β	ϵ_p	T_o–T_i	η_{th} (%)
1	GA	2000	1.0471	301.7647	33.4902	0.8873	17.796	33.6309
	PSO	2000	–	–	–	–	–	–
	SA	2000	–	–	–	–	–	22.6143
	TLBO	2000	1.3772	298.5773	21.6737	0.8934	15.3859	**35.4562**
	ETLBO	2000	1.4389	299.4785	24.6884	0.8882	15.4968	**35.5674**
2	GA	4000	2.1451	299.2549	24.9804	0.9339	12.7697	47.4312
	PSO	4000	–	–	–	–	–	–
	SA	4000	–	–	–	–	–	36.5246
	TLBO	4000	2.4436	295.7382	42.8322	0.9132	11.6380	**49.4673**
	ETLBO	4000	2.4958	296.5773	38.5994	0.9291	11.7847	**49.5893**
3	GA	6000	1.8392	293.8431	40.0784	0.8947	10.3696	55.2555
	PSO	6000	–	–	–	–	–	–
	SA	6000	–	–	–	–	–	45.6225
	TLBO	6000	1.5263	290.3843	65.3738	0.9028	9.3496	**56.9932**
	ETLBO	6000	1.4794	291.4885	66.5883	0.9084	9.5185	**57.1057**
4	GA	8000	2.0431	306.2353	61.2157	0.9233	8.4761	60.3607
	PSO	8000	–	–	–	–	–	–
	SA	8000	–	–	–	–	–	51.9252
	TLBO	8000	1.8127	302.6325	52.4771	0.8992	7.8913	**62.6748**
	ETLBO	8000	1.7389	300.5773	56.3885	0.8936	7.9947	**62.7739**
5	GA	10,000	1.6431	302.3137	4.6667	0.9139	7.1846	64.0505
	PSO	10,000	–	–	–	–	–	–
	SA	10,000	–	–	–	–	–	56.4899
	TLBO	10,000	1.2328	297.1183	21.3392	0.8846	6.9964	**65.8992**
	ETLBO	10,000	1.1840	299.6638	24.5883	0.8961	7.0748	**65.9834**
6	GA	12,000	1.0863	303.9608	56.549	0.8567	6.3845	66.5759
	PSO	12,000	–	–	–	–	–	–
	SA	12,000	–	–	–	–	–	60.0281
	TLBO	12,000	1.3823	299.7837	23.5732	0.8628	5.8732	**68.2184**
	ETLBO	12,000	1.3193	300.5732	25.6993	0.8692	5.9215	**68.3695**
7	GA	14,000	2.8745	296.9804	59.8431	0.8547	5.6448	68.6168
	PSO	14,000	–	–	–	–	–	–
	SA	14,000	–	–	–	–	–	62.9441
	TLBO	14,000	2.6122	299.3822	47.2685	0.9163	5.3181	**69.9247**
	ETLBO	14,000	2.5683	298.5710	42.5993	0.9038	5.4748	**69.9833**
8	GA	16,000	1.6118	295.9608	34.0392	0.879	5.0491	70.2132
	PSO	16,000	–	–	–	–	–	–
	SA	16,000	–	–	–	–	–	65.1525
	TLBO	16,000	1.2178	302.5732	21.4778	0.8927	4.8832	**71.4692**
	ETLBO	16,000	1.2499	299.0684	26.5935	0.8883	4.9385	**71.5831**

(continued)

Table 10.14 (continued)

S. No.	Algorithm	Re	v	T_a	β	ϵ_p	T_o–T_i	η_{th} (%)
9	GA	18,000	1.2353	301.6078	23.6078	0.8684	4.5681	71.4894
	PSO	18,000	–	–	–	–	–	–
	SA	18,000	–	–	–	–	–	67.0587
	TLBO	18,000	1.5721	304.9956	8.3772	0.9027	4.4616	**72.3785**
	ETLBO	18,000	1.4950	302.6884	11.9553	0.9065	4.5127	**72.4880**
10	GA	20,000	1.5961	292.8235	21.9608	0.8547	4.1981	72.5851
	PSO	20,000	–	–	–	–	–	–
	SA	20,000	–	–	–	–	–	68.5195
	TLBO	20,000	1.8973	290.3753	16.4672	0.8623	4.1109	**74.1743**
	ETLBO	20,000	1.7748	292.5630	18.4886	0.8748	4.1623	**74.2811**

The values corresponding to GA are taken from Varun and Siddhartha (2010). The values corresponding to PSO and SA are taken from Siddhartha et al. (2012). The values corresponding to TLBO are taken from Rao and Waghmare (2015)

Table 10.15 Set of optimum results at different Reynolds numbers ($N = 3$ and $S = 1200$ W/m^2)

S. No.	Algorithm	Re	v	T_a	β	ϵ_p	T_o–T_i	η_{th} (%)
1	GA	2000	1.2431	297.0588	61.4902	0.8602	21.3196	38.1579
	PSO	2000	–	–	–	–	–	–
	SA	2000	–	–	–	–	–	17.4604
	TLBO	2000	1.0151	294.3668	45.3568	0.8926	17.3942	**40.2748**
	ETLBO	2000	1.1748	295.7830	51.4885	0.9045	17.5385	**40.3629**
2	GA	4000	1.8784	301.6078	12.6275	0.8645	14.505	52.4193
	PSO	4000	–	–	–	–	–	–
	SA	4000	–	–	–	–	–	29.0595
	TLBO	4000	1.6714	297.6782	27.3564	0.8849	12.8242	**54.2748**
	ETLBO	4000	1.5639	299.5671	30.4885	0.8865	12.9584	**54.3702**
3	GA	6000	1.3216	294.4706	64.2353	0.8594	11.5417	59.8554
	PSO	6000	–	–	–	–	–	–
	SA	6000	–	–	–	–	–	37.6487
	TLBO	6000	1.7893	291.5683	56.7544	0.8941	10.5689	**61.4773**
	ETLBO	6000	1.8952	292.2589	56.3676	0.8836	10.8174	**61.5628**
4	GA	8000	2.7725	294.6275	32.6667	0.9473	9.1277	64.5362
	PSO	8000	–	–	–	–	–	–
	SA	8000	–	–	–	–	–	43.9187
	TLBO	8000	2.4844	298.4632	12.7348	0.9147	8.7635	**66.2726**
	ETLBO	8000	2.4199	300.5673	10.2219	0.9054	8.8493	**66.3953**

(continued)

Table 10.15 (continued)

S. No.	Algorithm	Re	v	T_a	β	ϵ_p	T_o–T_i	η_{th} (%)
5	GA	10,000	2.5608	302.4706	58.1961	0.8825	7.7548	67.592
	PSO	10,000	–	–	–	–	–	–
	SA	10,000	–	–	–	–	–	48.8642
	TLBO	10,000	2.8637	298.3570	63.4563	0.9123	7.1074	**69.3674**
	ETLBO	10,000	2.9478	299.5884	65.2054	0.9048	7.2358	**69.4883**
6	GA	12,000	2.9686	290.7059	43.9216	0.8908	6.7211	69.9637
	PSO	12,000	–	–	–	–	–	–
	SA	12,000	–	–	–	–	–	52.7259
	TLBO	12,000	2.8868	293.9874	56.7432	0.9038	6.3987	**71.2371**
	ETLBO	12,000	2.8199	295.7883	55.3786	0.9021	6.5949	**71.3394**
7	GA	14,000	1.2118	293.7647	57.098	0.8986	5.9539	71.6221
	PSO	14,000	–	–	–	–	–	–
	SA	14,000	–	–	–	–	–	55.7717
	TLBO	14,000	1.6585	291.3683	34.8643	0.9142	5.5853	**72.6738**
	ETLBO	14,000	1.6937	293.9503	31.4782	0.9067	5.7023	**72.6639**
8	GA	16,000	2.6314	309.2157	16.4706	0.8535	5.225	72.9702
	PSO	16,000	–	–	–	–	–	–
	SA	16,000	–	–	–	–	–	58.3572
	TLBO	16,000	2.8726	305.3783	31.4786	0.8736	4.9643	**73.9738**
	ETLBO	16,000	2.9588	306.4821	29.4383	0.8883	5.0867	**74.0927**
9	GA	18,000	2.7647	299.098	44.1961	0.9025	4.7531	74.0904
	PSO	18,000	–	–	–	–	–	–
	SA	18,000	–	–	–	–	–	60.3263
	TLBO	18,000	2.2169	303.4577	29.8642	0.8856	4.5775	**74.8992**
	ETLBO	18,000	2.1857	301.5684	27.5880	0.8754	4.6829	**74.9821**
10	GA	20,000	2.8353	306.7843	7.9608	0.9406	4.2964	74.969
	PSO	20,000	–	–	–	–	–	–
	SA	20,000	–	–	–	–	–	62.3454
	TLBO	20,000	2.7825	301.9751	24.5735	0.8821	4.2063	**76.0374**
	ETLBO	20,000	2.6793	300.5835	21.8832	0.8928	4.2338	**76.1432**

The values corresponding to GA are taken from Varun and Siddhartha (2010). The values corresponding to PSO and SA are taken from Siddhartha et al. (2012). The values corresponding to TLBO are taken from Rao and Waghmare (2015)

References

Deb, K., 2000. An efficient constraint handling method for genetic algorithm, Computer Methods in Applied Mechanics and Engineering 186, 311–338.

Klien, S.A., 1975. Calculation of flat plate loss coefficients. Solar Energy 17, 79–80.

Rao, R.V., Waghmare, G., 2015. Optimization of thermal performance of a smooth flat plate solar air heater using teaching-learning-based optimization algorithm. Cogent Engineering, doi:10.1080/23311916.2014.997421.

Siddhartha, Sharma, N., Varun, 2012. A particle swarm optimization algorithm for optimization of thermal performance of a smooth flat plate solar air heater. Energy 38, 406–413.

Varun, Siddhartha, 2010. Thermal performance optimization of a flat plate solar air heater using genetic algorithm. Applied Energy 87, 1793–1799.

Chapter 11
Design Optimization of a Robot Manipulator Using TLBO and ETLBO Algorithms

Abstract The effectiveness of the TLBO and ETLBO algorithms is verified for design optimization of a robot manipulator by considering four cases and imposing different conditions to demonstrate the efficiency of the design process. The workspace volume is considered as an objective function. The results of the TLBO and ETLBO algorithms are compared with the SQP, GA, DE, and PSO algorithms. The computational results show that for all the four cases the TLBO and ETLBO algorithms have obtained more accurate solutions than those obtained by the other optimization methods.

11.1 Design Optimization of Robot Manipulator

Over the past few decades, the interest of researchers is growing in the field of design optimization of robotic system in order to improve the system performance using advanced optimization techniques. In the present work, design optimization of a robot manipulator is considered. An industrial robot is a programmable, multifunctional manipulator designed to move materials, parts, tools, or special devices through variable programmed motions for a variety of tasks. Robots come in variety of sizes, shapes, and capabilities. Robots have four basic components namely a manipulator, an end effect which is a part of manipulator, computer controller, and a power supply. Robot anatomy is concerned with the physical construction of body, arm, and wrist of the machine. Relative movements between various components of the body, arm, and wrist are provided by a series of either sliding or rotating joints. The body, arm, and wrist assembly is called as manipulator. The manipulator is a mechanism consisting of the major linkages, minor linkages, and the end effector (gripper or tool). Robot is designed to reach a work piece within its work volume. Work volume is the term that refers to the space within which the robot can manipulate its wrist end. It is also called as work space. The surface of work space is termed as work envelop.

R.V. Rao, *Teaching Learning Based Optimization Algorithm*,
DOI 10.1007/978-3-319-22732-0_11

Several investigations had focused on the properties of the workspace of open chain robotics with the purpose of emphasizing its geometric and kinematic characteristics and to devise analytical algorithms and procedures for its design (Waghmare 2015). Ceccarelli (1996) presented algebraic formulation to determine the workspace of a revolution manipulator. The author obtained the workspace boundary from the envelope of a torus family which was traced by the parallel circles cut in the boundary of a revolving hyper-ring. The formulation was a function of the dimensional parameters in the manipulator chain and specifically of the last revolute joint angle, only. The formulation developed was used to obtain the equation of the family of plane curves that represents the workspace boundary. The workspace mathematically developed was of crucial importance; however, the manipulator's optimal design was not considered.

Abdel-Malek et al. (2000) proposed a generic formulation to determine voids in the workspace of serial manipulators. Lanni et al. (2002) investigated and solved the design of manipulators modeled as an optimization problem that takes into account the characteristics of the workspace. The authors had applied two different numerical techniques, the first using sequential quadratic programming (SQP) and the second involving a random search technique. These methodologies cannot be applied to calculate the workspace volume in case there is a ring void; however, it was an important initial study.

Some researchers had focused on determining the workspace boundary and on determining the presence of voids and singularities in the workspace. Saramago et al. (2002) proposed a form of characterizing the workspace boundary, formulating a general analytical condition to deduce the existence of cusp points at the internal and external boundaries of the workspace. Ceccarelli and Lanni (2004) presented a suitable formulation for the workspace that can be used in the design of manipulators which was formulated as a multiobjective optimization problem using the workspace volume and robot dimensions as objective functions. The authors had used volume maximization and dimensional minimization as objective functions.

One of the most commonly used methods to geometrically describe a general open chain 3R manipulator with three revolute joints is the one which uses the Hartenberg and Denavit (H–D) notation, whose scheme is exhibited in Fig. 11.1.

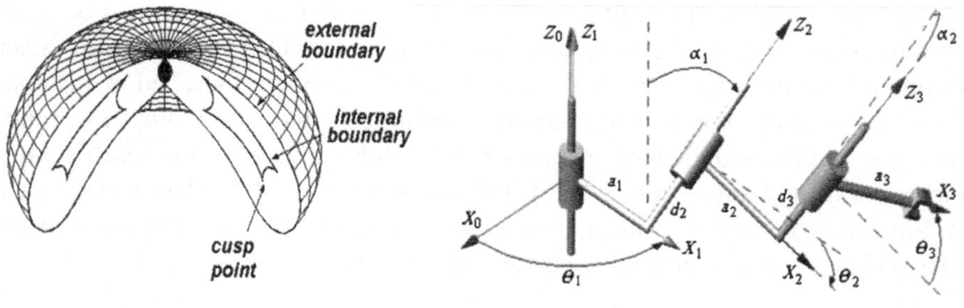

Fig. 11.1 The workspace for 3R manipulators and design parameters (Bergamaschi et al. 2008; Reprinted with permission from Elsevier)

The design parameters for the link size are represented as a_1, a_2, a_3, d_2, d_3, α_1, α_2 (d_1 is not meaningful, since it shifts the workspace up and down). Bergamaschi et al. (2008) optimized the design of manipulator with three revolute joints using an optimization problem that took into account the characteristics of the workspace.

The objective of the optimization problem is to maximize the workspace volume V, thereby obtaining the best dimensions. The optimization problem was defined as (Bergamaschi et al. 2008)

$$\max F_c = V(x), x_i^l \leq x \leq x_i^j, \quad i = 1, 2, \ldots, n \tag{11.1}$$

where $x = [a_1\ a_2\ a_3\ d_2\ d_3\ \alpha_1\ \alpha_2]$ and the side constraints were:

$$0.01 \leq a_i \leq a_i^u, \quad \text{for } i = 1, 2, 3, \tag{11.2}$$

$$0.01 \leq d_j \leq d_j^u, \quad \text{for } j = 2, 3, \tag{11.3}$$

$$0.01 \leq \alpha_k \leq 90°, \quad \text{for } k = 1, 2, \tag{11.4}$$

The optimization problem was subjected to certain constraints such as

$$g_1 = \frac{\partial^2 f}{\partial \theta_3^2} \neq 0, \tag{11.5}$$

$$g_2 = z > 0 \tag{11.6}$$

and

$$g_3 = r < 0.7\, L\text{max} \tag{11.7}$$

where

$$L\text{max} = \sqrt{(a_1^u)^2 + (d_2^u)^2} + \sqrt{(a_2^u)^2 + (d_3^u)^2} + a_3^u \tag{11.8}$$

The values a_i^u and d_j^u are the maximum values that the respective parameters a_i and d_j can assume during the optimization process, for i = 1, 2, 3 and j = 2, 3. Four cases are considered.

Case 1: Optimization problem given by Eqs. (11.1)–(11.4) considering $a_i^u = d_j^u = 1$ and the regularity constraint (11.5).

Case 2: Optimization problem given by Eqs. (11.1)–(11.4) considering $a_i^u = d_j^u = 1$ and the constraint (11.6).

Case 3: Optimization problem given by Eqs. (11.1)–(11.4) considering $a_i^u = d_j^u = 1$ and the constraints (11.6) and (11.7).

Case 4: Optimization problem given by Eqs. (11.1)–(11.4) considering $a_i^u = d_j^u = 1$ and without imposing any constraint.

11.2 Results and Discussion

To check the effectiveness of the TLBO and ETLBO algorithms, extensive computations are conducted on design of robot manipulator considered by Bergamaschi et al. (2008) and the results are compared. Bergamaschi et al. (2008) used the SQP, GA, DE, and PSO using 3000 function evaluations. Hence, to make fair comparison of results, the same number of function evaluation is considered. Hence, population size of 30 and maximum number of generations of 50 are considered (it may be mentioned here that the number of function evaluations in TLBO algorithm = 2 × population size × number of generations). As other optimization algorithms (e.g., PSO, ABC, ACO, etc.), TLBO algorithm also has no special mechanism to handle the constraints. So for constrained optimization problems it is necessary to incorporate any constraint handling techniques with the TLBO algorithm. In the present experiments, Deb's heuristic constrained handling method (Deb 2000) is used to handle the constraints. After making few trials, the population size of 30 and number of generations of 50 and elite size of 12 are considered for the ETLBO algorithm.

For case 1, it can be seen from Tables 11.1 and 11.2 that the TLBO and ETLBO algorithms perform better than the SQP, GA, DE, and PSO algorithms in terms of

Table 11.1 Results obtained with the six methods for case 1

	a_1 (u.m.)	a_2 (u.m.)	a_3 (u.m.)	d_2 (u.m.)	d_3 (u.m.)	α_1 (°)	α_2 (°)	Vol. (u.v.)
SQP	1.00	0.97	1.00	0.49	1.00	84.85	52.41	101.55
GA	1.00	1.00	1.00	0.48	1.00	86.50	56.09	105.28
DE	1.00	1.00	1.00	0.45	1.00	83.73	57.17	105.23
PSO	1.00	1.00	1.00	0.50	1.00	89.58	57.48	104.81
TLBO	1.00	1.00	1.00	0.47	1.00	87.54	56.87	**106.43**
ETLBO	1.00	1.00	1.00	0.48	1.00	87.69	57.11	**106.52**

The results of SQP, GA, DE, and PSO are from Bergamaschi et al. (2008)
u.m.: unit meter
u.v.: unit volume

Table 11.2 Results obtained with the six methods for case 1 for maximum, average, and minimum volume space

	Minimum volume space	Maximum volume space	Average volume space	Standard deviation
SQP	34.34	101.55	73.98	28.810
GA	99.84	105.28	103.45	1.616
DE	94.94	105.23	104.01	3.190
PSO	104.65	104.81	104.72	0.055
TLBO	105.56	**106.43**	105.79	**0.034**
ETLBO	105.67	**106.52**	105.86	**0.029**

The results of SQP, GA, DE, and PSO are from Bergamaschi et al. (2008)

workspace volume for the considered problem. From Table 11.2, it can be observed that the standard deviations achieved by the ETLBO and TLBO algorithms are less than those given by SQP, GA, DE, and PSO algorithms for the considered problem. This indicates the more robustness of the TLBO and ETLBO algorithms. The values shown bold in tables indicate the best values.

For case 2, it can be seen from Tables 11.3 and 11.4 that the TLBO and ETLBO algorithms perform better than the SQP, GA, DE, and PSO in terms of workspace volume for the considered problem. From Table 11.4, it can be observed that the standard deviation achieved is less than that obtained by the SQP, GA, DE, and PSO algorithms for the considered problem. This indicates the more robustness of the TLBO and ETLBO algorithms.

For case 3, it can be seen from Tables 11.5 and 11.6 that the TLBO and ETLBO algorithms perform better than the SQP, GA, DE, and PSO algorithms in terms of workspace volume for the considered problem. From Table 11.6 it can be observed that the standard deviation achieved is less than that obtained by the SQP and other metaheuristic methods for the considered problem. This also indicates the more robustness of the TLBO and ETLBO algorithms.

For case 4, it can be seen from Tables 11.7 and 11.8 that the TLBO and ETLBO algorithms perform better than the SQP, GA, DE, and PSO algorithms in terms of workspace volume for the considered problem. It can be observed from Table 11.8 that the standard deviation achieved is less than that of the SQP and other

Table 11.3 Results obtained with the six methods for case 2

	a_1 (u.m.)	a_2 (u.m.)	a_3 (u.m.)	d_2 (u.m.)	d_3 (u.m.)	α_1 (°)	α_2 (°)	Vol. (u.v.)
SQP	1.00	1.00	1.00	1.00	1.00	36.97	28.48	67.24
GA	1.00	1.00	1.00	1.00	1.00	37.30	27.12	66.81
DE	1.00	1.00	1.00	1.00	1.00	36.34	29.37	67.00
PSO	1.00	1.00	1.00	1.00	1.00	35.96	30.10	66.95
TLBO	1.00	1.00	1.00	1.00	1.00	36.65	28.27	**68.19**
ETLBO	1.00	1.00	1.00	1.00	1.00	36.71	28.37	**68.31**

The results of SQP, GA, DE, and PSO are from Bergamaschi et al. (2008)

Table 11.4 Results obtained with the six methods for case 2 for maximum, average, and minimum volume space

	Minimum volume space	Maximum volume space	Average volume space	Standard deviation
SQP	62.50	67.24	65.82	1.464
GA	61.30	66.81	64.33	1.851
DE	66.94	67.00	66.97	0.025
PSO	66.74	66.95	66.84	0.081
TLBO	66.99	**68.19**	67.34	**0.021**
ETLBO	67.17	**68.31**	67.56	**0.019**

The results of SQP, GA, DE, and PSO are from Bergamaschi et al. (2008)

Table 11.5 Results obtained with the six methods for case 3

	a_1 (u.m.)	a_2 (u.m.)	a_3 (u.m.)	d_2 (u.m.)	d_3 (u.m.)	α_1 (°)	α_2 (°)	Vol. (u.v.)
SQP	0.48	0.88	1.00	1.00	0.98	16.91	65.80	39.64
GA	0.82	0.60	0.84	1.00	1.00	26.59	61.12	37.27
DE	0.46	0.90	1.00	1.00	1.00	16.45	66.11	39.27
PSO	0.44	0.91	0.99	1.00	1.00	17.70	63.69	39.29
TLBO	0.47	0.92	0.99	1.00	1.00	17.56	64.41	**39.89**
ETLBO	0.46	0.91	1.00	1.00	1.00	17.67	65.33	**39.97**

The results of SQP, GA, DE, and PSO are from Bergamaschi et al. (2008)

Table 11.6 Results obtained with the six methods for case 3 for maximum, average, and minimum volume space

	Minimum volume space	Maximum volume space	Average volume space	Standard deviation
SQP	35.43	39.64	38.24	1.134
GA	30.48	37.27	33.74	2.145
DE	35.90	39.27	37.92	1.383
PSO	30.13	39.29	36.56	3.419
TLBO	36.32	**39.89**	38.29	**1.003**
ETLBO	36.41	**39.97**	38.49	**0.893**

The results of SQP, GA, DE, and PSO are from Bergamaschi et al. (2008)

Table 11.7 Results obtained with the six methods for case 4

	a_1 (u.m.)	a_2 (u.m.)	a_3 (u.m.)	d_2 (u.m.)	d_3 (u.m.)	α_1 (°)	α_2 (°)	Vol. (u.v.)
SQP	1.00	1.00	1.00	1.00	1.00	80.39	75.13	121.94
GA	1.00	1.00	1.00	1.00	1.00	85.10	75.16	122.04
DE	1.00	1.00	1.00	1.00	1.00	84.17	77.11	122.35
PSO	1.00	1.00	1.00	1.00	1.00	82.49	77.87	122.18
TLBO	1.00	1.00	1.00	1.00	1.00	84.51	77.48	**122.97**
ETLBO	1.00	1.00	1.00	1.00	1.00	84.82	77.61	**122.99**

The results of SQP, GA, DE, and PSO are from Bergamaschi et al. (2008)

Table 11.8 Results obtained with the six methods for case 4 for maximum, average, and minimum volume space

	Minimum volume space	Maximum volume space	Average volume space	Standard deviation
SQP	84.36	121.94	73.98	28.810
GA	121.90	122.04	103.45	1.616
DE	121.30	122.35	104.01	3.190
PSO	94.90	122.18	104.72	0.055
TLBO	121.89	**122.97**	110.34	**0.039**
ETLBO	121.95	**122.99**	111.48	**0.027**

The results of SQP, GA, DE, and PSO are from Bergamaschi et al. (2008)

metaheuristic methods for the considered problem. This indicates the more robustness of the TLBO and ETLBO algorithms. Also, it is observed that the concept of elitism enhances the performance of the TLBO algorithm for the considered design optimization of robot manipulator.

The next chapter presents the details of optimization of design and manufacturing tolerances using TLBO algorithm.

References

Abdel-Malek, K., Yeh, H-J., Othman, S., 2000. Understanding voids in the workspace of Serial Robot Manipulators, in: Proceedings of ASME Design Engineering Technical Conference, Baltimore, MD, 1–8.

Bergamaschi, P.R., Saramago, S., Coelho, L.,2008. Comparative study of SQP and metaheuristics for robotic manipulator design. Applied Numerical Mathematics 58, 1396–1412.

Ceccarelli, M., 1996. A formulation for the workspace boundary of general N-revolute manipulators, Mechanism and Machine Theroy 31(5), 637–646.

Ceccarelli, M., Lanni, C., 2004. A multi-objective optimum design of general 3R manipulators for prescribed workspace limits, Mechanism and Machine Theory 39, 119–132.

Deb, K., 2000. An efficient constraint handling method for genetic algorithm, Computer Methods in Applied Mechanics and Engineering 186, 311–338.

Lanni, C., Saramago, Ceccarelli, M., 2002. Optimal design of 3R manipulators using classical techniques and simulated annealing, Revista Brasileira de Ciencias Mecanicas 24 (4), 293–301.

Saramago, S.F.P., Ottaviano, E., Ceccarelli, M., 2002. A characterization of the workspace boundary of three-revolute manipulators, in: Proceedings of ASME Design Engineering Technical Conference, Montreal, Canada, 34342–34352.

Waghmare, G.G., 2015. Single and Multiobjective Design Optimization Using TLBO algorithm, PhD Thesis, S. V. National Institute of Technology, Surat, India (supervisor: Rao, R.V.).

Chapter 12
Multiobjective Optimization of Design and Manufacturing Tolerances Using TLBO Algorithm

Abstract Tolerance design has become a very responsive and key issue in product and process development because of an informal compromise between functionality, quality, and manufacturing cost. The problem formulation becomes more complex with simultaneous selection of design and manufacturing tolerances and the optimization problem is difficult to solve with the traditional optimization techniques. Rao and More (Prod Manuf Res Open Access J 2: 71–94, 2014) used TLBO algorithm for the optimal selection of design and manufacturing tolerances with an alternative manufacturing process to obtain the optimal solution. An example of knuckle joint assembly with three arms is presented in this chapter and it is found that the TLBO algorithm has produced better results when compared to those obtained using NSGA-II and MOPSO algorithms.

12.1 Optimization of Design and Manufacturing Tolerances

For manufacturing every dimension, the tolerance design problem becomes more intricate in the existence of different processes or machines. This is due to the fact that the manufacturing cost-tolerance features change from machine to machine and from process to process. The total costs incurred throughout a product's life cycle can be divided into two main categories: manufacturing cost, which occurs before the product reaches the customer; and quality loss, which occurs after the product is sold. A rigid tolerance (i.e., high manufacturing cost) indicates that the variability of product quality characteristics will be low (i.e., little quality loss). On the other hand, a slack tolerance (i.e., low manufacturing cost) indicates that the variability of product quality characteristics will be high (i.e., high quality loss). Therefore, there is a need to adjust the design tolerance between the quality loss and the manufacturing cost and to reach an economic balance during product tolerance design.

R.V. Rao, *Teaching Learning Based Optimization Algorithm*,
DOI 10.1007/978-3-319-22732-0_12

During the last few decades, researchers have focused their attention on obtaining the best tolerance allocation in such a way that the product design not only meets the efficient needs but also minimizes manufacturing cost. In order to solve the tolerance allocation problem, various optimization methods were employed in the past to deal with complicated computations associated with tolerance design models. Rao and More (2014) attempted the three case studies presented by Sivakumar et al. (2011) to demonstrate the applicability of the TLBO algorithm for the overrunning clutch assembly, knuckle joint with three arms, and the helical spring design problem to see if any further improvement could be obtained in the results and in the computational time. Out of these three problems, the problem of knuckle joint with three arms which is a five-objective optimization problem is presented in this chapter to demonstrate the applicability of the TLBO algorithm.

The roles of design and manufacturing tolerances, stock removal allowances, selection of manufacturing processes, manufacturing cost, and quality loss function are described in next subsections (Sivakumar et al. 2011; Rao and More 2014).

12.1.1 Design and Manufacturing Tolerances

In any product design, the proposed functions and assembly needs are distorted into a set of related tolerances and appropriate dimensions. The dimensions are known as the assembly dimensions and the related tolerances are known as assembly tolerances. The assembly tolerance is sensibly distributed among the part dimensions in particular dimension sequence considering the practical aspects. A tolerance specified in the design stage is further refined to suit the requirements of process planning for producing the constituent dimensions (Singh et al. 2005).

12.1.2 Stock Removal Allowances

The amount of stock removal allowance has very much effect on the manufacturing tolerance selection. The stock removal allowance is the layer of material to be removed from the surface of a work piece in manufacturing to obtain the required profile accuracy and surface quality through different machining processes. The stock removal allowance is very much affecting the quality and the production efficiency of the manufactured features. A disproportionate stock removal allowance will add to the consumption of material, machining time, tool, and power, and hence raise the manufacturing cost. On the other hand, with insufficient stock removal allowance, the faulty surface layer caused by the previous operation cannot be rectified. Variation in the stock removal is the sum of manufacturing tolerances in the prior and the current operations. An appropriate stock removal allowance is required for each successful manufacturing operation (Haq et al. 2005).

12.1.3 Selection of Machining Process

The selection of machining process is depending on the equipment precision, machining series, setup mode, and cutting parameters. The selection of machining process is strongly affected by the tolerance of the part to be machined. So it is vital to do a simultaneous selection of the best machining process while allocating the tolerance (Sivakumar et al. 2011).

12.1.4 Manufacturing Cost

For a given manufacturing process, the material, setup, tool, and inspection costs, in addition to overheads, comprise the fixed cost. The variable cost, which usually depends on the value of tolerance assigned, is mostly because of rework and/or rejection. The actual processing (labor) cost also accounts for the variable cost, though it is not predominant.

Manufacturing cost usually increases as the tolerance of quality characteristics in relation to the ideal value is reduced. On the other hand, large tolerances are having minimum cost to get as they want less specific manufacturing processes; but usually they provide poor result in performance, premature wear, increase in scrap, and part rejection.

The manufacturing cost function for the available manufacturing processes is assumed to be exponential (Singh et al. 2005).

$$C = c_0 * e^{-c_1 * t} + c_2 \tag{12.1}$$

where C is the manufacturing cost as a function of tolerance, c_0, c_1, and c_2 are the cost function values, and t is the design tolerance dimension.

12.1.5 Quality Loss Function

Variability in the production process is compulsory due to changeability in tool work piece, material, and process parameters. In this study it is referred as the quality loss function (Feng and Kusiak 1997). This loss function is a quadratic expression for measuring the cost in terms of financial loss due to product failure in the eyes of the consumers. The quality loss function (QL) is

$$\mathrm{QL} = \frac{A}{T^2}\sum_{i=1}^{J}\sigma_i^2 \tag{12.2}$$

$$\sigma_i = \frac{t_i}{3} \tag{12.3}$$

$$\mathrm{QL} = \frac{A}{9T^2}\sum_{i=1}^{J}t_i^2 \tag{12.4}$$

where T is the tolerance stack-up limit of the dimensional chain, A is the quality loss cost, i is the component tolerance index, j is the process index, t is the design tolerance dimension, and σ_i is the standard deviation of dimension i.

12.2 Example: Knuckle Joint with Three Arms

There are several dimensions in the Knuckle joint assembly shown in Fig. 12.1 (Sivakumar et al. 2011). There are three interrelated dimension chains corresponding to the respective design functions giving rise to three constraints. The permissible variation T associated with the respective assembly dimension Y is

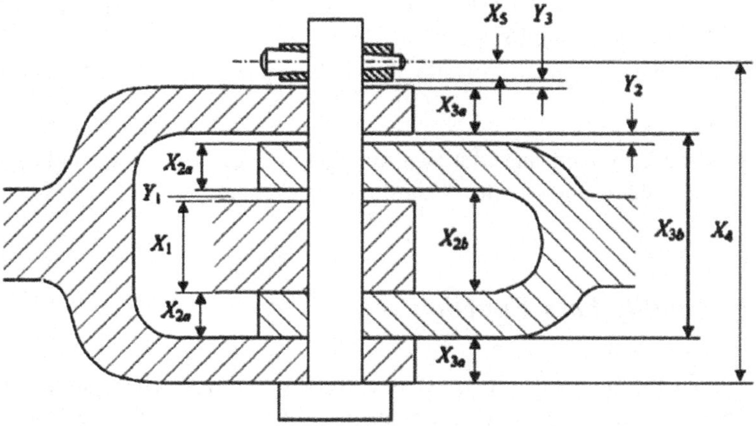

Fig. 12.1 Knuckle joint assembly with three arms (Sivakumar et al. 2011; Reprinted with permission from Elsevier). (X_i: the individual dimensions in product assembly; Y_1, Y_2, Y_3: design dimensions)

assumed to be equal to 0.2. All similar dimensions are manufactured by the same machine; hence, the design tolerance associated to X_{2a} and X_{2b} is t_2, and that associated to X_{3a} and X_{3b} is t_3. For each manufacturing process alternative machines are available with the details given in Table 12.1.

Design functions are given as follows:

$$Y_1 = X_{2b} - X_1 \tag{12.5}$$

$$Y_2 = X_{3b} - (2X_{2a} + X_{2b}) \tag{12.6}$$

$$Y_3 = X_4 - (2X_{3a} + X_{3b} + X_5) \tag{12.7}$$

where Y_1, Y_2, Y_3 are the design dimensions; X_1, X_{2a}, X_{2b}, X_{3a}, X_{3b}, X_4, X_5 are the individual dimensions in product assembly.

Sivakumar et al. (2011) used non-dominated sorting genetic algorithm-II (NSGA-II) and multiobjective particle swarm optimization algorithm (MOPSO) for simultaneous optimum selection of design and manufacturing tolerances with alternative process selection. The objective functions considered were minimum

Table 12.1 Manufacturing process characteristics for the knuckle joint assembly with three arms (Sivakumar et al. 2011; Reprinted with permission from Elsevier)

Dimensions	Process	Parameters of cost function			Minimum tolerance (mm)	Maximum tolerance (mm)
		C_0	C_1	C_2		
X_1	1	311.0	15.8	24.2	0.01	0.15
	2	280.0	14.0	19.8	0.01	0.15
	3	296.4	19.5	23.82	0.01	0.15
	4	331.2	17.64	20.0	0.01	0.15
X_{2a}, X_{2b}	1	311.0	15.8	24.2	0.01	0.15
	2	280.0	14.0	19.8	0.01	0.15
	3	296.4	19.5	23.82	0.01	0.15
	4	331.5	17.64	20.1	0.01	0.15
X_{3a}, X_{3b}	1	311.0	15.8	24.2	0.01	0.15
	2	280.0	14.0	19.8	0.01	0.15
	3	296.4	19.5	23.82	0.01	0.15
	4	331.5	17.64	20.0	0.01	0.15
X_4	1	92.84	82.45	32.5	0.02	0.20
	2	82.43	16.70	21.0	0.02	0.20
X_5	1	128.25	82.65	32.5	0.01	0.10
	2	160.43	86.70	29.2	0.01	0.10
	3	231.42	50.05	28.05	0.01	0.10
	4*a*	134.16	78.82	500.0	0.01	0.10

tolerances stack-up, minimum total manufacturing cost of the assembly, and minimum quality loss function. The objective functions considered by Rao and More (2014) are same as those considered by Sivakumar et al. (2011) and these are given below:

Minimize:

$$Z_1 = \Delta Y_1 = t_{1j} + t_{2j} \tag{12.8}$$

$$Z_2 = \Delta Y_2 = 3t_{2j} + t_{3j} \tag{12.9}$$

$$Z_3 = \Delta Y_3 = 3t_{3j} + t_{4j} + t_{5j} \tag{12.10}$$

$$Z_4 = C_{\text{asm}} = Cx\_\{1\} + 2Cx\_\{2a\} + Cx\_\{2b\} + 2Cx\_\{3a\} + Cx\_\{3b\} + Cx\_\{4\} + Cx\_\{5\} \tag{12.11}$$

$$Z_5 = \text{QL} = \frac{A}{9T^2}\sum_{i=1}^{J} t_{ij}^2 \tag{12.12}$$

where Z_1, Z_2, and Z_3 are the minimum tolerances stack-up; Z_4 is the minimum total manufacturing cost, and Z_5 is the minimum quality loss function.

The following tolerance stack-up constraints were considered:

$$t_{1j} + t_{2j} \le 0.2 \tag{12.13}$$

$$3t_{2j} + t_{3j} \le 0.2 \tag{12.14}$$

$$3t_{3j} + t_{4j} + t_{5j} \le 0.2 \tag{12.15}$$

Parameter cost functions and tolerance limits for the knuckle joint assembly with three arms are given in Table 12.1. Multiple objectives are combined into scalar objective using a weight vector (i.e., a priori approach is used). The combined objective function considered by Sivakumar et al. (2011) was used in the research work of Rao and More (201$) to make the comparison of results of optimization meaningful. Sivakumar et al. (2011) had combined the multiple objectives into scalar objective using the weight vector. Hence, the same approach is used in the present work for comparison purpose. For this problem, the combined objective function (f_c) is defined as follows (Sivakumar et al. 2011):

Minimize:

$$f_c = W_1 \times Z_1/N_1 + W_2 \times Z_2/N_2 + W_3 \times Z_3/N_3 + W_4 \times Z_4/N_4 + W_5 \times Z_5/N_5 \tag{12.16}$$

The values of $N_1 = N_2 = N_3 = N_5 = 1.0$ and $N_4 = 100$ are the normalizing parameters of the objective functions Z_1, Z_2, Z_3, Z_4, and Z_5, and W_1, W_2, W_3, W_4, and W_5 are the weights given to the objective functions 1, 2, 3, 4, and 5, respectively. $N_1 = 1.0$ means that it is the minimum value of Z_1 if only Z_1 is considered as a single objective function (by ignoring all the other objective functions). Similarly, $N_2 = 1.0$, $N_3 = 1.0$, $N_4 = 100$, and $N_5 = 1.0$ indicate the minimum values of Z_2, Z_3, Z_4, and Z_5, respectively, by considering them as single objective (by ignoring the other objective functions). The designer may give any weight to a particular objective function, but the summation of all weight values should be equal to 1. It means that the total weight should be 100 %.

Sivakumar et al. (2011) used the equal weights of $W_1 = W_2 = W_3 = W_4 = W_5 = 0.2$ and the same weights are used by Rao and More (2014) for comparison purpose. The calculated values of first, second, third, fourth, and fifth objective functions using TLBO algorithm are 0.186008, 0.191812, 0.196046, 1008.7, and 0.2466, respectively. To get the normalized values, the first, second, third, fourth, and fifth objective functions are divided by their individual average values (i.e., 1.0, 1.0, 1.0, 100, and 1.0), respectively. The normalized values of the first, second, third, fourth, and fifth objective functions were 0.186008, 0.191812, 0.196046, 10.087, and 0.2466, respectively.

Table 12.1 shows the manufacturing process characteristics for the knuckle joint assembly with three arms (Sivakumar et al. 2011) and Table 12.2 shows the optimum selection of the manufacturing process (machine), allocated tolerance value, the values of objective functions, and the combined objective function value obtained using various optimization techniques. For optimal result of the knuckle joint assembly with three arms, a population size of 50 and number of generations of 60 are used by the TLBO algorithm. As the evaluation is done in both the teacher and student phases, the number of function evaluations used in TLBO is calculated as 2 * Population size * number of generations. Thus the function evaluations are 6000 in the case of TLBO for the optimization of knuckle joint assembly with three arms.

It is observed that the TLBO algorithm has given better results than the NSGA-II and MOPSO algorithms for the multiobjective optimization problem. Each problem is run 30 times by Rao and More (2014) and the standard deviation is 0.0246 for the knuckle joint assembly. Figure 12.2 shows the convergence rate of the TLBO algorithm for the example considered. In this example of the knuckle joint assembly with three arms, the convergence has taken place after 12th iteration of the TLBO algorithm. It is to be noted that the TLBO algorithm has given the best results (i.e., minimum values of Z_1, Z_2, Z_3, and Z_5 for knuckle joint assembly) and the computational time to find the optimum solutions by TLBO algorithm is reported less than that of NSGA-II and MOPSO algorithms which implies that the TLBO algorithm is faster than the NSGA-II and MOPSO algorithms. Also, the

Table 12.2 Optimization results for the knuckle joint with three arms (Rao and More 2014)

Technique	Dimension	Machine	Tolerance notation	Tolerance value (mm)	Objective functions					Combined objective function (f_c)
					Z_1	Z_2	Z_3	Z_4	Z_5	
NSGA-II (Sivakumar et al. 2011)	X_1	3	tX_1	0.069665	0.119203	0.194848	0.179988	1076.895	0.124079	2.2774136
	X_{2a}, X_{2b}	3	tX_{2a}, tX_{2b}	0.049538						
	X_{3a}, X_{3b}	3	tX_{3a}, tX_{3b}	0.046234						
	X_4	2	tX_4	0.020007						
	X_5	2	tX_5	0.021279						
MOPSO (Sivakumar et al. 2011)	X_1	3	tX_1	0.087427	0.137138	0.197005	0.195297	1029.393	0.154244	2.1955228
	X_{2a}, X_{2b}	3	tX_{2a}, tX_{2b}	0.049711						
	X_{3a}, X_{3b}	3	tX_{3a}, tX_{3b}	0.047872						
	X_4	2	tX_4	0.027344						
	X_5	2	tX_5	0.024337						
TLBO (Rao and More 2014)	X_1	3	tX_1	0.137619	0.186008	0.191812	0.196046	1008.7	0.2466	**2.1814**
	X_{2a}, X_{2b}	3	tX_{2a}, tX_{2b}	0.048389						
	X_{3a}, X_{3b}	3	tX_{3a}, tX_{3b}	0.046645						
	X_4	2	tX_4	0.030979						
	X_5	2	tX_5	0.025132						

The number shown in bold indicates the better value

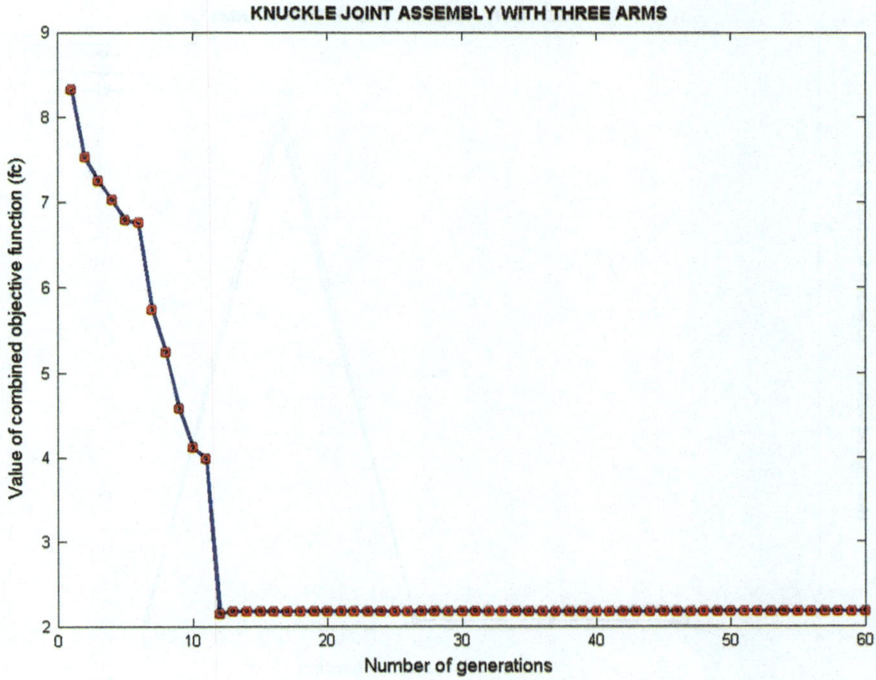

Fig. 12.2 Convergence of combined objective function obtained by TLBO algorithm for knuckle joint assembly with three arms

computational time to find the optimum solutions in TLBO is approximately half of that of NSGA-II and MOPSO algorithms.

Figure 12.3 shows the optimal solution trade-offs obtained from NSGA-II, MOPSO, and TLBO algorithms for the knuckle joint assembly (Rao and More 2014). The TLBO algorithm has given 4.2 and 0.62 % better value of the combined objective function than those given by NSGA-II and MOPSO algorithms, respectively.

The TLBO algorithm has shown its ability in solving multiobjective optimization problems using the normalized weighting objective function. The convergence behavior of the TLBO algorithm to a near global solution has been observed to be more effective than that of NSGA-II and MOPSO algorithms. Hence, the TLBO algorithm is proved better than the other optimization algorithms in terms of results and the convergence. The TLBO algorithm may be conveniently used for the optimal tolerance design of the other machine elements also.

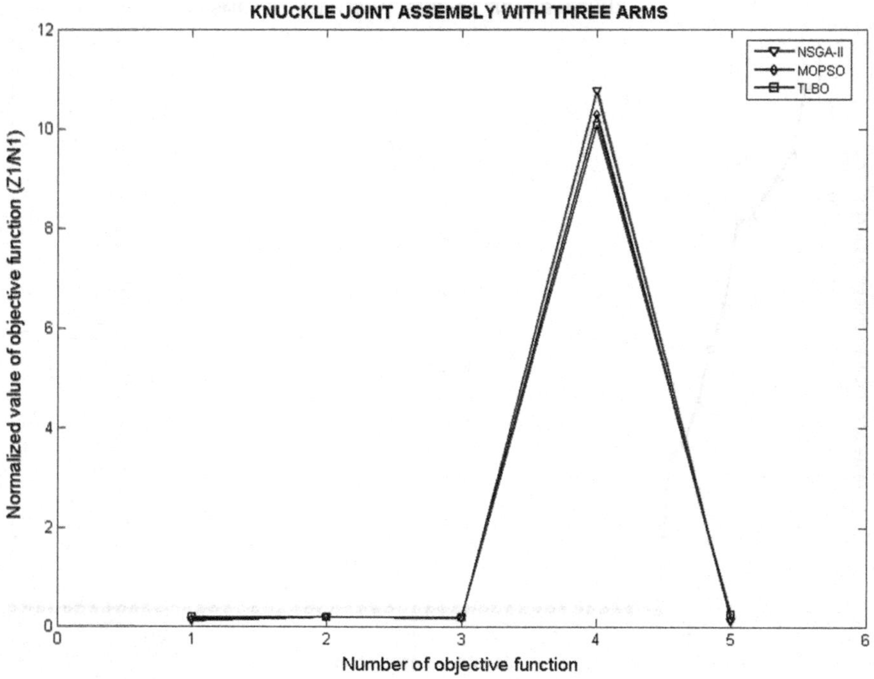

Fig. 12.3 Best solution trade-offs obtained from TLBO, NSGA-II, and MOPSO algorithms for the knuckle joint with three arms assembly

References

Feng, C-X., Kusiak, A., 1997. Robust tolerance design with the integer programming approach. Transactions of ASME Journal Manufacturing Science Engineering 119, 603–610.

Haq, A. N., Sivakumar, K., Saravanan, R., Muthiah, V., 2005. Tolerance design optimization of machine elements using genetic algorithm. International Journal of Advanced Manufacturing Technology 25, 385–391.

Rao, R.V., More, K.C., 2014. Advanced optimal tolerance design of machine elements using teachinglearning-based optimization algorithm Production & Manufacturing Research: An Open Access Journal 2(1), 71–94.

Singh, P. K., Jain, P. K., Jain, S. C., 2005. Advanced optimal tolerance design of mechanical assemblies with interrelated dimension chain and process precision limits. Computers in Industry 56, 179–194.

Sivakumar, K., Balamurugan, C., Ramabalan, S., 2011 .Simultaneous optimal selection of design and manufacturing tolerances with alternative manufacturing process selection. Computer-Aided Design 43, 207–218.

Chapter 13
Parameter Optimization of Machining Processes Using TLBO Algorithm

Abstract This chapter presents the optimization aspects of process parameters of an advanced machining process known as abrasive water jet machining process and an important conventional machining process known as milling. The TLBO algorithm is used to find the optimal combination of process parameters of the considered machining processes. The results obtained using TLBO algorithm are compared with those obtained using other advanced optimization techniques such as GA, SA, PSO, HS, and ABC algorithms. The results show better performance of the TLBO algorithm.

13.1 Parameter Optimization of Machining Processes

Determination of optimum parameters of any machining process is usually a difficult work where the following aspects are required: knowledge of machining process, empirical equations to develop realistic constraints, specification of machine tool capabilities, development of effective optimization criteria, and knowledge of mathematical and numerical optimization techniques. A human process planner selects proper machining process parameters using his own experience or from the handbooks. However, these parameters do not give optimal result. The selection of optimum process parameters plays a significant role to ensure quality of product, to reduce the machining cost, to increase productivity in computer-controlled machining processes and to assist in computer-aided process planning. Pawar and Rao (2013a, b) considered the TLBO algorithm for optimization of process parameters of selected machining processes and described the comparative performance of the TLBO algorithm with the other traditional and advanced algorithms in terms of its ability to find global optimum solution, accuracy of solution, and convergence rate.

R.V. Rao, *Teaching Learning Based Optimization Algorithm*,
DOI 10.1007/978-3-319-22732-0_13

13.1.1 Optimization of Abrasive Water Jet Machining (AWJM)

The AWJM process uses a high-velocity water jet in combination with abrasive particles for cutting different types of materials. A stream of small abrasive particles is introduced and entrained in the water jet in such a manner that the water jet's momentum is partly transferred to the abrasive particles. The role of carrier water is primarily to accelerate large quantities of abrasive particles to a high velocity and to produce a highly coherent jet. Important process parameters of abrasive water jet machining can be categorized as hydraulic parameters: water pressure, water flow rate, abrasive parameters (type, size, shape, and flow rate of abrasive particles), and cutting parameters (traverse rate and stand-off distance).

The model presented in this chapter is based on the analysis given by Hashish (1989). The five decision variables considered for this model are water jet pressure at the nozzle exit (P_w), diameter of abrasive water jet nozzle (d_{awn}), feed rate of nozzle (f_n), mass flow rate of water (M_w), and mass flow rate of abrasives (M_a). The objective function and constraint are discussed below:

Objective function:
The objective is to maximize the material removal rate (Z_1) as given by Eq. (13.1):

$$\text{Maximize } Z_1 = d_{\text{awn}} f_n (h_c + h_d) \tag{13.1}$$

where 'h_c' is the indentation depth due to cutting wear as given by Eqs. (13.2) and (13.3):

$$h_c = \left(\frac{1.028 \times 10^{4.5}\xi}{C_k \rho_a^{0.4}}\right)\left(\frac{d_{\text{awn}}^{0.2} M_a^{0.4}}{f_n^{0.4}}\right)\left(\frac{M_w P_w^{0.5}}{M_a + M_w}\right) - \left(\frac{18.48 K_a^{2/3} \xi^{1/3}}{C_k^{1/3} f_r^{0.4}}\right)\left(\frac{M_w P_w^{0.5}}{M_a + M_w}\right)^{1/3}; \text{ if } \alpha_t \leq \alpha_0. \tag{13.2}$$

$$h_c = 0, \text{ if } \alpha_t > \alpha_0. \tag{13.3}$$

'h_d' is the indentation depth due to deformation wear as given by Eq. (13.4):

$$h_d = \frac{\eta_a d_{\text{awn}} M_a \left[K_1 M_w P_w^{0.5} - (M_a + M_w) v_{\text{ac}}\right]^2}{(1570.8\sigma_{\text{fw}}) d_{\text{awn}}^2 f_n (M_a + M_w)^2 + (K_1 C_{\text{fw}} \eta_a)\left[K_1 M_w P_w^{0.5} - (M_a + M_w) v_{\text{ac}}\right] M_a M_w P_w^{0.5}} \tag{13.4}$$

$$\alpha_0 \approx \left(\frac{0.02164 C_K^{1/3} f_r^{0.4}}{K_a^{2/3} \xi^{1/3}}\right)\left(\frac{\dot{M}_a + \dot{M}_w}{\dot{M}_w P_w^{0.5}}\right)^{1/3} (^\circ). \tag{13.5}$$

$$\alpha_t \approx \left(\frac{0.389 \times 10^{-4.5} \rho_a^{0.4} C_K}{\xi}\right)\left(\frac{d_{\text{awn}}^{0.8} f_n^{0.4} (\dot{M}_a + \dot{M}_w)}{M_a^{0.4} \dot{M}_w P_w^{0.5}}\right) (°). \tag{13.6}$$

$$v_{\text{ac}} = 5\pi^2 \frac{\sigma_{\text{cw}}^{2.5}}{\rho_a^{0.5}} \left[\frac{1 - v_a^2}{E_{\text{Ya}}} + \frac{1 - v_w^2}{E_{\text{Yw}}}\right]^2 (\text{mm/s}) \tag{13.7}$$

$$K_1 = \sqrt{2} \times 10^{4.5} \xi. \tag{13.8}$$

$$C_K = \sqrt{3000 \sigma_{\text{fw}} f_r^{0.6} / \rho_a}\, (\text{mm/s}) \tag{13.9}$$

$$\text{K}_\text{a} = 3.$$

Constraint:

Constraint is on power consumption as given by Eq. (13.10):

$$1.0 - \frac{P_w M_w}{P_{\text{max}}} \geq 0 \tag{13.10}$$

Description of various symbols is provided in Table 13.1.

Table 13.1 Values of the constants and parameters for abrasive water jet machining process (Pawar and Rao 2013a, b; Reprinted with permission from Springer Science + Business Media)

Notation	Description	Unit	Value
ρ_a	Density of abrasive particles	Kg/mm^3	3.95×10^{-6}
v_a	Poisson ratio of abrasive particles		0.25
E_{Ya}	Modulus of elasticity of abrasive particles	MPa	350000
f_r	Roundness factor of abrasive particles		0.35
f_s	Sphericity factor of abrasive particles		0.78
η_a	Proportion of abrasive grains effectively participating in machining		0.07
v_w	Poisson ratio of work material		0.2
E_{Yw}	Modulus of elasticity of work material	MPa	114000
σ_{ew}	Elastic limit of work material	MPa	883
σ_{fw}	Flow stress of work material	MPa	8142
C_{fw}	Drag friction coefficient of work material		0.002
ξ	Mixing efficiency between abrasive and water		0.8
P_{max}	Allowable power consumption value	kW	56

Variable bounds:

The bounds for the five variables are given below:

$$50 \leq P_{\mathrm{w}} \leq 400 \text{ (MPa)} \quad (13.11)$$

$$0.5 \leq d_{\mathrm{awn}} \leq 5 \text{ (mm)} \quad (13.12)$$

$$0.2 \leq f_{\mathrm{n}} \leq 25 \text{ (mm/s)} \quad (13.13)$$

$$0.02 \leq M_{\mathrm{w}} \leq 0.2 \text{ (Kg/s)} \quad (13.14)$$

$$0.0003 \leq M_{\mathrm{a}} \leq 0.08 \text{ (Kg/s)} \quad (13.15)$$

As TLBO algorithm is an algorithm-specific parameter-less algorithm, only population size and number of generations need to be specified to run the algorithm. Based on trial runs, the population size decided for the present example is 20 and the number of generations is 50. The results of optimization of AWJM process using TLBO algorithm are presented in Table 13.2 along with those obtained using other optimization algorithms (i.e., GA and SA).

For abrasive water jet machining, if angle of impingement at the top of the machined surface 'α_t' exceeds the critical impact angle 'α_0' then no material removal is assumed to occur by cutting wear (i.e., $h_c = 0$) and the material removal occurs only due to the deformation wear (h_d), which results into relatively less material removal rate (Paul et al. 1998). As shown in Table 13.2, for the solution obtained using GA (Jain et al. 2007), as 'α_t' exceeds 'α_0,' indentation depth of cutting wear (h_c) becomes zero and hence results in very poor material removal rate as compared to the solution obtained using TLBO algorithm for which, as 'α_t' < 'α_0,' significant amount of material removal rate is contributed by cutting wear. Besides that, the optimum values of process variables obtained using TLBO algorithm also result in higher value of depth of deformation wear (h_d) than that obtained using genetic algorithm which further increases the material removal rate. The combined effect thus leads to the significant improvement in material removal rate from 90.257 mm^3/s to 239.54 mm^3/s. It can also be observed that TLBO algorithm provides better solution accuracy as compared to the solution obtained using SA algorithm. TLBO algorithm provides an improvement of about 9 % in objective function over that obtained using SA algorithm. It is observed that the algorithm required only 30 generations to achieve the global optimum solution.

Table 13.2 Results of optimization of AWJM process using TLBO (Pawar and Rao 2013a, b; Reprinted with permission from Springer Science + Business Media)

Method	d_{awn}	f_n	M_w	P_w	M_a	α_0	α_t	h_c	h_d	MRR	Power
	(mm)	(mm/s)	(Kg/s)	(MPa)	(Kg/s)	(°)	(°)	(mm)	(mm)	(mm^3/s)	(kW)
GA*	3.726	23.17	0.141	398.3	0.079	0.384	0.572	0	1.045	90.257	56
SA**	2.9	15	0.138	400	0.08	0.385	0.378	2.97	2.04	218.19	56
TLBO***	5	5.404	0.14	400	0.07	0.379	0.352	5.663	3.237	239.18	56

*Jain et al. (2007); ** Hashish (1989); ***Pawar and Rawar (2013a, b))

13.1.2 Optimization of Milling Process

The optimization model for milling process presented in this chapter is based on the analysis given by Sonmez et al. (1999). The decision variables considered for this model are feed per tooth (f_z), cutting speed (V), and depth of cut (a).

The objective function in this model is to minimize the production time (T_{pr}) as given by Eq. (13.16):

$$T_{\mathrm{Pr}} = \frac{T_s}{N_b} + T_L + N_p T_a + \sum_{i=1}^{N_p} \frac{\pi \mathrm{DL}}{f_{\mathrm{zi}} z 1000 V_i} + \frac{T_d \pi L V_i^{\left(\frac{1}{m}-1\right)} a_i^{e_v/m} f_{\mathrm{zi}}^{\left(\frac{u_v}{m}-1\right)} a_r^{r_v/m} z^{\left(\frac{n_v}{m}-1\right)} \lambda_s^{q_v/m}}{1000 C_v^{1/m} D^{\left(\frac{b_v}{m}-1\right)} \times \left(B_m B_h B_p B_t\right)^{1/m}} \tag{13.16}$$

where T_s is the setup time; N_b is the total number of components in batch; T_L is the loading and unloading time; N_p is the total number of passes and subscript 'i' denotes ith pass; T_a is the process adjusting and quick return time; T_d is the tool changing time; f_z is teh feed per tooth; z is the number of teeth on milling cutter; D is the cutter diameter; L is the length of the cut; a_r is the width of the cut; a is the depth of cut; V is the cutting speed; and B_m, B_k, B_p, B_t, m, e_v, u_v, r_v, n_v, q_v, C_v, b_v, C_{zp}, b_z, u_z, are the process constants.

Following three constraints are considered in this optimization model:

(a) Arbor strength:

$$\mathrm{F_s} - \mathrm{F_c} \geq 0 \tag{13.17}$$

$$\text{where, mean peripheral cutting force} = F_c = C_{zp} a_r z D^{b_z} a^{e_z} f_z^{u_z} \tag{13.18}$$

Permissible force for arbor strength (Kg) = F_s

$$F_s = \frac{0.1 k_b d_a^3}{0.08 L_a + 0.65 \sqrt{\left((0.25 L_a)^2 + (0.5 \alpha D)^2\right)}} \tag{13.19}$$

where k_b is the permissible bending strength of arbor; d_a is the arbor diameter of 27 mm; L_a is the arbor length between supports; $\alpha = k_b/(1.3\ k_t)$; and k_t is the permissible torsional strength of arbor.

(b) Arbor deflection:

$$F_d - F_c \geq 0 \tag{13.20}$$

$$\text{where, Permissible force for arbour deflection (Kg)} = F_d = \frac{4 E e d_a^4}{L_a^3} \tag{13.21}$$

where E is the modulus of elasticity of arbor material and e is the permissible value of arbor deflection. For roughing operation, 'e' is equal to 0.2 mm and for finishing operation 'e' = 0.05 mm.

(c) Power:===

$$P_c - \frac{F_c V}{6120} \geq 0 \tag{13.22}$$

where P_c is the cutting power (KW) = $P_m \times \eta$;

P_m is the nominal motor power; and η is the overall efficiency.

The three process variables and their bounds considered in this work are given below:

(a) Feed per tooth:

$$0.000875 \leq f_z \leq 3.571 \tag{13.23}$$

(b) Cutting speed:

$$6.234 \leq V \leq 395.84 \tag{13.24}$$

(c) Depth of cut:

$$0.5 < a \leq 4 \text{ (mm)} \tag{13.25}$$

Values of the constants and parameters considered in the present example are given below:

P_m = 5.5 kW, η = 0.7, d_a = 27 mm, L_a = 210 mm, k_b: 140 MPa, k_t: 120 MPa, E = 200 GPa, D = 63 mm, z = 8, L_a = 160 mm, a_r = 50 mm, a = 5 mm, T_L = 1.5 min, T_s = 10 min, T_c = 5 min, T_a = 0.1 (min/part), N_b = 100; Constants: B_m = 1, B_k = 1, B_p = 0.8, B_t = 0.8, m = 0.33, e_v = 0.3, u_v = 0.4, r_v = 0.1, n_v = 0.1, q_v = 0, C_v = 35.4, b_v = 0.45, C_{zp} = 68.2, b_z = –0.86, e_z = 0.86, and u_z = 0.72.

Results of optimization of milling process using TLBO algorithm are presented in Table 13.3 for the optimum cutting strategy indicating three rough passes each of 1.5 mm and one finishing pass of 0.5 mm. Table 13.4 provides the results of optimization of milling process obtained by various algorithms. As shown in Table 13.4, the results obtained using geometric programming (GP), genetic

Table 13.3 Results of optimization of milling process using TLBO (Pawar and Rao 2013a; Reprinted with permission from Springer Science + Business Media)

Cutting strategy	f_z (mm/tooth)	V (m/min)	T_2 (per pass) (min)	T_2 (min)	T_1 (min)	T_{pr} ($T_1 + T_2$) (min)
a_{rough} = 1.5	0.341	46.641	0.342	**1.237**	**2.0**	**3.237**
a_{rough} = 1.5	0.341	46.641	0.342			
a_{rough} = 1.5	0.341	46.641	0.342			
a_{finish} = 0.5	0.434	66.8576	0.211			

where a_{rough} = depth of cut for rough pass (mm); a_{finish} = depth of cut for finish pass (mm)

Table 13.4 Results of optimization of milling process using various optimization algorithms (Pawar and Rao 2013a; Reprinted with permission from Springer Science + Business Media)

Method	Cutting strategy	f_z (mm/tooth)	V (m/min)	SC	DC	PC	T_2 (min)	T_{pr} ($T_1 + T_2$) (min)
GP (Sonmez et al. 1999)	$a_{rough} = 3$	0.338	26.40	**–405**	24.92	**–0.08**	0.813	2.614
	$a_{finish} = 2$	0.570	25.16	**–430**	**–702**	0		
GA (Wang et al. 2005)	$a_{rough} = 3$	0.366	24.69	**–459**	**–28.81**	**–0.04**	0.8102	2.61
	$a_{finish} = 2$	0.5667	25.16	**–427**	**–698**	0		
PGSA (Wang et al. 2005)	$a_{rough} = 3$	0.3693	24.25	**–465**	**–35**	0.2	0.8	2.60
	$a_{finish} = 2$	0.5886	24.58	**–452**	**–74**	0		
Tribes (Onwubolu 2006)	$a_{rough} = 3$	0.587	36.27	**–8.50**	**–420**	**–4.18**	0.512	2.212
	$a_{finish} = 2$	0.902	30.16	**–797**	**–1069**	**–2.57**		
ABC (Rao and Pawar 2010)	$a_{rough} = 1.5$	0.337	46.982	4.708	435.02	0.0047	1.240	3.240
	$a_{rough} = 1.5$	0.337	46.982	4.708	435.02	0.0047		
	$a_{rough} = 1.5$	0.337	46.982	4.708	435.02	0.0047		
	$a_{finish} = 0.5$	0.432	64.410	271.97	1.131	1.400		
PSO (Rao and Pawar 2010)	$a_{rough} = 1.5$	0.340	46.610	1.5	431.9	0.01	1.240	3.240
	$a_{rough} = 1.5$	0.340	46.610	1.5	431.9	0.01		
	$a_{rough} = 1.5$	0.340	46.610	1.5	431.9	0.01		
	$a_{finish} = 0.5$	0.434	63.580	271.9	0.35	1.422		
SA (Rao and Pawar 2010)	$a_{rough} = 1.5$	0.336	44.633	5.779	436.09	0.204	1.263	3.263
	$a_{rough} = 1.5$	0.336	44.633	5.779	436.09	0.204		
	$a_{rough} = 1.5$	0.336	44.633	5.779	436.09	0.204		
	$a_{finish} = 0.5$	0.429	57.230	273.91	2.296	1.683		
TLBO (Pawar and Rao 2013a)	$a_{rough} = 1.5$	0.341	46.641	0.435	430.755	0.0001	1.237	3.237
	$a_{rough} = 1.5$	0.341	46.641	0.435	430.755	0.0001		
	$a_{rough} = 1.5$	0.341	46.641	0.435	430.755	0.0001		
	$a_{finish} = 0.5$	0.434	66.8576	271.975	0.355	1.297		

SC Arbor strength constraint, *DC* Arbor deflection constraint, and *PC* Power constraint

algorithm (GA), parallel genetic simulated annealing (PGSA), and Tribes are inappropriate, as these results violate the specified constraints.

$$T_1 = \frac{T_s}{N_b} + T_L + N_p T_a$$

$$T_2 = \sum_{i=1}^{N_p} \frac{\pi DL}{f_{zi} z 1000 V_i} + \frac{T_d \pi L V_i^{\left(\frac{1}{m}-1\right)} a_i^{e_v/m} f_{zi}^{\left(\frac{u_v}{m}-1\right)} a_r^{r_v/m} z^{\left(\frac{n_v}{m}-1\right)} \lambda_s^{q_v/m}}{1000 C_v^{1/m} D^{\left(\frac{b_v}{m}-1\right)} \times \left(B_m B_h B_p B_t\right)^{1/m}}$$

It can be seen from Table 13.2 that the solution obtained using TLBO algorithm is slightly better in terms of accuracy of solution as compared to ABC, PSO, and SA algorithms (Pawar and Rao 2013a, b).

Rao and Kalyankar (2013a) carried out the parameter optimization of a multi-pass turning operation using the TLBO algorithm. Two different examples were considered by the authors that were attempted previously by various researchers using different optimization techniques such as simulated annealing, genetic algorithm, ant colony algorithm, particle swarm optimization, etc. The first example was a multiobjective problem and the second example was a single objective multi-constrained problem with 20 constraints. The TLBO algorithm had proved its effectiveness over the other algorithms. The performance of the TLBO algorithm was studied in terms of the convergence rate and accuracy of the solution. The TLBO algorithm required less number of iterations for convergence to the optimal solution and the algorithm had shown its ability in handling multi-constrained problems.

Rao and Kalyankar (2013b) applied the TLBO algorithm for the process parameter optimization of selected modern machining processes. The important modern machining processes identified for the process parameters optimization were ultrasonic machining (USM), abrasive jet machining (AJM), and wire electrical discharge machining (WEDM) process. The examples considered for these processes were attempted previously by various researchers using different optimization techniques such as GA, SA, ABC, PSO, harmony search (HS), shuffled frog leaping (SFL), etc. The comparison between the results obtained by the TLBO algorithm and those obtained by different optimization algorithms showed the better performance of the TLBO algorithm. However, in the case of WEDM process of Example 3, the objective of the work was to maximize the cutting speed (V_m) by ensuring the constraint value of surface roughness (R_a) which should not exceed the permissible surface roughness (R_{per}) of 2.0 μm. The optimum process parameters setting obtained by the TLBO algorithm was given in Table 13.4 of their paper and the maximum cutting speed given by the TLBO algorithm was reported as 1.4287 mm/min. However, it has been observed that the corresponding set of process parameters leads to a slight variation of the surface roughness constraint by 0.0189 μm. Even though this difference is small, the TLBO algorithm is now rerun using the same number of 20 iterations and the population size of 10. The new result for V_m is 1.421034 mm/min and the corresponding R_a value is 1.999997 μm and this satisfies the constraint. The values given in Rao and Pawar (2010) and Rao (2011) are also relooked into and slight corrections are made. The corrected values are 1.420907 mm/min in the case of ABC of Rao and Pawar (2009), 1.420498 mm/min in the case of PSO, 1.414212 mm/min in the case of HS_M and SA, and 1.417831 mm/min in the case of SFL of Rao (2011). It can be observed that the maximum cutting speed (V_m) given by the TLBO algorithm is 1.421034 mm/min which is still better than the results given by all the other optimization algorithms used for the same model. ABC algorithm has given the next best result. However, the number of iterations used in the case of ABC algorithm was 150, whereas TLBO algorithm has given the result using only 20 iterations. Thus, in the case of WEDM process, the TLBO algorithm has proved its superiority and has given slight improvement in the result with less iterations compared to the other advanced optimization algorithms. Similarly, Tables 2 and 3 of Rao and Kalyankar (2013b) need to be corrected in the case of Example 2 keeping in view of the slight violation in the constraint value. The values given in Jain et al. (2007) and Rao et al. (2010) are also relooked into and slight corrections are now made. The mass flow rate of abrasive particles (kg/s) and velocity of abrasive particles (mm/s) in example 4.2.1 are 0.0005 and 315,772, respectively, in the case of TLBO algorithm with the optimum value of MRR (mm^3/s) as 8.2528. The optimum value of MRR (mm^3/s) is 8.2525 in the case of SA of Rao and Pawar (2010) and the optimum value of MRR (mm^3/s) is 8.242 in the case of Jain et al. (2007). Similarly, in the case of example 4.2.2, velocity of abrasive particles (mm/s) is recalculated as 333,600 in the case of TLBO of Rao and Kalyankar (2013) with the optimum value of MRR (mm^3/s) as 0.6056. The optimum value of MRR (mm^3/s) is 0.6035 in the case of GA of Jain et al. (2007). However, it can be observed that the results given by the TLBO algorithm are found still better than those given by GA and SA in this example.

The TLBO algorithm can also be easily modified to suit optimization of process parameters of other machining processes such as drilling, grinding, advanced machining processes, etc. Also, the TLBO algorithm can efficiently handle the multiobjective optimization models. The next chapter presents the application of the TLBO algorithm for the process parameter optimization of multiobjective machining processes using a posteriori approach.

References

Hashish, M., 1989. A model for abrasive water jet (AWJ) machining. Transactions of ASME: Journal of Engineering Materials and Technology 111, 154–162.

Jain, N.K., Jain, V.K., Deb, K., 2007. Optimization of process parameters of mechanical type advanced machining processes using genetic algorithm. International Journal of Machine Tools and Manufacture, 47, 900–919.

Onwubolu, G.C., 2006. Performance based optimization of multi-pass face milling operations using tribes. International Journal of Machine Tools and Manufacture 46, 717–727.

Paul, S., Hoogstrate, A.M., van Luttervelt, C.A., Kals, H.J.J., 1998. Analytical modeling of the total depth of cut in abrasive water jet machining of polycrystalline brittle materials. Journal of Material Processing Technology 73, 206–212.

Pawar, P.J., Rao, R.V., 2013a. Parameter optimization of machining processes using teaching-learning-based-optimization algorithm. International Journal of Advanced Manufacturing Technology 67, 995–1106.

Pawar, P.J., Rao, R.V., 2013b. Erratum to: Parameter optimization of machining processes using teaching-learning-based-optimization algorithm. International Journal of Advanced Manufacturing Technology 67, 1955.

Rao, R.V., 2011. Advanced Modeling and Optimization of Manufacturing Processes: International Research and Development. London: Springer-Verlag.

Rao, R.V., Kalyankar, V.D., 2013a. Multi-pass turning process parameter optimization using teaching–learning-based optimization algorithm. ScientiaIranica Transactions E: Industrial Engineering 20(3), 967–974.

Rao, R.V., Kalyankar, V.D., 2013b. Parameter optimization of modern machining processes using teaching–learning-based optimization algorithm. Engineering Applications of Artificial Intelligence 26, 524–531.

Rao, R.V., Pawar, P.J., 2010. Parameter optimization of a multi-pass milling process using non-traditional optimization algorithms. Applied Soft Computing 10(2), 445–456.

Sonmez, A.I., Baykasoglu, A., Dereli, T., Filiz, I.H., 1999. Dynamic optimization of multipass milling operations via geometric programming. International Journal of Machine Tools and Manufacture 39(2), 297–32.

Wang, Z.G,, Rahman, M., Wong, Y.S., Sun, J., 2005. Optimization of multi-pass milling using parallel genetic algorithm and par.llel genetic simulated annealing. International Journal of Machine Tools and Manufacture 45(15), 1726–1734.

Chapter 14
Multiobjective Optimization of Machining Processes Using NSTLBO Algorithm

Abstract Multiobjective optimization aspects of a traditional machining process namely surface grinding and five modern machining processes namely, wire-electro discharge machining process, micro-wire-electric discharge machining process, laser cutting process, electrochemical machining process, and electrochemical discharge machining process are presented in this chapter. A posteriori multiobjective optimization algorithm named as nondominated sorting teaching-learning-based optimization (NSTLBO) algorithm is proposed to solve the multiobjective optimization problems of the machining processes. A Pareto optimal set of solutions along with a Pareto front is presented for each of the considered machining processes.

14.1 Multiobjective Optimization of Machining Processes

Finding the optimum combination process parameters of any machining process requires comprehensive knowledge of the manufacturing process, empirical equations to develop realistic constraints, specification of machine tool capabilities, development of effective optimization criteria, and knowledge of mathematical and numerical optimization techniques. A human process planner selects proper machining process parameters using his own experience or machining tables. In most of the cases, the selected parameters are conservative and far from optimum. Selecting optimum combination of process parameters through experimentation is costly, time-consuming, and tedious. These factors have steered the researchers toward applying numerical and heuristics-based optimization techniques for process parameter optimization of machining processes.

In order to determine the optimum combination of process parameters, researchers had applied many traditional optimization algorithms such as geometric programming, nonlinear programming, sequential programming, goal programming, and dynamic programming (Mukherjee and Ray 2006). Although these methods had performed well in many practical cases, they may fail in complex

R.V. Rao, *Teaching Learning Based Optimization Algorithm*,
DOI 10.1007/978-3-319-22732-0_14

situations which involve complex functions with large number of independent variables. Most of the real-world machining process optimization problems involve complex functions and large number of process parameters. In such problems, traditional optimization techniques fail to provide a optimum solution as they may get caught into local optima. Moreover, traditional optimization techniques require an excellent initial guess of the optimal solutions and the results and the rate of convergence of the solution are very sensitive to these guesses.

In order to overcome these problems and to search a near optimum solution for complex problems, many population-based heuristic algorithms have been developed by researchers in the past two decades. In the field of machining also, many population-based algorithms have been applied by the researchers (Yusup et al. 2012; Rao and Kalyankar 2014). Researchers had also applied the TLBO algorithm to optimize the process parameters of machining processes (Rao and Kalyankar 2013, 2014; Pawar and Rao 2013), and it was reported that the TLBO algorithm was capable of solving complex machining optimization problems and had outperformed the other traditional and advanced optimization techniques in terms of objective function values, number of function evaluations required to achieve the optimum solution, and computational time. However, the most of the machining processes involve more than one machining process performance characteristic. This gives rise to the need to formulate and solve multiobjective optimization problems. The previous researchers had solved the multiobjective optimization problem of machining processes using the priori approach. In the priori approach, multiobjective optimization problem is transformed into a single-objective optimization problem by assigning an appropriate weight to each objective. This ultimately leads to a unique optimum solution. However, the solution obtained by this process depends largely on the weights assigned to various objective functions. This approach does not provide a dense spread of the Pareto points. Furthermore, in order to assign weights to each objective the process planner is required to precisely know the order of importance of each objective in advance which may be difficult in today's volatile market scenario. This drawback of the priori approach is eliminated in the posteriori approach, wherein it is not required to assign the weights to the objective functions prior to the simulation run. The posteriori approach does not lead to a unique optimum solution at the end but provides a dense spread of Pareto points (Pareto optimal solutions). The process planner can then select one solution from the set of Pareto optimal solutions based on the order of importance of objectives. Thus, the posteriori approach is very suitable for solving multiobjective optimization problems in machining processes, wherein the extremely volatile market leads to frequent change in the order of importance of objectives and determining the weights to be assigned to the objectives in advance is difficult.

In order to solve the multiobjective optimization problems and to successfully obtain the Pareto set without getting trapped into local optima two important aspects are required: balance between exploration and exploitation capability and good diversity ensuring mechanism in order to avoid stagnation at the local optima.

Researchers had already solved the multiobjective optimization problems in machining processes using well-known posteriori approaches such as NSGA, NSGA-II, MOGA, and MODE (Mitra and Gopinath 2004; Kuriakose and Shunmugam 2005; Konak et al. 2006; Mandal et al. 2007; Kodali et al. 2008; Kanagarajan et al. 2008; Palanikumar et al. 2009; Datta and Deb 2009; Yang and Natarajan 2010; Senthilkumar et al. 2010, 2011; Joshi and Pande 2011; Mitra 2009; Acharya et al. 2013). However, these algorithms require tuning of algorithm-specific parameters and improper tuning of algorithm-specific parameters may lead to non-Pareto optimal solutions. Thus, in this work, a parameter-less posteriori multiobjective optimization algorithm based on the TLBO algorithm is proposed and is named as "Non-dominated Sorting Teaching-Learning-Based Optimization (NSTLBO)" algorithm. In the NSTLBO algorithm, the teacher phase and learner phase maintain the vital balance between the exploration and exploitation capabilities and the teacher selection based on nondominance rank of the solutions and crowding distance computation mechanism ensures the selection process toward better solutions with diversity among the solutions in order to obtain a Pareto optimal set of solutions for multiobjective optimization problems in a single simulation run.

In order to demonstrate and validate the performance of the NSTLBO algorithm, it is applied to solve the multiobjective optimization problems of a traditional machining process namely surface grinding and turning and five modern machining processes such as wire-electric discharge machining (WEDM), micro-wire-electric discharge machining process, laser cutting, electrochemical machining (ECM), and electrochemical discharge machining (ECDM).

14.2 Examples

14.2.1 Optimization of Process Parameters of Surface Grinding Process

In the past, the researchers had attempted to optimize the process parameters of surface grinding process using QP, GA, DE, PSO, SA, HS, ABC, TLBO, etc. (Mukherjee and Ray 2006; Yusup et al. 2012; Wen et al. 1992; Rowe et al. 1994; Saravanan et al. 2002; Dhavalikar et al. 2003; Mitra and Gopinath 2004; Baskar et al. 2004; Krishna 2007; Pawar et al. 2010; Rao and Pawar 2010). Now, NSTLBO algorithm is applied to solve the mutiobjective optimization problem in surface grinding process. The optimization models formulated in this work are based on analysis given by Wen et al. (1992). The optimization aspects of rough grinding and finish grinding processes are considered separately with the help of two examples.

14.2.1.1 Optimization of Process Parameters of Rough Grinding Process

Objective functions:

The two objectives considered in case of rough grinding operation are as follows:

(a) Minimize production cost as given by Eq. (14.6).

$$C_T = \frac{M_c}{60p}\left(\frac{L_w + L_e}{V_w 1000}\right)\left(\frac{b_w + b_e}{f_b}\right)\left(\frac{a_w}{a_p} + S_p + \frac{a_w b_w L_w}{\pi D_e b_s a_p G}\right) + \frac{M_c}{60p}\left(\frac{S_d}{V_r} + t_1\right) + \frac{M_c t_{ch}}{60N_t} + \frac{M_c \pi b_s D_e}{60pN_d L V_s 1000} + C_s\left(\frac{a_w b_w L_w}{pG} + \frac{\pi(\text{doc}) b_s D_e}{pN_d}\right) + \frac{C_d}{pN_{td}} \quad (14.1)$$

where M_c is cost per hour labor and administration, L_w is length of workpiece, L_e is empty length of grinding, b_w is width of workpiece, b_e is empty width of grinding, f_b is cross feed rate, a_w is total thickness of cut, a_p is down feed of grinding, S_p is number of spark out grinding, D_e is diameter of wheel, b_s is width of wheel, G is grinding ratio, S_d is distance of wheel idling, p is number of workpieces loaded on table, V_r is speed of wheel idling, t_1 is time of loading and unloading workpieces, t_{ch} is time of adjusting machine tool, N_t is batch size of the workpieces, N_d is total number of workpieces to be ground between two dressing, N_{td} is total number of workpieces to be ground during life of dressing, and C_d is cost of dressing.

(b) Maximize of workpiece removal parameter (WRP) given by Eq. (14.2) as,

$$\text{WRP} = 94.4\frac{(1 + (2\text{doc}/3L))L^{11/19}(V_w/V_s)^{3/19} V_s}{D_e^{43/304}\text{VOL}^{0.47} d_g^{5/38} R_c^{27/19}} \quad (14.2)$$

where VOL is wheel bond percentage, d_g is grind size, R_c is workpiece hardness.

Constraints:

The following four constraints are considered.

(a) Thermal damage constraint

The grinding process requires very high energy per unit volume of material removed. The energy that is concentrated within the grinding zone is converted into heat. The high-thermal energy causes damage to the work piece and it leads to the reduced production rate. The specific energy U is calculated by Eq. (14.3).

$$U = 13.8 + \frac{9.64\times10^{-4} V_s}{a_p V_w} + \left(6.9\times10^{-3}\frac{2102.4 V_w}{D_e V_s}\right) \times \left(A_0 + \frac{K_u V_s L_w a_w}{V_w D_e^{1/2} a_p^{1/2}}\right)\frac{V_s D_e^{1/2}}{V_w a_p^{1/2}} \quad (14.3)$$

where K_u = wear constant.

The critical specific energy U^* at which the burning starts is expressed in terms of operating parameters as follows:

$$U^* = 6.2 + 1.76\left(\frac{D_e^{1/4}}{a_p^{3/4} V_w^{1/2}}\right) \tag{14.4}$$

The thermal damage constraint is then specified as follows:

$$U^* - U \geq 0 \tag{14.5}$$

(b) Wheel wear parameter constraint

Wheel wear parameter is related directly to the grinding conditions. For single-point diamond dressing, it is given by Eq. (14.6).

$$\text{WWP} = \left(\frac{k_p a_p d_g^{5/38} R_c^{27/29}}{D_c^{1.2/\text{VOL}-43/304} \text{VOL}^{0.38}}\right) \times \frac{(1 + (\text{doc}/L))L^{27/19}(V_s/V_w)^{3/19} V_w}{(1 + (2\text{doc}/3L))} \tag{14.6}$$

The wheel wear constraint is obtained as follows:

$$\frac{\text{WRP}}{\text{WWP}} - G \geq 0 \tag{14.7}$$

(c) Machine tool stiffness constraint

Chatter results in poorer surface quality and lowers the machining production rate. Chatter avoidance is a significant constraint in selection of machining parameters. The relationship between grinding stiffness K_c (in N/mm), wheel wear stiffness K_s (N/mm), and operating parameters during grinding is given below:

$$K_c = \frac{1000 V_w f_b}{\text{WRP}} \tag{14.8}$$

$$K_s = \frac{1000 V_s f_b}{\text{WWP}} \tag{14.9}$$

To avoid chatter during machining, the constraint given by Eq. (14.10) has to be fulfilled:

$$\text{MSC} - \frac{|R_{\text{em}}|}{K_m} \geq 0 \tag{14.10}$$

where

$$\text{MSC} = \frac{1}{2K_c}\left(1 + \frac{V_w}{V_s G}\right) + \frac{1}{K_s} \tag{14.11}$$

R_{em} is dynamic machine characteristics, K_m static machine stiffness.

(d) Surface roughness constraint

The surface roughness constraint is given by Eqs. (14.12)–(14.15) as,

$$R_a = 0.4587 T_{ave}^{0.30} \text{ for } 0 < T_{ave} < 0.254 \text{ else,} \tag{14.12}$$

$$R_a = 0.78667 T_{ave}^{0.72} \text{ for } 0.254 < T_{ave} < 2.54 \tag{14.13}$$

$$T_{ave} = 12.5 \times 10^3 \frac{d_g^{16/27} a_p^{19/27}}{D_e^{8/27}} \left(1 + \frac{doc}{L}\right) L^{16/27} \left(\frac{V_w}{V_s}\right)^{16/27} \tag{14.14}$$

$$R_a \le 1.8\,\mu\text{m} \tag{14.15}$$

Values of the constants and parameters considered in the present example are given below:

M_c = 30 \$/h, L_w = 300 mm, L_e = 150 mm, b_w = 60 mm, b_e = 25 mm, f_b = 2 mm/pass, a_w = 0.1 mm, a_p = 0.0505 mm/pass, S_p = 2, D_e = 355 mm, b_s = 25 mm, G = 60, S_d = 100 mm, p = 1, V_r = 254 mm/min, t_1 = 5 min, t_{ch} = 30 min, N_t = 12, N_d = 20, N_{td} = 2,000, C_d = 25 \$, VOL = 6.99 %, d_g = 0.3 mm, R_c = 58 HRC, K_u = 3.937 × 10^{-7} mm^{-1}, R_{em} = 1, K_m = 100,000 N/mm, K_a = 0.0869.

Parameter bounds:

$$\text{Wheel speed (m/min):} \quad 1000 \le V_s \le 2023 \tag{14.16}$$

$$\text{Workpiece speed (m/min):} \quad 10 \le V_w \le 27.7 \tag{14.17}$$

$$\text{Depth of dressing (mm):} \quad 0.01 \le V_s \le 0.137 \tag{14.18}$$

$$\text{Lead of dressing (mm/rev):} \quad 0.01 \le V_s \le 0.137 \tag{14.19}$$

The optimization problem for rough grinding process is solved using the NSTLBO algorithm and the nondominated set of solutions is obtained and reported in Table 14.1. The same problem was solved by Saravanan et al. (2002) using GA. They used 2000 (i.e., a population size of 20 and no. of generations equal to 100) function evaluations to obtain the optimum solution. Thus, in this work, the NSTLBO algorithm has considered the same number of function evaluations for fair comparison of results. The solutions suggested by other researchers using optimization techniques such as GA, SA, PSO, ABC, HS, and TLBO are reported in Table 14.2. Figure 14.1 shows the Pareto front obtained for rough grinding operation using NSTLBO algorithm and the solutions obtained by the above-mentioned advanced optimization algorithms. It is observed that all the solutions suggested by other researchers for rough grinding process are inferior to the nondominated set of solutions obtained using NSTLBO algorithm.

Table 14.1 Nondominated set of solutions for rough grinding process obtained using NSTLBO

Sr. no.	V_s (m/min)	V_w (m/min)	doc (mm)	L (mm/rev)	C_T ($/pc)	WRP (mm^3/min N)
1	2023	22.700	0.01098	0.1370	5.7834	20.0054
2	2023	22.700	0.01446	0.1370	5.7980	20.3273
3	2023	22.700	0.01848	0.1370	5.8148	20.6986
4	2023	22.700	0.02272	0.1370	5.8325	21.0901
5	2023	22.700	0.02558	0.1370	5.8445	21.3543
6	2023	22.700	0.03296	0.1370	5.8753	22.0362
7	2023	22.700	0.03902	0.1370	5.9007	22.5967
8	2023	22.700	0.04185	0.1370	5.9125	22.8578
9	2023	22.700	0.04487	0.1370	5.9252	23.1373
10	2023	22.700	0.04674	0.1370	5.9330	23.3101
11	2023	22.700	0.04878	0.1370	5.9415	23.4983
12	2023	22.700	0.05604	0.1370	5.9718	24.1690
13	2023	22.700	0.05864	0.1370	5.9827	24.4099
14	2023	22.700	0.06169	0.1370	5.9955	24.6913
15	2023	22.700	0.06514	0.1370	6.0099	25.0108
16	2023	22.700	0.06895	0.1370	6.0259	25.3628
17	2023	22.700	0.07277	0.1370	6.0419	25.7159
18	2023	22.700	0.07604	0.1370	6.0555	26.0173
19	2023	22.700	0.07867	0.1370	6.0665	26.2610
20	2023	22.700	0.08408	0.1370	6.0891	26.7603
21	2023	22.700	0.08723	0.1365	6.1023	27.0245
22	2023	22.700	0.08919	0.1370	6.1105	27.2328
23	2023	22.700	0.09176	0.1370	6.1212	27.4701
24	2023	22.700	0.08540	0.0100	6.1266	27.9317
25	2023	22.700	0.08840	0.0100	6.1391	28.7661
26	2023	22.700	0.09027	0.0100	6.1468	29.2368
27	2023	22.700	0.09273	0.0100	6.1572	29.9695
28	2023	22.700	0.09458	0.0100	6.1650	30.4859
29	2023	22.700	0.09652	0.0100	6.1731	31.0246
30	2023	22.700	0.09772	0.0100	6.1781	31.3579
31	2023	22.700	0.09980	0.0100	6.1868	31.9386
32	2023	22.698	0.10192	0.0100	6.1958	32.5263
33	2023	22.700	0.10360	0.0100	6.2026	32.9665
34	2023	22.700	0.10594	0.0100	6.2125	33.6444
35	2023	22.700	0.10713	0.0100	6.2175	33.9779
36	2023	22.700	0.10830	0.0100	6.2224	34.3007
37	2023	22.700	0.10975	0.0100	6.2284	34.7060
38	2023	22.700	0.11198	0.0100	6.2377	35.3245
39	2023	22.700	0.11502	0.0100	6.2505	36.1712
40	2023	22.700	0.11828	0.0100	6.2641	37.0794
41	2023	22.700	0.12032	0.0100	6.2726	37.6455

(continued)

Table 14.1 (continued)

Sr. no.	V_s (m/min)	V_w (m/min)	doc (mm)	L (mm/rev)	C_T ($/pc)	WRP (mm^3/min N)
42	2023	22.700	0.12244	0.0100	6.2815	38.2350
43	2023	22.696	0.12377	0.0100	6.2874	38.6052
44	2023	22.698	0.12703	0.0100	6.3008	39.5124
45	2023	22.700	0.12936	0.0100	6.3105	40.1613
46	2023	22.700	0.13082	0.0100	6.3165	40.5665
47	2023	22.700	0.13268	0.0100	6.3243	41.0846
48	2023	22.699	0.13429	0.0100	6.3311	41.5224
49	2023	22.700	0.13587	0.0100	6.3377	41.9711
50	2023	22.700	0.13700	0.0100	6.3424	42.2847

Table 14.2 Results of optimization for rough grinding process obtained using other techniques

Method	Author(s)	V_s	V_w	doc	L	C_T ($/pc)	WRP (mm^3/N)	R_a
QP	Wen et al. (1992)	2000	19.96	0.055	0.044	6.2	17.47	1.74
GA	Saravanan et al. (2002)	1998	11.30	0.101	0.065	7.1	21.68	1.79
ACO	Baskar et al. (2004)	2010	10.19	0.118	0.081	7.5	24.20	1.798
PSO	Rao and Pawar (2010)	2023	10.00	0.110	0.137	8.33	25.63	1.768
SA	Rao and Pawar (2010)	2023	11.48	0.089	0.137	7.755	24.45	1.789
HS	Rao and Pawar (2010)	2019.35	12.455	0.079	0.136	7.455	23.89	1.796
ABC	Rao and Pawar (2010)	2023	10.973	0.097	0.137	7.942	25.00	1.80
TLBO	Pawar and Rao (2013)	2023	11.537	0.0899	0.137	7.742	24.551	1.798

14.2.1.2 Optimization of Process Parameters of Finish Grinding Process

Objective functions:

The objectives considered in the case of finish grinding process are described as follows:

(a) Minimize production cost as given by Eq. (14.1).
(b) Minimization of surface roughness given by Eqs. (14.12) and (14.13).

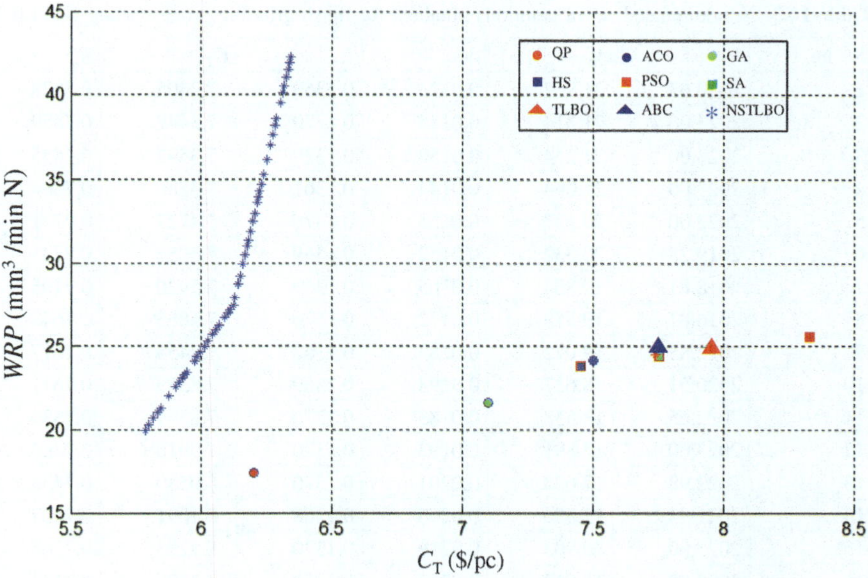

Fig. 14.1 Pareto front for rough grinding process obtained using NSTLBO algorithm

Constraints:

The thermal damage constraint, wheel wear parameter constraint, and machine tool stiffness constraint remain the same. In addition, the constraint on the WRP is expressed as follows:

$$\mathrm{WRP} \geq 20 \tag{14.20}$$

The parameter bounds in the case of finish grinding process are same as those considered for rough grinding process. The values of constants considered in the case of finish grinding process are same as those considered in rough grinding process except the values of a_w = 0.055 mm and a_p = 0.0105 mm/pass are considered in the case of finish grinding process.

The nondominated set of solutions obtained for the finish grinding process using NSTLBO algorithm is reported in Table 14.3. The solutions suggested by other researchers using optimization algorithms such as QP, GA, ACO, and PSO are reported in Table 14.4. Figure 14.2 shows the Pareto front obtained for the finish grinding process using NSTLBO algorithm. It is observed that solutions obtained using QP and GA are inferior to the solutions obtained using NSTLBO, whereas the solutions obtained by ACO and PSO do not dominate any solution obtained using the NSTLBO algorithm.

Table 14.3 Nondominated set of solutions obtained for finish grinding process using NSTLBO

Sr. no.	V_s	V_w	doc	L	C_T	R_a
1	2020.63	21.952	0.0123	0.1369	7.2405	0.7878
2	2023.00	21.314	0.0145	0.1370	7.3447	0.7859
3	2023.00	21.254	0.0158	0.1339	7.3593	0.7835
4	2023.00	21.094	0.0144	0.1361	7.3786	0.7804
5	2023.00	20.537	0.0155	0.1370	7.4727	0.7771
6	2019.28	20.332	0.0152	0.1369	7.5056	0.7733
7	2018.81	19.864	0.0165	0.1369	7.5920	0.7705
8	2016.87	19.358	0.0172	0.1370	7.6865	0.7648
9	2022.83	19.072	0.0183	0.1362	7.7454	0.7617
10	2020.51	18.637	0.0198	0.1369	7.8369	0.7611
11	2022.65	18.538	0.0183	0.1370	7.8509	0.7539
12	2023.00	17.891	0.0193	0.1370	7.9910	0.7460
13	2022.98	17.625	0.0201	0.1370	8.0530	0.7438
14	2021.94	17.386	0.0205	0.1368	8.1091	0.7407
15	2023.00	16.903	0.0219	0.1370	8.2295	0.7366
16	2022.99	16.813	0.0217	0.1370	8.2508	0.7343
17	2023.00	16.401	0.0226	0.1370	8.3587	0.7295
18	2023.00	16.223	0.0230	0.1370	8.4072	0.7276
19	2022.99	15.946	0.0246	0.1357	8.4883	0.7252
20	2023.00	15.625	0.0246	0.1370	8.5786	0.7212
21	2022.99	15.479	0.0251	0.1369	8.6225	0.7198
22	2022.46	15.391	0.0255	0.1370	8.6500	0.7194
23	2022.71	15.257	0.0265	0.1358	8.6938	0.7178
24	2023.00	14.748	0.0273	0.1364	8.8554	0.7112
25	2023.00	14.645	0.0269	0.1370	8.8866	0.7085
26	2022.63	14.475	0.0272	0.1369	8.9439	0.7059
27	2023.00	14.279	0.0277	0.1368	9.0120	0.7031
28	2023.00	14.153	0.0278	0.1370	9.0562	0.7010
29	2023.00	13.988	0.0284	0.1370	9.1169	0.6995
30	2022.23	13.871	0.0287	0.1370	9.1601	0.6980
31	2023.00	13.744	0.0289	0.1370	9.2075	0.6959
32	2022.97	13.277	0.0315	0.1363	9.3970	0.6923
33	2022.82	13.134	0.0312	0.1370	9.4527	0.6891
34	2023.00	13.042	0.0309	0.1370	9.4888	0.6860
35	2023.00	12.730	0.0326	0.1370	9.6271	0.6841
36	2018.98	12.610	0.0326	0.1370	9.6790	0.6817
37	2023.00	12.532	0.0327	0.1370	9.7141	0.6799
38	2023.00	12.365	0.0328	0.1370	9.7899	0.6763
39	2023.00	12.255	0.0331	0.1370	9.8420	0.6747
40	2023.00	11.974	0.0343	0.1370	9.9804	0.6714
41	2023.00	11.880	0.0345	0.1370	10.026	0.6694

(continued)

Table 14.3 (continued)

Sr. no.	V_s	V_w	doc	L	C_T	R_a
42	2023.00	11.722	0.0360	0.1355	10.112	0.6680
43	2023.00	11.398	0.0367	0.1370	10.284	0.6640
44	2023.00	11.235	0.0376	0.1367	10.377	0.6619
45	2023.00	11.036	0.0379	0.1368	10.490	0.6580
46	2022.93	10.629	0.0409	0.1355	10.745	0.6539
47	2022.88	10.479	0.0419	0.1351	10.842	0.6518
48	2023.00	10.436	0.0429	0.1320	10.874	0.6496
49	2014.26	10.203	0.0410	0.1370	11.019	0.6457
50	2023.00	10.000	0.0408	0.1370	11.157	0.6385

Table 14.4 Results of optimization for finish grinding process obtained using other techniques

Method	Author(s)	V_s	V_w	doc	L	C_T	WRP	R_a
QP	Wen et al. (1992)	2000	19.99	0.052	0.091	7.7	20.00	0.83
GA	Saravanan et al. 2002	1986	21.40	0.024	0.136	7.37	20.08	0.827
ACO	Baskar et al. (2004)	2023	*19.36*	*0.019*	*0.134*	7.69	20.01	0.764
PSO	Rao and Pawar (2010)	2023	22.7	0.01	0.137	7.11	20.01	0.79

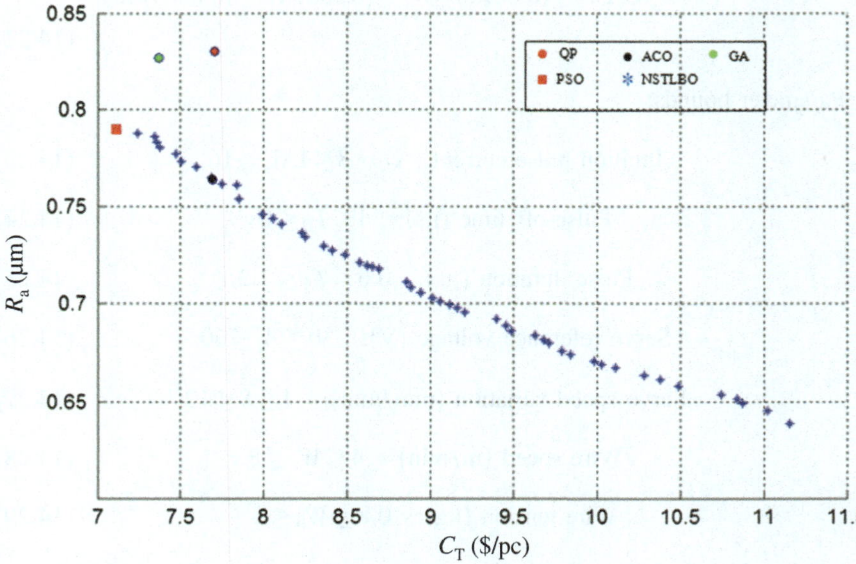

Fig. 14.2 Pareto front for finish grinding process obtained using NSTLBO algorithm

14.2.2 Optimization of Process Parameters of Wire-Electric Discharge Machining Process

In the past, researchers had determined the optimum combination of process parameters for WEDM using techniques such as GA, SA, ABC, NSGA, and TLBO (Yusup et al. 2012; Rao and Kalyankar 2014; Ramakrishnan and Karunamoorthy 2006, 2008; Chiang and Chang 2006; Pradhan et al. 2009; Satishkumar et al. 2011; Sadeghi et al. 2011; Mukherjee et al. 2012; Sharma et al. 2013; Rajyalakshmi and Ramaiah 2013; Garg et al. 2014). Now, the NSTLBO algorithm is applied to solve the multiobjective optimization problem of WEDM. The optimization model formulated in this work is based on the analysis given by Kuriakose and Shunmugam (2005). The process parameters considered for this model were applied voltage 'V,' ignition pulse current 'IAL,' pulse-off time 'T_B,' pulse duration 'T_A,' servo reference mean voltage 'A_j,' servo speed variation 'S,' wire speed 'W_s,' wire tension 'W_b' and injection pressure 'Inj' while cutting velocity 'CV' (mm/min) and surface roughness 'R_a' (μm) are considered performance measures.

Objective functions:

The objective functions are expressed by Eqs. (5.1) and (5.2) as follows:

$$\begin{aligned}\text{Maximize }(\text{CV}) = {} & 1.662 + (0.002375)\text{IAL} - (0.0639)T_B + (0.628)T_A - (0.01441)A_j \\ & + (0.008313)S - (0.001792)W_S - (0.673)W_b - (0.0294)\text{Inj}\end{aligned} \tag{14.21}$$

$$\begin{aligned}\text{Minimize }(R_a) = {} & 2.017 - (0.01236)\text{IAL} + (0.0075)T_B + (1.792)T_A - (0.006056)A_j \\ & + (0.01)S - (0.009583)W_s + (0.258)W_b - (0.0683)\text{Inj}\end{aligned} \tag{14.22}$$

Parameter bounds:

$$\text{Ignition pulse current (A):}\quad 8 \le \text{IAL} \le 16 \tag{14.23}$$

$$\text{Pulse-off time }(\mu\text{s}):\quad 4 \le T_B \le 8 \tag{14.24}$$

$$\text{Pulse duration }(\mu\text{s}):\quad 0.6 \le T_A \le 1.2 \tag{14.25}$$

$$\text{Servo reference voltage (V):}\quad 30 \le A_{\text{j}} \le 60 \tag{14.26}$$

$$\text{Servo speed variation (mm/min):}\quad 4 \le S \le 12 \tag{14.27}$$

$$\text{Wire speed (m/min):}\quad 4 \le W_{\text{s}} \le 8 \tag{14.28}$$

$$\text{Wire tension (kg):}\quad 0.8 \le W_{\text{b}} \le 1 \tag{14.29}$$

$$\text{Injection pressure (bar):} \quad 2 \le \text{Inj} \le 4 \qquad (14.30)$$

Now the multiobjective optimization problem in WEDM process is solved using the NSTLBO algorithm. The same problem was attempted by Kuriakose and Shanmugam (2005) using NSGA. They used 12,500 number of function evaluations (i.e., population size of 50 and no. of generations equal to 250). Hence, for fair comparison, NSTLBO algorithm has considered the same number of function evaluations.

The nondominated set of solutions obtained using NSTLBO algorithm for WEDM process is reported in Table 14.5. Figure 14.3 shows the comparison between the Pareto fronts obtained using NSGA and NSTLBO algorithm for WEDM process. It is observed that solutions obtained using NSGA are inferior to those obtained using NSTLBO in terms of cutting velocity and surface roughness.

14.2.3 Optimization of Process Parameters of Micro-Wire-Electric Discharge Machining Process

Several attempts had been made to study and optimize the process parameters of micro-WEDM process using techniques such as GA, SA, artificial neural networks, and fuzzy logic (Yusup et al. 2012; Rao and Kalyankar 2014; Yan and Fang 2008; Vijayaraj and Gowri 2010; Somashekhar et al. 2012a, b; Sivaprakasam et al. 2013; Kuriachen et al. 2015; Kovacevic et al. 2014). Now, the multiobjective optimization problem for micro-WEDM is solved using NSTLBO algorithm.

The optimization models considered in this work are based on the analysis given by Somashekhar et al. (2012a, b). The process parameters considered are gap voltage, capacitance, and feed rate. Material removal rate 'MRR' (mm^3/min), Overcut 'OC' (μm), and surface roughness 'R_a' (μm) are considered as performance measures.

Objective functions:

The objective functions based on coded levels of process parameters are expressed by Eqs. (14.31)–(14.33). The factors and levels of process parameters are given in Table 14.6.

$$\begin{aligned}\text{Minimize Overcut} = {} & 42.4 + 1.25A + 10.58B + 2.41C + 2.17AB \\ & + 0.73AC + 4.8BC - 12.93A^2 - 5.23B^2 + 1.62C^2\end{aligned} \qquad (14.31)$$

$$\begin{aligned}\text{Maximize MRR} = {} & 0.022 + 1.92 \times 10^{-3}A + 1.752 \times 10^{-3}B + 7.62 \times 10^{-3}C \\ & - 1.499 \times 10^{-3}AB - 4.159 \times 10^{-4}BC - 4.914 \times 10^{-3}A^2 \\ & + 5.833 \times 10^{-3}B^2 - 4.313 \times 10^{-3}C^2\end{aligned} \qquad (14.32)$$

Table 14.5 Nondominated set of solutions for WEDM process obtained using NSTLBO

Sr. no.	IAL	T_B	T_A	A_j	$S \times 7.32$	W_s	W_b	Inj	CV	R_a
1	16.00	4.00	0.600	59.94	4.000	8.000	0.800	4.000	0.320	2.4579
2	16.00	4.00	0.600	57.13	4.000	8.000	0.800	4.000	0.360	2.4749
3	16.00	4.00	0.600	54.37	4.000	8.000	0.800	4.000	0.400	2.4916
4	16.00	4.00	0.600	52.16	4.000	8.000	0.800	4.000	0.432	2.5050
5	16.00	4.00	0.600	50.73	4.000	8.000	0.800	4.000	0.453	2.5137
6	16.00	4.00	0.600	49.34	4.000	8.000	0.800	4.000	0.473	2.5221
7	16.00	4.00	0.600	47.67	4.000	8.000	0.800	4.000	0.497	2.5322
8	16.00	4.00	0.600	45.50	4.000	8.000	0.800	4.000	0.528	2.5454
9	16.00	4.00	0.600	42.79	4.000	8.000	0.800	4.000	0.567	2.5618
10	16.00	4.00	0.600	41.02	4.000	7.890	0.800	4.000	0.593	2.5735
11	16.00	4.00	0.600	38.10	4.000	8.000	0.800	4.000	0.634	2.5901
12	16.00	4.00	0.600	35.32	4.000	8.000	0.800	4.000	0.675	2.6070
13	16.00	4.00	0.600	34.17	4.000	7.861	0.800	4.000	0.691	2.6153
14	16.00	4.00	0.600	32.26	4.000	8.000	0.800	4.000	0.719	2.6255
15	16.00	4.00	0.600	30.00	4.000	8.000	0.800	3.884	0.755	2.6471
16	15.97	4.00	0.600	30.04	6.727	7.092	0.800	4.000	0.775	2.6752
17	16.00	4.00	0.600	30.00	9.005	5.205	0.800	4.000	0.798	2.7161
18	16.00	4.00	0.600	30.00	10.92	4.000	0.800	4.000	0.816	2.7468
19	15.93	4.00	0.600	30.00	12.00	7.498	0.800	3.329	0.838	2.7707
20	15.90	4.00	0.600	30.00	12.00	7.488	0.800	2.804	0.854	2.8070
21	16.00	4.00	0.600	30.00	11.91	5.241	0.800	2.000	0.881	2.8815
22	16.00	4.00	0.628	30.00	11.86	8.000	0.800	2.000	0.893	2.9052
23	16.00	4.00	0.647	30.00	11.39	7.359	0.800	2.000	0.902	2.9401
24	16.00	4.00	0.655	30.00	11.89	4.970	0.800	2.000	0.916	2.9832
25	16.00	4.00	0.689	30.00	12.00	8.000	0.800	2.000	0.933	3.0160
26	15.93	4.00	0.709	30.00	12.00	7.760	0.801	2.000	0.944	3.0549
27	15.96	4.00	0.736	30.00	12.00	8.000	0.800	2.000	0.962	3.1015
28	16.00	4.00	0.747	30.00	12.00	7.743	0.802	2.000	0.968	3.1229
29	16.00	4.00	0.767	30.00	12.00	8.000	0.800	2.000	0.982	3.1562
30	15.77	4.00	0.782	30.00	12.00	7.997	0.800	2.000	0.991	3.1863
31	16.00	4.00	0.796	30.00	12.00	5.307	0.801	2.000	1.004	3.2346
32	15.95	4.00	0.837	30.00	11.98	6.918	0.803	2.360	1.014	3.2678
33	15.62	4.00	0.845	30.00	12.00	7.996	0.800	2.000	1.030	3.3006
34	16.00	4.00	0.855	30.00	12.00	4.680	0.802	2.000	1.041	3.3454
35	15.77	4.00	0.891	30.00	11.86	6.109	0.805	2.265	1.050	3.3818
36	15.92	4.00	0.914	30.00	12.00	5.673	0.802	2.101	1.073	3.4354
37	14.91	4.00	0.928	30.00	12.00	6.007	0.800	2.000	1.084	3.4769
38	16.00	4.00	0.936	30.00	12.00	4.000	0.802	2.000	1.093	3.4985
39	15.81	4.00	0.986	30.00	11.92	6.084	0.803	2.167	1.114	3.5579
40	16.00	4.00	1.001	30.00	11.85	5.768	0.803	2.000	1.129	3.5959
41	15.90	4.00	1.012	30.00	12.00	5.956	0.800	2.000	1.138	3.6151

(continued)

Table 14.5 (continued)

Sr. no.	IAL	T_B	T_A	A_j	$S \times 7.32$	W_s	W_b	Inj	CV	R_a
42	15.95	4.00	1.047	30.00	11.43	6.667	0.801	2.000	1.155	3.6663
43	16.00	4.00	1.073	30.00	11.27	7.249	0.801	2.000	1.168	3.7048
44	15.97	4.00	1.095	30.00	11.97	7.936	0.801	2.000	1.186	3.7443
45	15.76	4.00	1.112	30.00	11.94	7.206	0.800	2.200	1.192	3.7702
46	15.94	4.00	1.126	30.00	11.98	6.662	0.801	2.000	1.208	3.8126
47	15.11	4.00	1.134	30.00	12.00	6.016	0.801	2.000	1.213	3.8433
48	15.15	4.03	1.163	30.00	12.00	6.375	0.801	2.000	1.228	3.8922
49	16.00	4.00	1.200	30.00	11.73	7.907	0.800	2.000	1.251	3.9294
50	16.00	4.00	1.200	30.00	12.00	4.319	0.800	2.000	1.260	3.9665

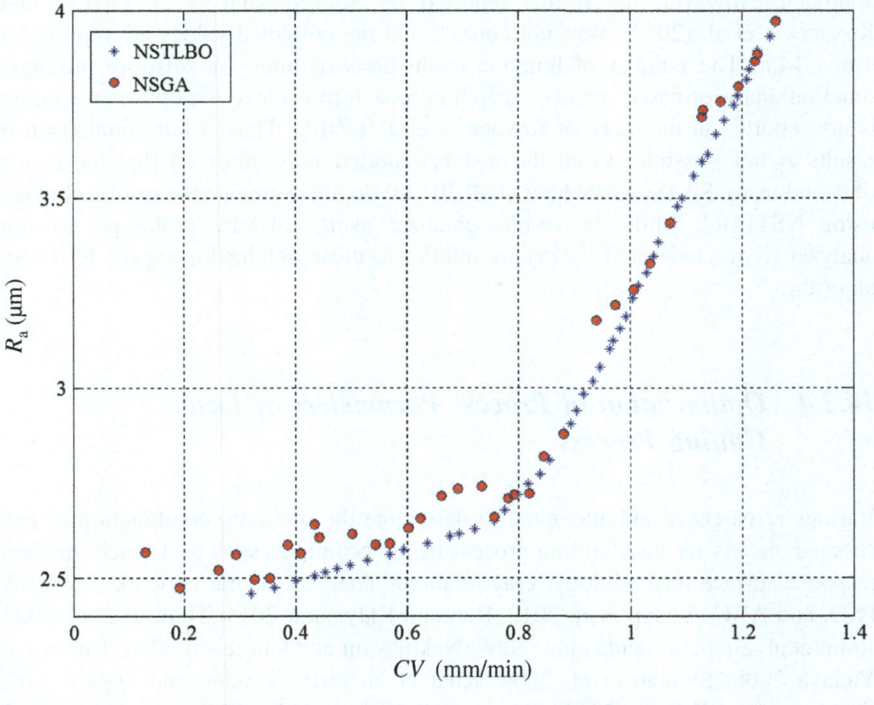

Fig. 14.3 Pareto fronts for WEDM process obtained using NSTLBO and NSGA

Table 14.6 Factors and levels for process parameters of micro-WEDM process (Somashekhar et al. 2012a, b)

Factor	Parameter	Low level (–1)	Middle level (0)	High level (1)
A	Gap voltage (V)	80	115	150
B	Capacitance (μF)	0.01	0.1([a]0.205)	0.4
C	Feed rate (μm/s)	1	6.0([a]5.5)	10

[a]Corrected values

$$\text{Minimize } \mathrm{Ln}(R_a) = 0.41 - 0.04A - 0.081B - 0.076C - 0.072AB + 0.076AC + 0.039BC - 0.11A^2 + 0.54B^2 + 0.079C^2 \quad (14.33)$$

Now, the NSTLBO algorithm is applied to solve the multiobjective optimization problem in micro-WEDM process. Somashekhar et al. (2012b) used function evaluations of 11,200. Hence, for a fair comparison of results, the NSTLBO algorithm has considered the same number of function evaluations. The Pareto optimal set of solution obtained using NSTLBO algorithm is reported in Table 14.7, while Fig. 14.4 shows the obtained Pareto front. Somashekhar et al. (2012b) solved the same problem using SA. The same problem was also attempted by Kovacevic et al. (2014) using a software prototype function analyzer-based iterative search approach. However, the results reported by Somashekhar et al. (2012b) and Kovacevic et al. (2014) were not correct, and the corrected values are reported in Table 14.8. The number of function evaluations required by software prototype function analyzer-based iterative search approach to achieve the optimum solution is not reported in the work of Kovacevic et al. (2014). Thus, a fair comparison of results is not possible. From the results reported it is observed that the results obtained using SA (Somashekhar et al. 2012b) do not dominate the results obtained using NSTLBO, while the results obtained using software prototype function analyzer (Kovacevic et al. 2014) are inferior to those obtained using the NSTLBO algorithm.

14.2.4 Optimization of Process Parameters of Laser Cutting Process

Various researchers had attempted to determine the optimum combination of process parameters for laser cutting process using techniques such as Taguchi method, response surface methodology, gray-relational analysis, neural networks, GA, SA, PSO, and ABC (Yusup et al. 2012; Rao and Kalyankar 2014; Thawari et al. 2005; Jimin et al. 2006; Almeida et al. 2006; Nakhjavani and Ghoreishi 2006; Dubey and Yadava 2008; Sivarao et al. 2009; Kuar et al. 2010; Pandey and Dubey 2012; Kondayya and Krishna 2013; Sharma and Yadava 2013; Mukherjee et al. 2013; Tamrin et al. 2015). Now, NSTLBO algorithm is applied to solve the multiobjective optimization problem for laser cutting process.

The optimization model formulated in this work is based on the analysis given by Pandey and Dubey (2012). The quality characteristics selected for analysis were surface roughness 'R_a' (μm) and kerf taper 'K_t' (°) considering gas pressure, pulse width, pulse frequency, and cutting speed as process parameters. The objective functions are mathematically expressed as given below.

Table 14.7 Nondominated set of solutions for micro-WEDM process obtained using NSTLBO

Sr. no.	Gap voltage	Capacitance	Feed rate	MRR	OC	R_a
1	80.000	0.08529	7.7806	0.0182	20.8607	1.6163
2	80.000	0.11597	9.2716	0.0184	24.0000	1.4475
3	80.000	0.23150	8.1783	0.0187	31.2284	1.3356
4	80.000	0.08921	8.9104	0.0188	21.4265	1.5657
5	80.000	0.07209	8.4281	0.0189	19.7086	1.6723
6	150.00	0.22961	5.5134	0.0191	32.2573	1.2831
7	80.000	0.07364	9.8792	0.0192	20.2037	1.6363
8	80.000	0.06110	8.4082	0.0192	18.6393	1.7452
9	80.000	0.06396	8.9960	0.0194	18.9877	1.7103
10	81.656	0.08313	10.000	0.0196	22.4460	1.6050
11	80.000	0.03065	7.5562	0.0197	15.6653	2.0356
12	80.000	0.05261	10.000	0.0198	18.0028	1.7741
13	80.000	0.03738	10.000	0.0203	16.3014	1.8968
14	150.00	0.16926	6.7149	0.0209	28.9360	1.3634
15	80.000	0.01302	9.9341	0.0213	13.4428	2.1411
16	80.000	0.01000	10.000	0.0215	13.0800	2.1749
17	150.00	0.10249	6.2411	0.0216	22.7184	1.6297
18	149.92	0.02864	4.7866	0.0222	15.2053	2.3353
19	150.00	0.16989	8.6065	0.0224	30.5988	1.4019
20	150.00	0.24660	8.7895	0.0225	37.1120	1.3502
21	140.21	0.15546	10.000	0.0244	36.5044	1.5535
22	111.25	0.17244	8.9632	0.0249	42.3818	1.5179
23	132.20	0.22613	8.5498	0.0251	44.0648	1.4484
24	132.06	0.24380	8.8298	0.0254	45.6859	1.4664
25	113.57	0.10127	8.3555	0.0258	35.8093	1.7757
26	118.40	0.23746	9.5117	0.0259	48.2622	1.5110
27	115.39	0.08983	8.2191	0.0262	34.6656	1.8509
28	119.45	0.27876	9.9664	0.0267	51.5901	1.6046
29	150.00	0.40000	7.0675	0.0271	41.2024	1.9546
30	148.42	0.39976	6.7541	0.0271	41.5517	1.9673
31	99.770	0.01000	9.3248	0.0274	23.4111	2.5281
32	101.23	0.01000	10.000	0.0277	23.8943	2.5485
33	145.02	0.38219	8.2825	0.0281	46.1384	1.9277
34	150.00	0.01000	8.8585	0.0282	12.4034	2.6324
35	150.00	0.01000	9.3768	0.0283	12.5122	2.6592
36	121.696	0.32494	10.000	0.0283	54.0721	1.7965
37	145.29	0.40000	7.6421	0.0287	45.1217	2.0569
38	121.88	0.03250	8.4326	0.0288	27.9031	2.4009
39	124.09	0.33852	10.000	0.0288	54.5366	1.8711
40	130.27	0.02781	8.5876	0.0292	25.3361	2.4671
41	142.82	0.39685	7.8350	0.0292	46.7079	2.0720

(continued)

Table 14.7 (continued)

Sr. no.	Gap voltage	Capacitance	Feed rate	MRR	OC	R_a
42	134.49	0.01859	9.0681	0.0296	22.6484	2.6018
43	120.30	0.01000	9.9942	0.0302	25.4929	2.7422
44	139.56	0.40000	9.0747	0.0304	50.9404	2.2235
45	96.901	0.40000	8.2239	0.0308	47.2535	2.4115
46	130.38	0.40000	7.3822	0.0310	50.1899	2.2498
47	119.99	0.38400	8.5262	0.0314	53.2909	2.2019
48	114.69	0.38809	8.2149	0.0316	52.4526	2.2603
49	118.42	0.40000	7.2536	0.0317	51.0445	2.3571
50	116.09	0.40000	10.000	0.0324	56.6973	2.4837

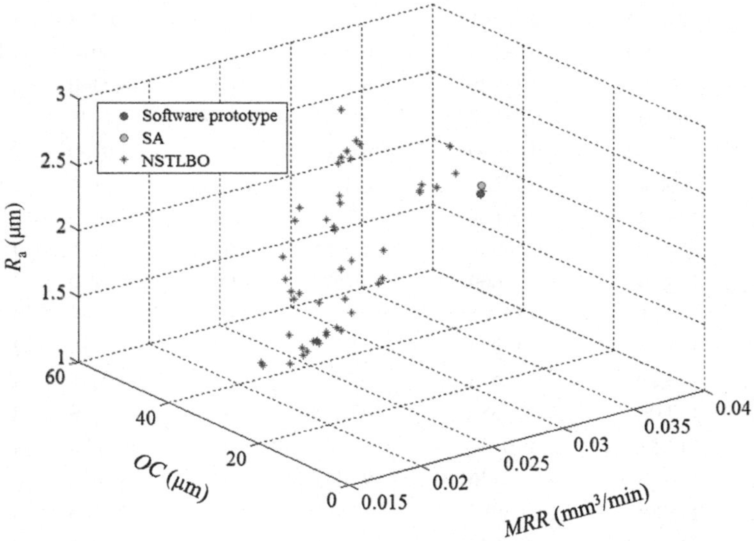

Fig. 14.4 Pareto optimal solutions for micro-WEDM obtained using NSTLBO algorithm

Table 14.8 Results of optimization of micro-WEDM obtained using SA (Somashekhar et al. 2012a, b) and Software prototype (Kovacevic et al. 2014)

Parameters and objective functions	SA algorithm	Software prototype (step 0.01)
Gap voltage (V)	150	150
Capacitance (μm)	0.01	0.01
Feed rate (μm/s)	9	10
Objective function value, Z	47.8([a]50.52)	46.6493([a]50.73)
MRR (mm^3/min)	0.024([a]0.0282)	0.0320([a]0.0283)
Overcut (μm)	5.24([a]12.4289)	12.7
R_a (μm)	0.9([a]2.6391)	2.6993

[a]Corrected values

$$R_a = -33.4550 + 7.2650x_1 + 12.1910x_2 + 1.8114x_3 - 0.2813x_2^2 - 0.0371x_3^2 - 0.7193x_1x_2 + 0.0108x_3x_4 + 0.0752x_1x_2 \quad (14.34)$$

$$K_t = -8.567 - 2.528x1 + 0.2093x_1^2 + 2.1318x_2^2 - 0.0371x_3^2 - 0.7193x_1x_2 + 0.0108x_3x_4 + 0.0752x_1x_3 \quad (14.35)$$

Parameter bounds:

$$\text{Gas pressure (kg/cm}^2) \; : \; 5 \leq x_1 \leq 9 \quad (14.36)$$

$$\text{Pulse width (ms)} \; : \; 1.4 \leq x_2 \leq 2.2 \quad (14.37)$$

$$\text{Pulse frequency (Hz)} \; : \; 6 \leq x_3 \leq 14 \quad (14.38)$$

$$\text{Cutting speed (mm/min)} \; : \; 15 \leq x_4 \leq 25 \quad (14.39)$$

Now the multiobjective optimization problem in laser cutting is solved using NSTLBO algorithm. The same problem was attempted by Pandey and Dubey (2012) using GA, and they used a number of function evaluations of 40,000. Hence, for fair comparison of results, NSTLBO algorithm has considered the same number of function evaluations. The nondominated set of solutions obtained using NSTLBO algorithm is reported in Table 14.9.

The same problem was also attempted by Kovacevic et al. (2014) using a software prototype function analyzer based on iterative search. However, fair comparison of results reported by Kovacevic et al. (2014) with those obtained using NSTLBO is not possible because the number of generations required by software prototype function analyzer to achieve the optimum solution was not given by Kovacevic et al. (2014). Figure 14.7 shows the comparison between the Pareto fronts obtained using NSTLBO algorithm, GA, and software prototype function analyzer. It is observed from Fig. 14.5 that the Pareto front obtained using NSTLBO algorithm completely dominates the Pareto front obtained using GA and is not inferior (if not superior) to the Pareto front obtained using the software prototype function analyzer of Kovacevic et al. (2014).

14.2.5 *Optimization of Parameters of Electrochemical Machining Process*

In order to improve, the performance of ECM process various researchers hade attempted to optimize the parameters of ECM process using techniques such as partial differentiation, goal programming, GA, PSO, ABC, NSGA, fuzzy logic, and

Table 14.9 Nondominated set of solutions for laser cutting obtained using NSTLBO

Sr. no.	x_1 (kg/cm^2)	x_2 (ms)	x_3 (Hz)	x_4 (mm/min)	K_t (°)	R_a (μm)
1	5.915	1.4	14	15.000	0.3822	11.9632
2	5.766	1.4	14	15.000	0.3877	11.7390
3	5.654	1.4	14	15.000	0.3980	11.5635
4	5.420	1.4	14	15.000	0.4364	11.1732
5	5.290	1.4	14	15.000	0.4678	10.9423
6	5.205	1.4	14	15.000	0.4922	10.7861
7	5.071	1.4	14	15.000	0.5366	10.5325
8	5.000	1.4	14	15.084	0.5758	10.3801
9	5.000	1.4	14	15.566	0.6487	10.2989
10	5.000	1.4	14	16.117	0.7319	10.2029
11	5.000	1.4	14	16.687	0.8181	10.1001
12	5.000	1.4	14	16.947	0.8575	10.0518
13	5.000	1.4	14	17.249	0.9032	9.99507
14	5.000	1.4	14	17.598	0.9559	9.92834
15	5.000	1.4	14	18.036	1.0221	9.84251
16	5.000	1.4	14	18.657	1.1161	9.71705
17	5.000	1.4	14	19.016	1.1704	9.64267
18	5.000	1.4	14	19.656	1.2671	9.50663
19	6.219	1.4	6.0	15.188	1.3100	9.47394
20	6.318	1.4	6.0	16.435	1.3457	9.38040
21	6.735	1.4	6.0	19.242	1.3809	9.29362
22	6.597	1.4	6.0	19.337	1.4276	9.13070
23	6.511	1.4	6.0	19.647	1.4770	8.96957
24	6.567	1.4	6.0	20.545	1.5158	8.83290
25	6.469	1.4	6.0	20.505	1.5476	8.73379
26	6.649	1.4	6.0	21.947	1.5805	8.59323
27	6.560	1.4	6.0	22.035	1.6145	8.47707
28	6.269	1.4	6.0	21.282	1.6815	8.31490
29	6.974	1.4	6.0	25.000	1.7027	8.1146
30	6.658	1.4	6.0	25.000	1.7757	7.8138
31	6.489	1.4	6.0	25.000	1.8317	7.6309
32	6.399	1.4	6.0	25.000	1.8665	7.5263
33	6.298	1.4	6.0	25.000	1.9096	7.4038
34	6.192	1.4	6.0	25.000	1.9593	7.2692
35	6.076	1.4	6.0	25.000	2.0192	7.1145
36	6.003	1.4	6.0	25.000	2.0597	7.0134
37	5.939	1.4	6.0	25.000	2.0969	6.9227
38	5.797	1.4	6.0	25.000	2.1862	6.7116
39	5.728	1.4	6.0	25.000	2.2323	6.6057
40	5.629	1.4	6.0	25.000	2.3028	6.4471

(continued)

Table 14.9 (continued)

Sr. no.	x_1 (kg/cm^2)	x_2 (ms)	x_3 (Hz)	x_4 (mm/min)	K_t (°)	R_a (μm)
41	5.520	1.4	6.0	25.000	2.3844	6.2681
42	5.471	1.4	6.0	25.000	2.4224	6.1858
43	5.351	1.4	6.0	25.000	2.5205	5.9773
44	5.276	1.4	6.0	25.000	2.5856	5.8413
45	5.216	1.4	6.0	25.000	2.6387	5.7317
46	5.161	1.4	6.0	25.000	2.6888	5.6292
47	5.124	1.4	6.0	25.000	2.7232	5.5594
48	5.068	1.4	6.0	25.000	2.7766	5.4518
49	5.028	1.4	6.0	25.000	2.8159	5.3731
50	5.000	1.4	6.0	25.000	2.8431	5.3189

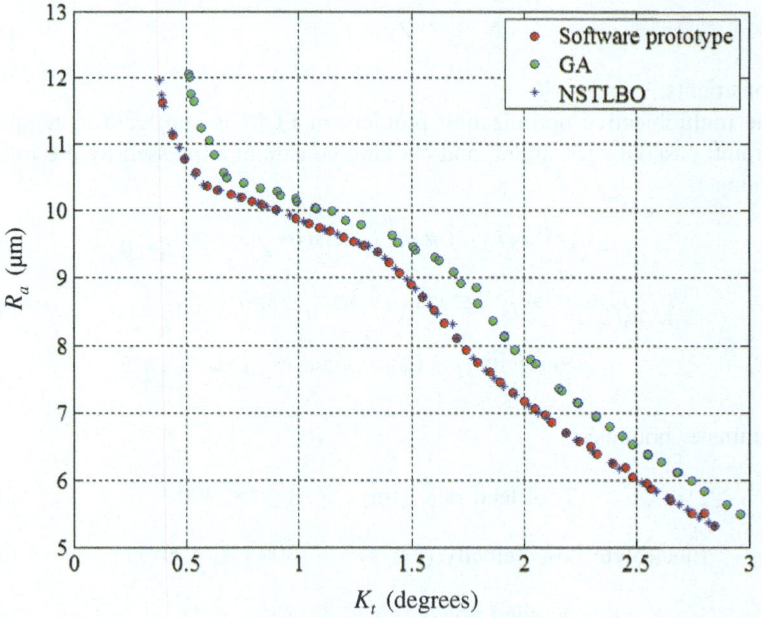

Fig. 14.5 Pareto fronts obtained using NSTLBO algorithm, GA and software prototype function analyzer for laser cutting process

neural networks (Yusup et al. 2012; Rao and Kalyankar 2014; Acharya et al. 1986; Choobineh and Jain 1993; Jain and Jain 2007; Rao et al. 2008; Samanta and Chakraborty 2011). Now, the NSTLBO algorithm is applied to solve the multi-objective optimization problem for electrochemical machining process.

The optimization model formulated in this work is based on the analysis given by Acharya et al. (1986). The three decision variables considered for this model are tool feed rate, electrolyte flow velocity, and applied voltage while material removal rate, dimensional accuracy and tool life are considered as performance measures.

Objective functions:

The objective functions are mathematically expressed by Eqs. (14.40)–(14.42).

(a) Minimizing of dimensional inaccuracy (Z_1)

$$Z_1 = f^{0.381067}\, U^{-0.372623}\, V^{3.155414}\, e^{-3.128926} \quad (14.40)$$

(b) Maximizing the tool life by minimizing the number of sparks per millimeter (Z_2)

$$Z_2 = f^{3.528345}\, U^{0.000742}\, V^{-2.5225}\, e^{0.391436} \quad (14.41)$$

(c) Maximizing the material removal rate (Z_3)

$$Z_3 = f \quad (14.42)$$

Constraints:

The multiobjective optimization problem in ECM is subjected to temperature constraint, passivity constraint, and choking constraint expressed by the following equations.

$$1 - (f^{2.133007}\, U^{-1.088937}\, V^{-0.351436}\, e^{0.321968}) \geq 0 \quad (14.43)$$

$$(f^{-0.844369}\, U^{-2.526076}\, V^{1.546257}\, e^{12.57697}) - 1 \geq 0 \quad (14.44)$$

$$1 - (f^{0.075213}\, U^{-2.488362}\, V^{0.240542}\, e^{11.75651}) \geq 0 \quad (14.45)$$

Parameter bounds:

$$\text{Tool feed rate } (\mu\text{m}) \quad : \quad 8 \leq f \leq 200 \quad (14.46)$$

$$\text{Electrolyte flow velocity (cm/s)} \quad : \quad 300 \leq U \leq 5000 \quad (14.47)$$

$$\text{Applied voltage (V)} \quad : \quad 3 \leq V \leq 21 \quad (14.48)$$

Now, the NSTLBO algorithm is applied to solve the multiobjective optimization problem in ECM. Jain and Jain (2007) solved the same problem and the number of function evaluations used by them was 2500. Thus, NSTLBO algorithm has used the same number of function evaluations for the purpose of fair comparison. Table 14.10 gives the results of optimization for ECM obtained by other researchers using geometric programming (GP), fuzzy sets, GA, and PSO. The nondominated set of solution obtained using NSTLBO algorithm is reported in Table 14.11. Figure 14.6 gives the Pareto front for ECM process obtained using NSTLBO algorithm. It is observed that the results of optimization in ECM obtained using

Table 14.10 Results of optimization for ECM obtained using other techniques

Method	Author(s)	f (μm)	U (cm/s)	V (volts)	Z_1	Z_2	Z_3	Z
GP	Acharya et al. (1986)	18.96	179	15	100	51.79	18.96	18.22
Fuzzy set	Choobineh and Jain (1993)	12.75	400	21	181.1	5.47	12.75	5.47
GA	Jain and Jain (2007)	8	2978.45	16.5	33.62	1.94	8	1.23
PSO	Rao et al. (2008)	8	300	13.225	39.34	3.39	8	1.811

Z = normalized combined objective function

NSTLBO algorithm are not inferior to those obtained using GP, fuzzy sets and PSO.

14.3 Optimization of Process Parameters of Electrochemical Discharge Machining Process

Electrochemical discharge machining is a hybrid process which combines the ECM and EDM processes. Several researchers in the past had attempted to determine the optimum combination of process parameters for ECDM process using techniques such as response surface methodology, gray-relational analysis, GA, ABC, and neural networks (Yusup et al. 2012; Rao and Kalyankar 2014; Sarkar et al. 2006; Bhuyan and Yadava 2013, 2014; Mallick et al. 2014).

Now, the multiobjective optimization problem in ECDM process is solved using NSTLBO algorithm. The optimization models formulated in this work are based on the analysis given by Sarkar et al. (2006). The objectives are to maximize material removal rate 'MRR' (mg/h), minimize radial overcut 'ROC' (mm), and minimize heat affected zone 'HAZ' (mm). The applied voltage, electrolyte concentration, and inter-electrode gap are considered as process parameters.

Objective functions:

The objective functions based on the coded levels of process parameters are expressed by Eqs. (14.49)–(14.51). The list of actual and corresponding coded values for each parameter is given in Table 14.12.

$$Y_u(\mathrm{MRR}) = 0.60266 + 0.16049X_1 - 0.04044X_2 - 0.03481X_3 + 0.08781X_1^2 - 0.03060X_2^2 + 0.01358X_3^2 - 0.065X_1X_2 - 0.0375X_1X_3 + 0.045X_2X_3 \quad (14.49)$$

Table 14.11 Nondominated set of solutions for ECM process obtained using NSTLBO algorithm

Sr. no.	*f*	*U*	*V*	Z_1	Z_2	Z_3
1	8.0000	300	15.9583	72.1369	2.1069	8.0000
2	8.1877	300	16.3320	78.2917	2.1569	8.1877
3	8.0245	300	16.7428	84.0277	1.8870	8.0245
4	8.0000	300	17.1612	90.7278	1.7540	8.0000
5	8.8744	300	17.1388	93.9977	2.5376	8.8744
6	8.4802	300	17.5382	99.3513	2.0396	8.4802
7	9.2351	300	17.5406	102.676	2.7547	9.2351
8	8.0000	300	17.9524	104.594	1.5655	8.0000
9	8.1595	300	18.0193	106.628	1.6628	8.1595
10	9.0414	300	17.9522	109.584	2.4110	9.0414
11	9.4633	300	17.9243	110.958	2.8430	9.4633
12	9.9229	300	17.9393	113.282	3.3537	9.9229
13	9.2704	300	18.2916	117.368	2.5119	9.2704
14	8.7463	300	18.5380	119.745	1.9777	8.7463
15	9.2300	300	18.5364	122.193	2.3919	9.2300
16	9.5400	300	18.6630	126.426	2.6418	9.5400
17	9.1242	300	19.0403	132.401	2.1463	9.1242
18	8.0000	300	19.4883	135.520	1.2726	8.0000
19	10.491	300	19.0310	139.424	3.5173	10.491
20	10.676	300	19.0641	141.125	3.7244	10.676
21	9.8364	300	19.3261	142.807	2.6949	9.8364
22	10.586	300	19.2055	143.987	3.5476	10.586
23	10.299	300	19.4162	147.477	3.1323	10.299
24	11.218	300	19.3897	151.707	4.2499	11.218
25	11.178	300	19.5914	156.529	4.0880	11.178
26	10.427	300	19.8786	159.596	3.0829	10.427
27	11.126	300	19.8093	161.803	3.9113	11.126
28	11.639	300	19.7710	163.607	4.6084	11.639
29	10.585	300	20.1392	167.249	3.1460	10.585
30	8.5504	300	20.7571	169.610	1.3727	8.5504
31	11.244	300	20.1835	172.335	3.8720	11.244
32	11.767	300	20.2222	176.408	4.5234	11.767
33	8.9871	300	20.9627	178.319	1.5963	8.9871
34	12.118	300	20.3085	180.806	4.9640	12.118
35	12.414	300	20.3132	182.614	5.4030	12.414
36	12.253	300	20.3919	183.939	5.1095	12.253
37	12.176	300	20.4727	185.800	4.9468	12.176
38	12.042	300	20.5522	187.294	4.7113	12.042
39	11.414	300	20.7177	188.212	3.8219	11.414
40	11.085	300	20.8692	190.453	3.3841	11.085
41	11.860	300	20.7361	191.522	4.3666	11.860

(continued)

Table 14.11 (continued)

Sr. no.	f	U	V	Z_1	Z_2	Z_3
42	11.005	300	21.0000	193.711	3.2475	11.005
43	11.480	300	21.0000	196.858	3.7700	11.480
44	11.987	300	21.0000	200.126	4.3909	11.987
45	12.347	300	21.0000	202.397	4.8745	12.347
46	12.739	300	20.9433	203.081	5.4791	12.739
47	12.708	300	21.0000	204.628	5.3952	12.708
48	12.975	300	21.0000	206.258	5.8065	12.975
49	13.121	300	21.0000	207.138	6.0400	13.121
50	13.198	300	21.0000	207.599	6.1655	13.197

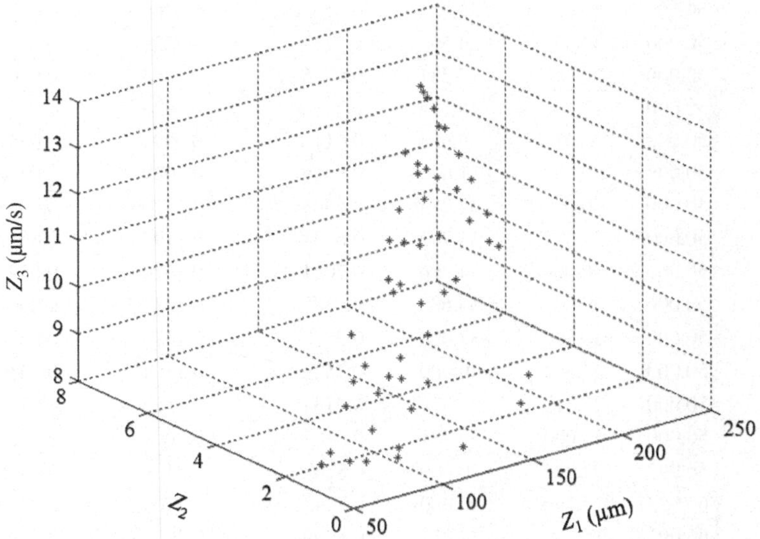

Fig. 14.6 Pareto front for ECM process obtained using NSTLBO algorithm

$$Y_u(\mathrm{ROC}) = 0.16114 + 0.05333X_1 - 0.01017X_2 - 0.00716X_3 + 0.02454X_1^2 + 0.01727X_2^2 + 0.00598X_3^2 + 0.02603X_1X_2 - 0.00940X_1X_3 + 0.01493X_2X_3 \quad (14.50)$$

$$Y_u(\mathrm{HAZ}) = 0.07835 + 0.01583X_1 - 0.00418X_2 - 0.00599X_3 + 0.00523X_1^2 + 0.00857X_2^2 + 0.00061X_3^2 + 0.00905X_1X_2 - 0.00060X_1X_3 + 0.00382X_2X_3 \quad (14.51)$$

where X_1, X_2, and X_3 are the coded values for the process parameters of applied voltage, electrolyte concentration, and inter-electrode gap, respectively.

Table 14.12 Actual and corresponding coded values for each parameter of ECDM (Sarkar et al. 2006)

	Different levels of parameters				
	–1.682	–1	0	+1	+1.682
Applied voltage (V)	50	54	60	66	70
Electrolyte concentration (wt%)	10	14	20	26	30
Inter-electrode gap (mm)	20	24	30	36	40

Table 14.13 Nondomiated set of solution for ECDM obtained using NSTLBO

Sr. no.	x_1 (*V*)	x_2 (wt%)	x_3 (mm)	MRR (mg/h)	ROC (mm)	HAZ (mm)
1	50.000	30.000	20.000	0.4740	0.0591	0.05740
2	50.000	29.648	20.000	0.4800	0.0614	0.05745
3	50.000	30.000	23.628	0.5135	0.0704	0.05742
4	50.000	30.000	24.569	0.5254	0.0738	0.05744
5	50.000	28.750	27.281	0.5735	0.0865	0.05649
6	50.000	27.707	28.235	0.5918	0.0925	0.05596
7	50.000	27.311	29.654	0.6116	0.0992	0.05576
8	50.000	30.000	31.676	0.6369	0.1089	0.05862
9	50.000	30.000	32.284	0.6483	0.1127	0.05880
10	50.000	27.311	33.742	0.6731	0.1197	0.05577
11	50.000	30.000	34.996	0.7021	0.1310	0.05976
12	50.000	30.000	36.688	0.7386	0.1437	0.06048
13	50.000	30.000	37.098	0.7477	0.1470	0.06067
14	50.000	24.504	40.000	0.7529	0.1594	0.05502
15	50.000	27.310	39.374	0.7784	0.1571	0.05671
16	50.000	30.000	39.781	0.8107	0.1694	0.06205
17	66.993	12.723	40.000	0.8208	0.2138	0.09161
18	66.923	14.760	40.000	0.8243	0.2160	0.08964
19	66.091	15.309	34.311	0.8499	0.2211	0.09456
20	67.217	14.778	38.678	0.8590	0.2224	0.09237
21	65.159	15.222	29.596	0.8698	0.2282	0.09903
22	68.644	10.000	40.000	0.9192	0.2319	0.10071
23	68.474	13.690	39.116	0.9357	0.2354	0.09648
24	67.409	16.042	32.883	0.9509	0.2450	0.10068
25	69.447	10.000	39.840	0.9856	0.2418	0.10309
26	70.000	11.784	40.000	1.0332	0.2495	0.10172
27	70.000	11.050	39.424	1.0430	0.2509	0.10363
28	69.428	12.862	36.435	1.0556	0.2577	0.10471
29	69.455	12.851	35.872	1.0690	0.2606	0.10577
30	70.000	10.000	37.505	1.0827	0.2604	0.10914
31	70.000	10.884	33,767	1.1707	0.2813	0.11468

(continued)

Table 14.13 (continued)

Sr. no.	x_1 (V)	x_2 (wt%)	x_3 (mm)	MRR (mg/h)	ROC (mm)	HAZ (mm)
32	70.000	10.000	33.763	1.1742	0.2832	0.11675
33	70.000	10.000	32.975	1.1948	0.2886	0.11842
34	70.000	10.000	31.555	1.2332	0.2988	0.12147
35	70.000	10.000	30.809	1.2539	0.3044	0.12310
36	70.000	10.000	30.518	1.2622	0.3067	0.12374
37	70.000	10.000	29.801	1.2826	0.3123	0.12533
38	70.000	10.000	28.707	1.3147	0.3213	0.12780
39	70.000	12.430	25.889	1.3649	0.3351	0.12752
40	70.000	10.000	26.094	1.3949	0.3443	0.13385
41	70.000	11.017	25.099	1.4117	0.3480	0.13293
42	70.000	12.742	23.233	1.4376	0.3570	0.13248
43	70.000	10.000	24.183	1.4567	0.3627	0.13841
44	70.000	12.743	22.056	1.4738	0.3679	0.13505
45	70.000	14.114	20.000	1.5039	0.3813	0.13640
46	70.000	10.510	22.268	1.5124	0.3786	0.14126
47	70.000	10.000	22.039	1.5294	0.3846	0.14368
48	70.000	10.000	21.419	1.5511	0.3912	0.14524
49	70.000	10.000	21.193	1.5591	0.3937	0.14581
50	70.000	10.844	20.000	1.5841	0.4003	0.14563

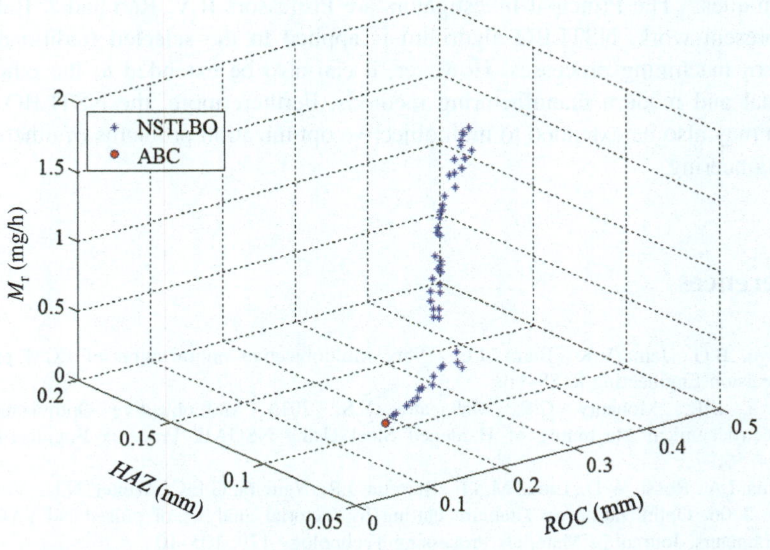

Fig. 14.7 Pareto optimal solutions for ECDM process obtained using NSTLBO algorithm

Table 14.14 Results of multiobjective of optimization of ECDM obtained using ABC (Samanta and Chakraborty 2011)

Parameters and objective functions	Value
Applied voltage (V)	50
Electrolyte concentration (wt%)	30
Inter-electrode gap (mm)	20
Metal removal rate (mg/h)	1.48603([a]0.4740)
Radial overcut (mm)	0.05912([a]0.0591)
HAZ thickness (mm)	0.056905([a]0.0574)

[a]Corrected value

In this work, the multiobjective optimization problem in ECDM process is solved using the NSTLBO algorithm. The same problem was solved by Samanta and Chakraborty (2011) using ABC algorithm with a number of function evaluations of 10,000. Hence, for a fair comparison of results NSTLBO algorithm has considered a same number of function evaluations. The nondominated set of solutions obtained using NSTLBO is reported in Table 14.13. Figure 14.7 shows the Pareto front obtained using NSTLBO algorithm. The values of objective functions reported by Samanta and Chakraborty (2011) were found not correct and the corrected values are reported in Table 14.14.

The present work is done under a bilateral research project supported by the Department of Science and Technology (DST), Ministry of Science and Technology of the Republic of India and the Slovenian Research Agency (ARRS), Ministry of Education, Science and Sport of the Republic of Slovenia for the project entitled "Optimization of Sustainable Manufacturing Processes using Advanced Techniques." The Principal Investigators are Professors R.V. Rao and J. Balic. In the present work, NSTLBO algorithm is applied to the selected traditional and modern machining processes. However, it can also be extended to the other traditional and modern manufacturing methods. Furtherermore, the NSTLBO algorithm may also be extended to multiobjective optimization problems of other fields of engineering.

References

Acharya, B.G., Jain, V.K., Batra, J.L., 1986. Multiobjective optimization of ECM process. Precision Engineering 8, 88–96.

Acharya, B.R., Mohanty, C.P., Mahapatra, S.S., 2013. Multiobjective Optimization of Electrochemical Machining of Hardened Steel Using NSGAII. Procedia Engineering 51, 554–560.

Almeida, I.A., Rossi, W.D., Lima, M.S.F., Berretta, J.R., Ngueira, G.E.C., Wetter, N.U., Vieira, N. D., 2006. Optimization of Titanium cutting by factorial analysis of pulsed Nd:YAG laser parameters. Journal of Materials Processing Technology 179, 105–10.

Baskar, N., Saravanan, R., Asokan, P., Prabhaharan, G., 2004. Ants colony algorithm approach for multiobjective optimization of surface grinding operations. International Journal of Advanced Manufacturing Technology 23, 311–317.

Bhuyan, B.K., Yadava, V., 2013. Experimental modeling and multiobjective optimization of traveling wire electrochemical spark machining (TW-ECSM) process. Journal of Mechanical Science and Technology 27(8), 2467–2476.

Bhuyan, B.K., Yadava, V., 2014. Experimental modelling and multi-response optimization of travelling wire electrochemical spark machining of pyrex glass. Journal of Engineering Manufacture 228(8), 902–916.

Chiang, K.T., Chang, F.P., 2006. Optimization of the WEDM process of particle-reinforced material with multiple performance characteristics using grey relational analysis. Journal of Materials Processing Technology 180, 96–101.

Choobineh, F., Jain, V.K., 1993. A fuzzy sets approach for selecting optimum parameters of an ECM process. Processing of Advanced Materials 3, 225–232.

Datta, R., Deb, K., 2009. A classical cum evolutionary multiobjective optimization for optimal machining parameters, in: Proceeding of World Congress on Nature and Biologically Inspired Computing, 607–612.

Dhavalikar, M.N., Kulkarni, M.S., Mariappan, V., 2003. Combined taguchi and dual response method for optimization of a centerless grinding operation. Journal of Materials Processing Technology 132, 90–94.

Dubey, A.K., Yadava, V., 2008. Robust parameter design and multiobjective optimization of laser beam cutting for aluminium alloy sheet. International Journal of Advanced Manufacturing Technology 38, 268–277.

Garg, M.P., Jain, A., Bhushan, G., 2014. Multiobjective optimization of process parameters in wire electrical discharge machining of Ti-6-2-4-2 Alloy. International Journal of Advanced Manufacturing Technology 39, 1465–1476.

Jain, N.K., Jain, V.K., 2007. Optimization of electrochemical machining process parameters using genetic algorithm. Machining Science and Technology 11, 235–258.

Jimin, C., Jianhua, Y., Shuai, Z., Tiechuan, Z., Dixin, G., 2006. Parametric optimization of non vertical laser cutting. International Journal of Advanced Manufacturing Technology 33, 469–73.

Joshi, S.N., Pande, S.S., 2011. Intelligent process modeling and optimization of die sink electric-discharge machining. Applied soft computing 11, 2743–2755.

Kanagarajan, D., Karthikeyan, R., Palanikumar, K., Davim, J.P., 2008. Optimization of electric discharge machining characteristics of WC/Co composites using nondominated sorting genetic algorithm (NSGA-II). International Journal of Advanced Manufacturing Technology 36, 1124–1132.

Konak, A., Coit, D.W., Smith, A.E., 2006. Multiobjective optimization using genetic algorithms: A tutorial. Reliability Engineering and System Safety 91, 992–1007.

Kodali, S.P., Kudikala, R., Deb, K., 2008. Multiobjective optimization of surface grinding process using NSGA-II. in: Proceedings of First International Conference on Emerging trends in Engineering and Technology, Washington, DC, 763–767.

Kondayya, D., Krishna, A.G., 2013. An integrated evolutionary approach for modeling and optimization of laser beam cutting process. International Journal of Advanced Manufacturing Technology 65, 259–274.

Kovacevic, M., Madic, M., Radovanovic, M., Rancic, D., 2014. Software prototype for solving multiobjective machining optimization problems: Application in non conventional machining processes. Expert Systems with Applications 41, 5657–5668.

Krishna, A.G., 2007. Optimization of surface grinding operations using a differential evolution approach. Journal of Materials Processing Technology 183, 202–209.

Kuar, A.S., Dhara, S.K., Mitra, S., 2010. Multi-response optimization of Nd:YAG laser micro-machining of die steel using response surface methodology. International Journal of Manufacturing Technology and Management 21(1–2), 17–29.

Kuriachen, B., Somashekhar, K.P., Mathew, Jose., 2015. Multiresponse optimization of micro-wire electric discharge machining process. International Journal of Advanced Manufacturing Technology 1–4, 91–104.

Kuriakose, S., Shunmugam, M.S., 2005. Multiobjective optimization of wire-electro discharge machining process by non-dominated sorting genetic algorithm. Journal of Materials Processing Technology 170, 133–141.

Mallick, B., Sarkar, B.R., Doloi, B., Bhattacharyya, B., 2014. Multi criteria optimization of electrochemical discharge micro-machining process during micro-channel generation on glass. Applied Mechanics and Materials 592–594, 525-526.

Mandal, D., Pal, S.K., Saha, P., 2007. Modeling of electric discharge machining process using back propagation neural network and multiobjective optimization using non-dominated sorting genetic algorithm-II. Journal of Materials Processing and Technology186, 154–162.

Mitra, K., 2009. Multiobjective optimization of industrial grinding operation under uncertainty. Chemical Engineering Science 64, 5043–5056.

Mitra, K., Gopinath, R., 2004. Multiobjective optimization of industrial grinding operation using elitist nondominated sorting genetic algorithm. Chemical Engineering Science 59, 385–396.

Mukherjee, I., Ray, P.K., 2006. A review of optimization techniques in metal cutting processes. Computers and Industrial Engineering 50, 15–34.

Mukherjee, R., Chakraborty, S., Samantha, S., 2012. Selection of wire electrical discharge machining process parameters using nontraditional optimization algorithms. Applied Soft Computing 12, 2506–2516.

Mukherjee, R., Goswami, D., Chakraborty, S., 2013. Parametric optimization of Nd:YAG laser beam machining process using artificial bee colony algorithm. Journal of Industrial Engineering. doi:10.1155/2013/570250.

Nakhjavani, O.B., Ghoreishi, M., 2006. Multi criteria optimization of laser percussion drilling process using artificial neural network model combined with genetic algorithm. Materials and Manufacturing Processes 21, 11–18.

Palanikumar, K., Latha, B., Senthilkumar, V.S., Karthikeyan, R., 2009. Multiple performance optimization in machining of GFRP composites by a PCD tool using non-dominated sorting genetic algorithm (NSGA-II). Metals and Materials International 15(2), 249–258.

Pandey, A.K., Dubey, A.K., 2012. Simultaneous optimization of multiple quality characteristics in laser cutting of titanium alloy sheet. Optics and Laser Technology 44, 1858–1865.

Pawar, P.J., Rao, R.V., Davim, J.P., 2010. Multiobjective optimization of grinding process parameters using particle swarm optimization algorithm. Materials and Manufacturing Process 25(6), 424–431.

Pawar, P.J., Rao, R.V., 2013. Parameter optimization of machining processes using teaching learning based optimization algorithm. International Journal of Advanced Manufacturing Technology 67, 995–1006.

Pradhan, B.B., Masanta, M., Sarkar, B.R., Bhattacharyya, B., 2009. Investigation of electro-discharge micro-machining of titanium super alloy. International Journal of Advanced Manufacturing Technology 41, 1094–1106.

Rajyalakshmi, G., Ramaiah, P.V., 2013. Multiple process parameters optimization of wire electrical discharge machining on Inconel 825 using Taguchi grey relational analysis. International Journal of Advanced Manufacturing Technology 69, 1249–1262.

Ramakrishnan, R., Karunamoorthy, L., 2006. Multi response optimization of wire EDM operations using robust design of experiments. International Journal of Advanced Manufacturing Technology 29, 105–112.

Ramakrishnan, R., Karunamoorthy, L., 2008. Modeling and multiresponse optimization of Inconel 718 on machining of CNC WEDM process. Journal of Materials Processing Technology 207, 343–349.

Rao, R.V., Pawar, P.J., 2010. Grinding process parameter optimization using non-traditional optimization algorithms. Journal of Engineering Manufacture 224(6), 887–898.

Rao, R.V., Kalyankar, V.D., 2013. Parameters optimization of modern machining processes using teaching learning based optimization algorithm. Engineering Applications of Artificial Intelligence 26, 524–531.

Rao, R.V., Kalyankar, V.D., 2014. Optimization of modern machining processes using advanced optimization techniques: a review. International Journal of Advanced Manufacturing Technology 73, 1159–1188.

Rao, R.V., Pawar, P.J., Shankar, R., 2008. Multiobjective optimization of electrochemical machining process parameters using a particle swarm optimization algorithm. Journal of Engineering Manufacture 222(8), 949–958.

Rowe, W.B., Yan, L., Inasaki, I., Malkin, S., 1994. Application of artificial intelligence in grinding. CIRP Annals-Manufacturing Technology 43, 521–531.

Sadeghi, M., Razavi, H., Esmaeilzadeh, A., Kolahan, F., 2011. Optimization of cutting conditions in WEDM process using regression modelling and tabu search algorithm. Proceedings of the Institution of Mechanical Engineers, Part B: Journal of Engineering Manufacture 225(10), 1825–1834.

Samanta, S., Chakraborty, S., 2011. Parametric optimization of some non-traditional machining processes using artificial bee colony algorithm. Engineering Applications of Artificial Intelligence 24, 946–957.

Saravanan, R., Asokan, P., Sachidanandam, M., 2002. A multiobjective genetic algorithm approach for optimization of surface grinding operations. International Journal of Machine Tools and Manufacture 42, 1327–1334.

Sarkar, B.R., Doloi, B., Bhattacharyya, B., 2006. Parametric analysis on electrochemical discharge machining of silicon nitride ceramics. International Journal of Advanced Manufacturing Technology 28, 873–881.

Satishkumar, D., Kanthababu, M., Vajjiravelu, M., Anburaj, R., Sundarrajan, N.T., Arul, H., 2011. Investigation of wire electrical discharge machining characteristics of Al6063/SiCp composites. International Journal of Advanced Manufacturing Technology 56, 975–986.

Senthilkumar, C., Ganesan, G., Karthikeyan, R., 2010. Bi-performance optimization of electrochemical machining characteristics of Al/20%SiCp composite using NSGA-II. Journal of Engineering Manufacture 224(9), 1399–1407.

Senthilkumar, C., Ganesan, G., Karthikeyan, R., 2011. Parametric optimization of electrochemical machining of Al/15%SiC_p composites using NSGA-II. Transactions of Non-ferrous Metals Society of China 21, 2294–2300.

Sharma, A., Yadava, V., 2013. Modelling and optimization of cut quality during pulsed Nd:YAG laser cutting of thin Al-alloy sheet for curved profile. Optics and Lasers in Engineering 51, 77–88.

Sharma, N., Khanna, R.R., Gupta, R.D., 2013. Sharma R Modeling and multiresponse optimization on WEDM for HSLA by RSM. International Journal of Advanced Manufacturing Technology 67, 2269–2281.

Sivaprakasam, P., Hariharan, P., Gowri, S., 2013. Optimization of micro-WEDM process of aluminum matrix composite (A413-B_4C): A response surface approach. Materials and Manufacturing Processes 28, 1340–1347.

Sivarao, P., Brevern., N.S.M., Tayeb, Vengkatesh, V.C., 2009. Modeling, testing and experimental validation of laser machining micro quality response by artificial neural network. International Journal of Engineering and Technology 9:161–166.

Somashekhar, K.P., Mathew, J., Ramachandran, N., 2012a. Multiobjective optimization of micro wire electric discharge machining parameters using grey relational analysis with Taguchi method. Journal of Mechanical Engineering and Science 225(7), 1742–1753.

Somashekhar, K.P., Mathew, J., Ramachandran, N., 2012b. A feasibility approach by simulated annealing on optimization of micro-wire electric discharge machining parameters. International Journal of Advanced Manufacturing Technology 61, 1209–1213.

Tamrin, K.F., Nukman, Y., Choudhury, I.A., Shirley, S., 2015. Multiple-objective optimization in precision laser cutting of different thermoplastics. Optics and Lasers in Engineering 67, 57–65.

Thawari, G., Sundar, J.K.S., Sundararajan, G., Joshi, S.V., 2005. Influence of process parameters during pulsed ND:YAG laser cutting of nickel-base superalloys. Journal of Materials Processing Technology 170, 229–239.

Vijayaraj, R., Gowri, S., 2010. Study on parametric influence, optimization and modeling in micro-WEDM of Al alloy. International Journal of Abrasive Technology 3(2), 157–164.

Wen, X.M., Tay, A.A.O., Nee, A.Y.C., 1992. Microcomputer based optimization of the surface grinding process. Journal of Materials Processing Technology 29, 75–90.

Yan, M.T., Fang, C.C., 2008. Application of genetic algorithm based fuzzy logic control in wire transport system of wire-EDM machine. Journal of Materials Processing Technology 205, 128–137.

Yang, S.H., Natarajan, U., 2010. Multiobjective optimization of cutting parameters in turning process using differential evolution and non-dominated sorting genetic algorithm-II approaches. International Journal of Advanced Manufacturing Technology 49, 773–784.

Yusup, N., Zain, A.M., Hashim, S.Z.M, 2012. Evolutionary techniques in optimizing machining parameters: Review and recent applications. Expert Systems and Applications 39, 9909–9927.

Chapter 15
Applications of TLBO Algorithm and Its Modifications to Different Engineering and Science Disciplines

Abstract After its introduction in 2011 by Rao et al. (Computer-Aided Design 43:303–315, 2011), the TLBO algorithm is finding a large number of applications in different fields of engineering and science. The major applications, as of April 2015, are found in *electrical engineering, mechanical design, thermal engineering, manufacturing engineering, civil engineering, structural engineering, computer engineering, electronics engineering, physics, chemistry, biotechnology, and economics.* This chapter presents an overview of the *year-wise applications* of the TLBO algorithm and its modifications since 2011.

15.1 Overview of the Applications of TLBO Algorithm and Its Modifications (Year-Wise)

15.1.1 Publications in the Year 2011

Rao et al. (2011) presented the TLBO algorithm based on the philosophy of the teaching-learning process and its performance was checked by experimenting with different benchmark problems with different characteristics. The effectiveness of TLBO algorithm was also checked for different performance criteria such as success rate, mean solution, average number of function evaluations required, and convergence rate. The results showed the better performance of TLBO over other nature-inspired optimization methods for the constrained benchmark functions and mechanical design problems considered. Also, the TLBO algorithm showed better performance with less computational effort for large-scale problems, i.e., problems of high dimensionality.

Rao and Savsani (2011a, b) presented the design optimization formulation of a robot gripper. Two different objective functions were considered for the optimization: the difference between maximum and minimum gripping forces for the assumed range of the gripper displacement and the force transmission ratio between the gripper actuator and the gripper ends. Geometrical dimensions were considered

R.V. Rao, *Teaching Learning Based Optimization Algorithm*,
DOI 10.1007/978-3-319-22732-0_15

as the design variables and geometric limitations were considered as the constraints. Optimization was carried out using the TLBO algorithm and the results were compared with the results presented by the previous researchers.

15.1.2 Publications in the Year 2012

Rao et al. (2012a) checked the performance of the TLBO algorithm with the well-known optimization algorithms such as GA, ABC, PSO, HS, DE, and Hybrid PSO by experimenting with different benchmark problems with different characteristics like multimodality, separability, regularity, and dimensionality. The effectiveness of TLBO method was also checked for different performance criteria like success rate, mean solution, average function evaluations required, convergence rate, etc. The results showed better performance of TLBO algorithm over other natured inspired optimization methods for the benchmark functions considered. Also, the TLBO method showed better performance with less computational efforts for the large-scale problems, i.e., problems with high dimensions. In another work, Rao et al. (2012b) attempted unconstrained and constrained real-parameter optimization problems using TLBO algorithm. In general, to maintain the consistency in comparison, the number of function evaluations is to be maintained same for all the optimization algorithms including TLBO algorithm for all the benchmark functions considered. However, it may be mentioned here that, in general, the algorithm which requires fewer number of function evaluations to get the same best solution can be considered as a better algorithm compared with the other algorithms. If an algorithm gives global optimum solution within certain number of function evaluations, then consideration of more number of function evaluations will go on giving the same best result. Rao et al. (2011, 2012a) showed that TLBO requires fewer number of function evaluations as compared to the other optimization algorithms. Even though certain experiments were not conducted by Rao et al. (2011, 2012a) in the same settings, but better test conditions (i.e., comparatively less number of function evaluations) were chosen by them which proved the better performance of the TLBO algorithm. There was no need for TLBO algorithm to go to the high settings followed by other researchers who themselves had used different number of function evaluations for the benchmark functions considered. The stopping conditions used by Rao et al. (2011, 2012a) in certain benchmark functions with 30 runs each time were better than those used by other researchers.

Rao and Patel (2012) introduced the concept of elitism in the TLBO algorithm and investigated its effect on the performance of the algorithm for the constrained optimization problems. Moreover, the effect of common controlling parameters (i.e., population size, elite size, and number of generations) on the performance of TLBO algorithm was also investigated by considering different combinations of common controlling parameters. The algorithm was implemented on 35 well-defined constrained optimization problems having different characteristics to identify the effect of elitism and common controlling parameters. The results showed that for many functions the strategy with elitism consideration produced better results than that without elitism consideration. Also, in general, the strategy with higher population size had produced better results than that with smaller population size for the same number of function evaluations. The results obtained using TLBO algorithm were compared with the other optimization algorithms available in the literature such as PSO, DE, ABC, and EP for the benchmark problems considered. Results had shown the satisfactory performance of TLBO algorithm for the constrained optimization problems. The computational experiments were conducted for the same number of function evaluations used by the other algorithms in all the 76 unconstrained optimization problems.

At this point, it is important to clarify that in the TLBO algorithm, the solution is updated in the teacher phase as well as in the learner phase. Hence, the total number of function evaluations in the TLBO algorithm can be computed as (2 × population size × number of generations). Rao and Patel (2012) computed the total number of function evaluations in the TLBO algorithm as = {(2 × population size × number of generations) + (function evaluations required for duplicate elimination)}, and they had used this formula to count the number of function evaluations while conducting experiments with TLBO algorithm. Since the 'function evaluations required for duplication removal' were not clearly known, experiments were conducted with different population sizes and based on the experiments the 'function evaluations required for the duplication removal' were reasonably concluded for different population sizes. The statement of Mernik et al. (2015) and their team mates like Črepinšek et al. (2015) that this way of deciding the 'function evaluations required for the duplication removal' as imprecise seems to be biased and not valid. In general, the researchers who have used the advanced optimization algorithms like GA, SA, ACO, PSO, DE, ABC, DE, ES, NSGA, NSGA-II, VEGA, SPEA, etc., have hardly talked about the 'function evaluations required for the duplication removal' while using those algorithms for solving different optimization problems. This concept of 'function evaluations required for the duplication removal' was not even used by the previous researchers while comparing the performance of different algorithms on various benchmark functions. In fact, there may not be any such need in actual practice. It is true that to maintain the consistency in comparison, the number of function evaluations is to be maintained same for all the optimization algorithms including TLBO algorithm for all the benchmark functions considered while making comparisons. However, as mentioned earlier, in general, the algorithm, which requires fewer number of function evaluations to get the same best solution can be considered better as compared with the other algorithms. If an algorithm gives global optimum solution within certain number of function evaluations, then consideration of more number of function evaluations will normally go on giving the same best result. In such situations, thinking about the 'function evaluations required for duplication removal' is not so meaningful. In fact, it is surprising that even though Mernik et al. (2015) described certain misconceptions when comparing different variants of the ABC algorithm, they had not computed any 'function evaluations required for duplication removal' in ABC algorithm and its variants! Thus, the statement of Mernik et al. (2015) on the way of deciding the function evaluations required for the duplication removal by Rao and Patel (2012) is not valid. Keeping in view of the works of researchers in the optimization field, the total number of function evaluations in the TLBO algorithm can be simply computed as (2 × population size × number of generations). The factor "2" is used in the computation because the solution is updated in the teacher phase as well as in the learner phase of the TLBO algorithm. It may be mentioned here that, in general, the total number of function evaluations required by many of the widely used optimization algorithms is computed by the researchers as (population size × number of generations). No separate provision was made in any of the widely used algorithms for computing the function evaluations required for the duplication removal and hence in TLBO also it is not considered and it may not be considered. Unnecessary emphasis on 'function evaluations required for the duplication removal' by Mernik et al. (2015) and Črepinšek et al. (2015) is not justified.

Rao et al. (2012a, b) tested the TLBO algorithm on 25 different unconstrained benchmark functions and 35 constrained benchmark functions with different characteristics. For the constrained benchmark functions, TLBO was tested with different constraint handling techniques such as superiority of feasible solutions, self-adaptive penalty, ϵ-constraint, stochastic ranking and ensemble of constraints. The codes for the elitist TLBO algorithm were also given in the Appendix of their paper. The performance of the TLBO algorithm was compared with that of other optimization algorithms and the results showed the better performance of the TLBO algorithm.

Kundu et al. (2012) proposed a swarm-based niching technique to enhance the diversity in the TLBO algorithm that adapts to the local neighborhood by controlled exploitation. The algorithm imitates the local explorative swarm behavior to hover around local sites in groups, exploiting the peaks with high degree of accuracy. In order to prove the effectiveness of the TLBO with local diversification strategy, it was applied to 18 multimodal benchmark functions. In addition to the teacher phase and the learner phase, a local diversification strategy was introduced. In the local diversification strategy, a distance-based matrix holding the members θ_j sorted in ascending order with respect to member θ_i $(i \neq j)$ was created. A local colony, comprising of member θ_i and its nearest k-best neighbors, was generated and the TLBO algorithm was applied to this colony, and it returned the fittest member X_{best} in the colony along with the updated k-best colony X. The results obtained were compared with those obtained using other algorithms such as CMA-ES with self-adaptive niche radius (SCMA-ES), sharing Differential Evolution (ShDE), crowding differential evolution (CDE), speciation-based differential evolution (SDE), speciation-based Particle Swarm Optimization (SPSO), fitness-Euclidean distance ratio particle swarm optimization (FERPSO) and lbest-PSO variants with ring-topology namely r2pso, r3pso, r2pso-lhc, and r3pso-lhc. The results proved the effectiveness in maintaining population diversity with a high success rate on 17 out of 18 cases.

Mohapatra et al. (2012) proposed a new learner stage based on the mutation strategy. A new set of learners was produced using differential random vectors and also the difference vector with respect to the best vector at that point. The algorithm was applied for optimal placement of capacitors in a distribution network. The main objective of the capacitor placement problem was to minimize the annual cost of the distribution system, considering other operational constraints of the system like, voltage limits at the distribution buses and the current limit of the lines. The objective function was formulated to minimize the annual cost of distribution system which the distribution company had to bear in order to place the capacitors in the distribution network. As the power loss depends on the load pattern, three load patterns for different durations in a year had been assumed. The light load condition for 1000 h, medium load for 6760 h, and heavy load for 10,000 h were assumed for the analysis purpose in a year. Only the fixed cost and the capacitor bank installation cost had been taken into account, whereas operation cost, maintenance cost, and depreciation cost had been neglected. The radial distribution system was considered to be balanced. The modified TLBO algorithm was applied to solve the capacitor placement problem for a 85 bus radial distribution system with one supply point as a single feeder distribution system having no laterals and the system voltage was 11 kV. Details of the feeder and the loads were adopted from the load data at different buses and were assumed to be medium loading condition for the test case. The second test case consisted of the same 85 bus system but with different loading conditions of various duration throughout a year. 80 % of the loading was assumed as light loading and 120 % of the loading was assumed to be heavy loading condition for the simulation over 1 year for a time-varying load pattern. The modified TLBO algorithm had achieved better results than those obtained using PSO algorithm and Plant Growth Algorithm (PGA). The percentage loss reduction was found better than both the approaches, i.e., PSO and PGA.

Niknam et al. (2012a) proposed a modified TLBO algorithm. In the modified algorithm, teacher and learner phases were modified. The performance of the modified TLBO algorithm was tested on a 70-bus distribution network in order to solve the optimal location of automatic voltage regulators (AVRs) in distribution systems with the presence of distributed generators (DGs). The objectives were the minimization of energy generation cost, minimization of electrical energy losses, and minimization of voltage deviation. The problem of optimal location of AVRs in the distribution system was subjected to the constraints of active power, AVRs tap position, and bus voltage magnitude. The first modification incorporated in the modified TLBO algorithm was related to the teaching factor T_f in teacher phase, and it was expressed as a vector which included n teaching factors for each elements of the control variable. The second modification was introduction of a cogent mutation strategy to diversify the TLBO population and to improve the TLBO's performance in preventing premature convergence to local minima.

Niknam et al. (2012b) proposed a stochastic model for optimal energy management with the goal of cost and emission minimization. In this model, the uncertainties related to the forecasted values for load demand, available output power of wind and photovoltaic units, and market price were modeled by a scenario-based stochastic programming. The scenarios were generated by a roulette wheel mechanism based on probability distribution functions of the input random variables. Through this method, the inherent stochastic nature of the proposed problem was released and the problem was decomposed into a deterministic problem. An improved multiobjective TLBO algorithm was implemented to yield the best expected Pareto optimal front. A novel self-adaptive probabilistic modification strategy was offered to improve the performance of the algorithm. Also, a set of nondominated solutions were stored in a repository during the simulation process. The size of the repository was controlled by usage of a fuzzy-based clustering technique. The best expected compromise solution stored in the repository was selected via the niching mechanism in a way that solutions were encouraged to seek the lesser explored regions. The framework was applied in a typical grid-connected micro grid in order to verify its efficiency and feasibility.

Rajasehkar et al. (2012) proposed the elitist teaching-learning oppositional-based (ETLOBA) algorithm in order to enhance the accuracy of the TLBO algorithm. The ETLOBA was empowered with two mechanisms, i.e., elitism to strengthen the exploitation capability of the optimization method by retaining the best solutions obtained so far and opposition-based optimization to strengthen the exploration capability of the optimization method. In order to test the performance of the ETLOBA, it was applied to solve a set of benchmark test functions. The results obtained using ETLOBA were compared with those obtained using HS, improved bees algorithm (IBA) and ABC algorithms in terms of mean and standard deviation for the five benchmark test functions. The results showed that the presence of the two robust mechanisms had improved the convergence rate of traditional TLBO method and had also shown better performance than the other algorithms such as HS, IBA, and ABC in terms of optimal values.

Rasoul et al. (2012) proposed a modified TLBO algorithm to solve the multi-objective wind-thermal economic emission dispatch problem based on point estimated method and a set of nondominated solutions was obtained. The efficiency and feasibility of the algorithm was test on three test systems. The first system consisted of 6 units neglecting power losses. The second system included 14 units and the power losses were considered. The third system comprised 40 units in which 5 out of 40 exhibited prohibited zones and the effect of power losses was neglected.

Satapathy and Naik (2012) proposed a modified TLBO algorithm by incorporating a random weighted differential vector. The performance of the algorithm was tested on separable, nonseparable, unimodal, and multimodal benchmark test functions. The performance of the modified TLBO algorithm on the standard benchmark test functions was also compared with that of OEA, HPSO-TVAC, CLPSO, APSO, and variants of DE such as JADE, jDE and SaDE, and the results proved that the modified TLBO algorithm outperformed the other algorithms. It had also outperformed different variants of ABC algorithm such as CABC, GABC, RABC, and IABC in terms of mean and standard deviations obtained for the benchmark test functions.

Toğan (2012) applied the TLBO algorithm for the design optimization of planar steel frames. The design algorithm aimed to obtain minimum weight frames subjected to strength and displacement requirements imposed by the American Institute for Steel Construction (AISC) load and resistance factor design (LRFD). Designs were obtained by selecting appropriate W-shaped sections from a standard set of steel sections specified by the AISC. Several frame examples from the literature were examined to verify the suitability of the design procedure and to demonstrate the effectiveness and robustness of the TLBO in creation of an optimal design for frame structures. The results of the TLBO were compared to those of the GA, ACO, harmony search (HS), and the improved ant colony optimization (IACO) and it was observed that the TLBO algorithm was a powerful search and applicable optimization method for the problem of engineering design applications.

15.1.3 Publications in the Year 2013

Waghmare (2013) made comments on a note published by Črepinšek et al. (2012) who tried to invalidate the performance supremacy of the TLBO algorithm. The views and the experimental results presented by Črepinšek et al. (2012) were questionable and hence Waghmare (2013) re-examined the experimental results and presented the correct understanding of the TLBO algorithm in an objective manner. The latest literature on TLBO algorithm was also presented and the algorithm-specific parameter-less concept of TLBO was explained. The results of the work demonstrated that the TLBO algorithm performs well on the problems where the fitness-distance correlations are low by proper tuning of the common control parameters of the algorithm. Yu et al. (2014) commented that the claim made by Črepinšek et al. (2012) in another study that Waghmare (2013) used different success rates was unsuitable. Furthermore, the comparisons of evolutionary algorithms conducted by Veček et al. (2014) (with Črepinšek as a co-author) attempted to cast the TLBO algorithm in a poor light, although this attempt may also be seen as not meaningful as the findings of Veček et al. (2014) were simply comparisons of the basic TLBO algorithm with different modified versions of DE and did not consider comparisons with other important algorithms such as the GA, SA, PSO, and ACO. It is to emphasize here that there is no origin bias in the TLBO algorithm.

Rao and Patel (2013a) investigated the effect of elitism on the performance of the TLBO algorithm for the unconstrained optimization problems. Furthermore, the effect of common controlling parameters on the performance of TLBO algorithm was also investigated by considering different combinations of common controlling parameters. The proposed algorithm was implemented on 76 unconstrained optimization problems having different characteristics to identify the effect of elitism and common controlling parameters. The results had shown that the strategy with elitism consideration produced better results for some functions than that without elitism consideration. The results obtained using TLBO algorithm were compared with the other optimization algorithms available in the literature such as GA, PSO, DE, ABC of Karaboga and Akay (2009), CES, FES, ESLAT, and CMA-ES for the considered benchmark problems. Results had shown the satisfactory performance of TLBO algorithm for the unconstrained optimization problems. The computational experiments were conducted for the same number of function evaluations used by the other algorithms in all the 76 unconstrained optimization problems.

Rao and Patel (2013b) modified the basic TLBO algorithm by introducing the concept of number of teachers and adaptive teaching factor. The presented modifications speeded up the convergence rate of the basic TLBO algorithm. The modified TLBO algorithm was applied successfully to the multiobjective optimization of heat exchangers considering two conflicting objectives: effectiveness and total cost. Two different heat exchangers namely plate-fin heat exchanger and shell and tube heat exchanger were investigated for the optimization. The ability of the proposed algorithm was demonstrated using case studies and the performance of the modified TLBO algorithm was compared with the performance of GA presented by previous researchers. Improvement in the results was observed using the modified TLBO algorithm as compared to the GA approach showing the improvement potential of the algorithm for such thermodynamic optimization. In addition, the competence of the modified TLBO algorithm was verified by changing the cost function as well as by changing the objective functions. The TLBO algorithm responded correctly in all the cases.

Rao and Patel (2013c) modified the basic TLBO algorithm by introducing the concept of number of teachers, adaptive teaching factor, and self-motivated learning. The modified TLBO algorithm was applied successfully to the multiobjective optimization of a two-stage thermoelectric cooler (TEC) considering two conflicting objectives: cooling capacity and COP. Two different configurations of TECs, i.e., electrically separated and electrically connected in series were investigated for the optimization. Moreover, the contact and spreading resistance of TEC were also considered. The ability of the algorithm was demonstrated using an example, and the performance of the modified TLBO algorithm was compared with the performance of basic TLBO and GA. Improvements in the results were observed using the basic TLBO and the modified TLBO algorithms as compared to the GA approach showing the improvement potential of the proposed algorithm for such thermodynamic optimization.

Rao and Patel (2013d) improved the basic TLBO algorithm to enhance its exploration and exploitation capacities by introducing the concept of number of teachers, adaptive teaching factor, tutorial training, and self-motivated learning. Performance of the improved TLBO algorithm was assessed by implementing it on a range of standard unconstrained benchmark functions having different characteristics. The results of optimization obtained using the improved TLBO algorithm were validated by comparing them with those obtained using the basic TLBO and other optimization algorithms available in the literature.

Rao and Kalyankar (2013a) carried out the parameter optimization of a multi-pass turning operation using the TLBO algorithm. Two different examples were considered that were attempted previously by various researchers using different optimization techniques, such as SA, GA, ACO, and PSO. The first example was a multiobjective problem and the second example was a single-objective multi-constrained problem with 20 constraints. The performance of the TLBO algorithm was studied in terms of the convergence rate and accuracy of the solution. The TLBO algorithm required a lower number of iterations for convergence to the optimal solution. The algorithm had shown its ability in handling multi-constrained problems.

Rao and Kalyankar (2013b) applied the TLBO algorithm for the process parameter optimization of selected modern machining processes. The important modern machining processes identified for the process parameters optimization were ultrasonic machining (USM), abrasive jet machining (AJM), and wire electrical discharge machining (WEDM) process. The examples considered for these processes were attempted previously by various researchers using different optimization techniques such as GA, SA, ABC, PSO, harmony search (HS), and shuffled frog leaping (SFL), etc. The comparison between the results obtained by the TLBO algorithm and those obtained by different optimization algorithms showed the better performance of the TLBO algorithm. However, in the case of WEDM process of Example 3, the objective of the work was to maximize the cutting speed (V_m) by ensuring the constraint value of surface roughness (R_a) which should not exceed the permissible surface roughness (R_{per}) of 2.0 μm. The optimum process parameters setting obtained by the TLBO algorithm was given in Table 4 of their paper, and the maximum cutting speed given by the TLBO algorithm was reported as 1.4287 mm/min. However, it has been observed that the corresponding set of process parameters lead to a slight variation of the surface roughness constraint by 0.0189 μm. Even though this difference is small, the TLBO algorithm is now rerun using the same number of 20 iterations and the population size of 10. The new result for V_m is 1.421034 mm/min and the corresponding R_a value is 1.999997 μm and this satisfies the constraint. The values given in Rao and Pawar (2010) and Rao (2011) are also relooked into and slight corrections are made. The corrected values are 1.420907 mm/min in the case of ABC of Rao and Pawar (2010), 1.420498 mm/min in the case of PSO, 1.414212 mm/min in the case of HS_M and SA, and 1.417831 mm/min in the case of SFL of Rao (2011). It can be observed that the maximum cutting speed (V_m) given by the TLBO algorithm is 1.421034 mm/min which is still better than the results given by all the other optimization algorithms used for the same model. ABC algorithm has given the next best result. However, the number of iterations used in the case of ABC algorithm was 150, whereas TLBO algorithm has given the result using only 20 iterations. Thus, in the case of WEDM process, the TLBO algorithm has proved its superiority and has given slight improvement in the result with less iterations compared to the other advanced optimization algorithms. Similarly, Tables 2 and 3 of Rao and Kalyankar (2013b) need to be corrected in the case of Example 2 keeping in view of the slight violation in the constraint value. The values given in Jain et al. (2007) are also relooked into and slight corrections are now made. The mass flow rate of abrasive particles (kg/s) and velocity of abrasive particles (mm/s) in Example 4.2.1 are 0.0005 and 315,772, respectively, in the case of TLBO algorithm with the optimum value of MRR (mm^3/s) as 8.2528. The optimum value of MRR (mm^3/s) is 8.2525 in the case of SA of Rao and Pawar (2010) and the optimum value of MRR (mm^3/s) is 8.242 in the case of Jain et al. (2007). Similarly, in the case of Example 4.2.2, velocity of abrasive particles (mm/s) is recalculated as 333,600 in the case of TLBO of Rao and Kalyankar (2013a, b) with the optimum value of MRR (mm^3/s) as 0.6056. The optimum value of MRR (mm^3/s) is 0.6035 in the case of GA of Jain et al. (2007). However, it can be observed that the results given by the TLBO algorithm are found still better than those given by GA and SA in this example.

Dede (2013) used TLBO algorithm for the optimum design of grillage systems based on the LRFD-AISC (Load and Resistance Factor Design-American Institute of Steel Construction). Cross-sectional area of W-shapes were considered as discrete design variables. Various grillage structures were designed to show the efficiency of the TLBO algorithm. The results obtained from the study were compared with those reported in the literature. It was concluded that the TLBO algorithm could be effectively used in the optimal design of grillage structures.

Degertekin and Hayalioglu (2013) applied the TLBO algorithm for optimization of truss structures. The validity of the method was demonstrated by the four design examples. Results obtained for the design examples revealed that although the TLBO developed slightly heavier designs than the other metaheuristic methods in a few cases, it obtained results as good as or better than the other metaheuristic optimization methods in terms of both the optimum solutions and the convergence capability in most cases.

García and Mena (2013) proposed TLBO algorithm to determine the optimal placement and size of distributed generation (DG) units in distribution systems. The objective function considered was to minimize total electrical power losses, although the problem could be easily configured as multiobjective, where the optimal location of DG systems, along with their sizes, was simultaneously obtained. The optimal DG site and size problem was modeled as a mixed integer nonlinear programming problem. Evolutionary methods were used by researchers in the past to solve this problem because of their independence from type of the objective function and constraints. The authors had modified the TLBO algorithm to find the best sites to connect DG systems in a distribution network, choosing among a large number of potential combinations. A comparison between the TLBO algorithm and a brute force method was performed. Besides this, a comparison using several results available in other articles published by other authors was carried out.

Jiang and Zhou (2013) proposed a hybrid differential evolution and TLBO (hDE-TLBO) algorithm to solve the multiobjective short-term optimal hydro-thermal scheduling (MOSOHTS) considering the total fuel cost and emission effects as two conflict objectives. The water transport delay between reservoirs and the valve-point effect of thermal units were also taken into consideration. The authors had opined that the teachers' knowledge needs to be refreshed and this was considered as the teachers refresher process in the hybrid version. The DE algorithm played the role of teacher refresher and enhanced the overall efficiency of the algorithm. The effectiveness of the algorithm was tested on three scenarios of a hydro-thermal power system with four cascaded hydro plants and three thermal units. The fuel cost of each thermal unit with the consideration of valve-point effects was expressed as the sum of a quadratic function and a sinusoidal function. The emission of each thermal unit was given as the sum of a quadratic and an exponential function of its power output. The multiobjective optimal short-term hydro-thermal scheduling problem was subjected to the constraints of system power balance, generation capacity, hydroelectric plant discharge, reservoir storage volume, initial and terminal reservoir storage volumes, and continuity equation for the hydro reservoir network. On comparing Pareto optimal fronts obtained by hDE-TLBO with those obtained using multiobjective differential evolution (MODE) and multiobjective TLBO (MOTLBO), it was found that in the three cases, the nondominated schemes obtained by the hDE-TLBO method were widely and uniformly distributed in the objective space. The results verified that the hDE-TLBO had an enhanced overall searchability and could be a viable alternative to generate optimal non-dominated scheduling schemes for the MOSOHTS problem.

Krishanand et al. (2013) proposed a hybrid self-evolving algorithm with its application to a nonlinear optimal power flow problem with the focus on the minimization of the fuel costs of the thermal units while maintaining the voltage stability at each of the load buses with restrictions on acceptable voltage levels, capacitance levels of shunt compensation devices, and transformer taps. The authors had proposed a self-evolving brain storming inclusive teaching-learning-based (SEBI-TLB) algorithm which was a combination of the learning principles from brain storming optimization algorithm and the TLBO algorithm, along with a self-evolving principle applied to the control parameter. The strategies used in the SEBI-TLB algorithm made it self-adaptive in performing the search over the multidimensional problem domain. The SEBI-TLB algorithm was tested so as to obtain optimal power flow solution for an IEEE 30-Bus and 6 generator system. Both objectives were individually optimized first and were then used as seeds during the consequent initialization. The results obtained using SEBI-TLB algorithm were compared with those obtained using the gradient-based approach, improved GA-based approach, PS0-based approach, DE approach, and adaptive Genetic Algorithm with adjusting population size (AGAPOP). It was observed that the SEBI-TLB algorithm had shown better or competitive performance.

Li et al. (2013) proposed an ameliorated TLBO algorithm (A-TLBO) in order to improve the solution quality and to quicken the convergence speed of TLBO. The A-TLBO was adopted to adjust the hyper-parameters of least squares support vector machine (LS-SVM) in order to build NO_x emissions model of a 330 MW coal-fired boiler. In LS-SVM, the regularization parameter and kernel parameter directly affect the generalization ability and regression accuracy. So the selection of appropriate hyper-parameters is a crucial step in obtaining well-adjusted LS-SVM. The A-TLBO algorithm was applied to tune the hyper-parameters of LS-SVM. During the selection of hyper-parameters, each learner represented a feasible solution consisted of a vector. Here, the fitness function was defined as leave-one-out cross-validation and the objective was to minimize the fitness value, so the learner with the minimal fitness value would outperform others and reserved during the optimization process. In the A-TLBO, in order to quicken the process of 'teaching' and 'learning' process, the elitist strategy to replace the greedy selection mechanism, inertia weight, and acceleration coefficients were introduced. The tuned model by A-TLBO showed better identification and generalization abilities under various operating conditions than the ones by the gravitational search algorithm (GSA), ABC, coupled simulated annealing (CSA), and TLBO.

Mandal and Roy (2013) incorporated a quasi-opposition-based learning concept in original TLBO algorithm in order to accelerate the convergence speed and to improve solution quality. The proposed algorithm was applied to solve the multi-objective optimal reactive power dispatch problem by minimizing real-power loss, voltage deviation, and voltage stability index. The quasi oppositional TLBO (QOTLBO) approach was applied to the standard IEEE 30-bus and IEEE 118-bus test systems, and the results obtained were compared with those obtained using other algorithms such as PSO, fully informed Particle Swarm Optimization (FIPS), quantum inspired evolutionary algorithm (QEA), ACS, and DE. The results showed that the QOTLBO algorithm was able to produce better voltage stability condition, voltage deviation, and transmission loss as compared to the other algorithms.

Mardaneh and Golstaneh (2014) proposed a harmonic elimination technique for online reducing harmonics in voltage source inverters. In the presented method, in order to provide required fundamental voltage and eliminate specified harmonics simultaneously, the values of the switching angles were determined using a modified version of TLBO. The efficiency of the presented strategy was validated on three-phase seven-level diode-clamped inverter, while other structures of multilevel inverters (MLIs) could be employed. The aim was to reduce the total harmonic distortion (THD) in multilevel converter. The modified version of the TLBO algorithm was successfully implemented to achieve optimal switching angles in order to harmonic elimination in MLIs. A three-phase seven-level diode-clamped MLI was selected as a test case to validate the effectiveness of the method. The simulation results demonstrated that the proposed method not only was in preference to the GA and PSO for including no user-defined parameters and simple implementation, but also for providing higher quality solutions. Moreover, in order to verify the superiority of the proposed algorithm over conventional methods for harmonic elimination, the Newton-Raphson (NR) method was implemented to find optimal witching angles. The results obtained by the NR method were compared with those of the modified TLBO algorithm, and the superiority of the algorithm in terms of the both computational time and the resulted %THD was concluded. As a whole, the results showed that the algorithm was expected to become widespread in power systems where online updating harmonics is needed since it provides accurate and high-quality solutions in extremely short time.

Medina et al. (2013) proposed a multiobjective teaching-learning algorithm based on decomposition (MOTLA/D). The performance of the MOTLA/D was tested on three systems, and the results were compared with those obtained using other multiobjective algorithm based on decomposition. The MOTLA/D utilizes the Tchebycheff's approach, to decompose the multiobjective optimization problem into scalar optimization subproblems. The reactive power handling problem consisted mainly of two objectives to be minimized, i.e., reactive power loss and the voltage stability index L_{index}. The reactive power handling problem was associated with the equality constraints and inequality constraints. The effectiveness and performance of MOTLA/D was compared to that of the multiobjective evolutionary algorithm based on decomposition (MOEA/D). The results showed that the MOTLA/D outperformed MOEA/D in all cases.

Niknam et al. (2013) introduced a modified phase based on the self-adaptive mechanism in the TLBO algorithm in order to achieve a balance between the exploration and exploitation capabilities of the TLBO algorithm. The modified TLBO algorithm was applied to solve the reserve constrain dynamic economic dispatch (RCDED) problem for thermal units. Minimization of fuel cost was the main objective in the RCDED problem. The constraints associated with the RCDED problem were related to power balance, power loss, up/down ramp rate limits, generation limits, and spinning reserve requirements. In the modified TLBO algorithm, the teacher phase and the learner phase were similar to that in the TLBO algorithm. However, a modified phase based on the self-adaptive learning strategy was introduced. The basic idea behind the modified TLBO approach was to simultaneously select adaptively multiple effective strategies from the candidate strategy pool on the basis of their previous experiences in the generated promising solutions and applied to perform the mutation operation. It means that at different steps of the optimization procedure, multiple strategies may be assigned a different probability based on their capability in generating improved solutions. Accordingly, during the evolution process, with respect to each target solution in the current population which is extracted from the second phase (learner phase), one method will be selected from the strategy pool based on its probability. The more successfully one mutation method behaved in previous iterations to generate promising solutions, the more probably it will be chosen in the current iteration to produce solutions. Niknam et al. (2013) suggested four mutation strategies to be implemented in the modified TLBO algorithm and the algorithm was applied to four case studies to investigate the RCDED problem. In the first case, five thermal units with transmission losses were considered, in the second case ten-unit network was investigated with and without transmission losses, the third case was formulated by tripling the number of units in case 2, and the fourth case was the scalability study to test the performance of the modified TLBO algorithm. The results obtained by applying the modified TLBO algorithm to cases 1 and 2 were compared with those obtained using hybrid evolutionary programming-sequential quadratic programming (EP-SQP), simulated annealing (SA), adaptive Particle Swarm Optimization (APSO), improved particle swarm optimization (IPSO), artificial immune system (AIS), and basic TLBO. The results proved that the proposed modified TLBO algorithm had a good global searching capability.

Pawar and Rao (2013a, b) presented the optimization aspects of process parameters of three machining processes including an advanced machining process known as abrasive water jet machining process and two important conventional machining processes namely grinding and milling. The TLBO algorithm was to find the optimal combination of process parameters of the considered machining processes. The results obtained using TLBO algorithm were compared with those obtained using other advanced optimization techniques such as GA, SA, PSO, HS, and ABC algorithm. The results showed better performance of the TLBO algorithm.

Singh and Verma (2014) presented the TLBO algorithm to solve parameter identification problems in the designing of a digital infinite impulse response (IIR) filter. TLBO-based filter modeling was applied to calculate the parameters of unknown plant in simulations. Unlike other heuristic search algorithms, Big Bang–Big Crunch (BB–BC) optimization and PSO algorithms were also applied to the filter design for comparison. Unknown filter parameters were considered as a vector to be optimized by these algorithms. Experimental results showed that the TLBO was more accurate to estimate the filter parameters than the BB–BC optimization algorithm and had shown faster convergence rate when compared to the PSO algorithm.

Rao and Waghmare (2013) solved complex composite test functions using TLBO algorithm. Satapathy et al. (2013) proposed an orthogonal design-based TLBO (OTLBO) algorithm. The OTLBO algorithm was incorporated with the orthogonal design and thus generated an optimal offspring by a statistical optimal method. OTLBO was also incorporated with a new selection strategy to decrease the number of generations and to make the algorithm converge faster. The performance of the OTLBO algorithm was tested on 20 unimodal, multimodal, separable, and non-separable benchmark functions. Similar experimentation was conducted to compare the performance of OTLBO algorithm with ortogonal design-based constrained evolutionary algorithm (OEA), hierarchical particle swarm optimizer with time-varying acceleration coefficients (HPSO-TVAC), CLPSO, and adaptive particle swarm optimization (APSO) algorithms, and the OTLBO algorithm was ranked first based on the Friedman's test. The performance of the OTLBO algorithm was also compared with JADE, jDE, self-adaptive Differential Evolution (SaDE), CoDE, and EPSDE algorithms wherein the OTLBO algorithm was ranked first based on the Friedman's test.

Roy et al. (2013) applied a modified version of TLBO (called QTLBO) for solving the short-term hydro-thermal scheduling (HTS) problem with highly complex and nonlinear relationship of the problem variables and cascading nature of hydroreservoirs, water transport delay, and scheduling time linkage as constraints. The main objective of HTS problem was to minimize the thermal production cost over the scheduling period while satisfying various operational constraints. The nonlinear constrained HTS problem was subjected to various system constraints such as the varying load demand, the cascading nature of the hydroreservoirs, the time-varying hourly reservoir in flowrate, the maximum and minimum power generation limits of the thermal plant and hydro units, limits of water discharge rate, storage capacity limits of hydroreservoirs, initial and final reservoir storage limits, and hydraulic continuity constraints. The results obtained using QTLBO algorithm were compared with those obtained using two-phase neural networks, augmented lagrange method, PSO, ISA, PSO, and basic TLBO algorithms.

Shabanpour-Haghighi et al. (2014a, b) proposed a modified TLBO algorithm with self-adaptive wavelet mutation strategy and fuzzy clustering technique to control the size of the repository and smart population selection for the next iteration. The modified TLBO algorithm was applied to solve multiobjective optimal power flow problem considering the total fuel cost and total emissions of the generating units. The modified TLBO algorithm used an external archive called repository to save the non-dominated solutions found. In each iteration of the algorithm, besides the optimization process, solutions found by each phase of the algorithm were compared with the repository members. The new non-dominated solutions were stored in the repository. Also the algorithm removed the dominated members of the repository. The repository size of most optimization problems could be increased extremely large during the optimization algorithm. It is obvious that the large number of individuals in the repository may lead to more computation burden and even the memory constraints. Thus in order to limit the size of the repository without losing the characteristic and quality of the Pareto-optimal front, a repository with a determined size was used and the new solution was added to it if one of the four conditions was satisfied. Whenever the repository was full, in order to find out whether the new non-dominated solution should be replaced by one of the repository members or not, a fuzzy decision making strategy was used. To find out the best compromise solution among the final repository members, the fuzzy membership functions of all objectives were extracted separately and the fuzzy solution was calculated.

Singh et al. (2013) discussed the application of TLBO algorithm for optimal coordination of DOCR relays in a looped power system. Combination of primary and backup relay was chosen using Far vector of LINKNET structure to avoid mis-coordination of relays. Coordination of DOCR was tested for IEEE 3, 4, and 6 bus systems using the TLBO. Also, the objective function was modified to optimize the operating time between backup and primary relays. The results were compared with the optimized values of time dial setting and plug setting values obtained from modified DE algorithm. The proposed algorithm TLBO gave optimal coordination margin between 0.3 and 0.8 s and no mis-coordination between primary and backup pairs. Results were also verified using Digsilent powerfactory simulation software.

Tang et al. (2013) proposed an improved TLBO algorithm with Memetic method (ITLBO-M). The memetic method improved the global exploring ability, whereas the one-to-one teaching improved the local searchability and the effectiveness of the method was tested on eight benchmark test functions. The results obtained using the ITLBO-M algorithm were compared with those obtained using PSO, SFL, DE, and basic TLBO algorithm. The experimental results showed that ITLBO-M algorithm was efficient and robust in the case of unimodal and multimodal functions.

Theja and Rajasekhar (2013) applied a modified version of TLBO algorithm for designing proportional-integral (PI) controller-based power system stabilizer (PSS) for a single machine infinite bus power system equipped with thyristor-controlled series compensator (TCSC). In order to enhance the performance of the basic TLBO algorithm, it was incorporated with the concept of opposite number. The design of coordinated controller was done based on minimizing the objective function considered such that the power system oscillations after a disturbance were effectively damped out so as to improve the stability. The objective function was formulated in such way that the rotor speed deviation was minimized.

Tuo et al. (2013) proposed an Improved Harmony Search-based teaching-learning (HSTL) optimization algorithm in order to maintain a balance between the convergence speed and the population diversity. The HSTL algorithm was incorporated with four modifications (i.e., harmony memory consideration, teaching-learning strategy, local pitch adjusting, and random mutation). The associated control parameter values were dynamically changed according to the process of evolution. To demonstrate the robustness and convergence, the success rate and convergence analysis were also studied on 31 complex benchmark functions. The experimental results for 31 complex benchmark functions demonstrated the strong convergence capability and robustness of the HSTL algorithm. The results proved that the HSTL algorithm had better balance capacity of space exploration and local exploitation on high-dimension complex optimization problems.

Yildiz (2013) integrated Taguchi method with the TLBO algorithm known as the hybrid robust TLBO (HRTLBO) algorithm. The HRTLBO was applied to determine the optimum machining parameters for the multipass turning operation. The objective was to minimize the unit production cost. The results obtained by the HRTLBO algorithm for rough machining and finish machining were compared to those obtained using PSO algorithm, hybrid Genetic Algorithm (HRGA), scatter search (SS) algorithm, float encoding genetic algorithm (FEGA), and integration of simulated annealing and Hooke–Jeeves patter search (SA/PS) algorithm. It was observed that better results were achieved for the multipass turning optimization problem using the HRTLBO algorithm compared to the PSO, HRGA, SS, FEGA, and SA/PS algorithms.

Zou et al. (2013a) proposed a multiswarm TLBO algorithm for optimization in dynamic environment. In this method, all learners were divided up into several subswarms so as to track multiple peaks in the fitness landscape. Each learner learns from the teacher and the mean of his or her corresponding subswarm instead of the teacher and the mean of the class in teaching phase, learners also learn from interaction among themselves in their corresponding subswarm in learning phase. All subswarms are regrouped periodically so that the information exchange was made with all the learners in the class to achieve proper exploration ability. The performance of the algorithm was evaluated on moving peaks benchmark problem in dynamic environments. The moving peak benchmark problem (MPB) proposed by Branke (1999) is widely used as dynamic benchmark problem in the literature. Within the MPB problem, the optima can be varied by three features, i.e., the location, height, and width of peaks. The multidimensional problem space of the moving peaks function contains several peaks of variable height, width, and shape. These move around with height and width changing periodically. The MPB problem was solved and the results obtained were compared with those obtained using Multiswarm PSO (mPSO) algorithm, clustering quasi settler object selection (CQSO) algorithm and clustering PSO (CPSO) algorithm. The algorithm showed a much better performance than the other algorithms. The results proved that the algorithm was robust with number of peaks and could achieve acceptable results when the number of peaks were low or when the number of peaks were high.

Zou et al. (2013b) proposed a TLBO algorithm for multiobjective optimization problems (MOPs) by adopting the nondominated sorting concept and the mechanism of crowding distance computation. The teacher of the learners was selected from among current nondominated solutions with the highest crowding distance values and the centroid of the nondominated solutions from current archive was selected as the mean of the learners. The performance of multiobjective TLBO algorithm was investigated on the two bar truss design problem and the I-beam design problem. An external archive was used to keep the best solutions generated so far. To select the individuals in better fronts and in order to push the population toward the Pareto front, the non-dominated sorting concept and the mechanism of crowding distance computation were incorporated into the algorithm specifically on teacher selection and in the deletion method of population including an external archive of the current best solutions. Initially NP (the number of individuals of initial population) solutions were sorted on the basis of nondomination rank and crowding distance rank and added to the external archive. As the evolution progressed, the algorithm applied the teacher phase and the learner phase of the TLBO algorithm to create NP new solutions. The two population sets (i.e., current and external archive) were then combined so that the total size of the set after combination becomes 2NP. Thereafter, NP solutions were selected on the basis of nondomination rank and crowding distance rank for the next generation from 2NP solutions. Those solutions which were in the most crowded areas were most likely to be selected so that the diversity among the stored solutions in the archive was promoted. The nondominated set of solutions in the external archive was used as the teacher of the learners. However, in order to ensure that the learners in the population move toward the sparse regions of the search space the teacher of the learners was selected from among those nondominated solutions with the highest crowding distance values. Selecting different teachers for each learner in a specified top part of the external archive based on a decreasing crowding distance allowed the learners in the primary population to move toward those nondominated solutions in the external archive, which were in the least crowded area in the objective space. The centroid of the nondominated solutions from current archive was considered as the mean of the learners. In this approach, each solution was compared with every other solution in the population to find if it was dominated in order to sort a population according to the level of nondomination. For each solution i of a solutions set, two entities were calculated, i.e., n_i (the number of solutions which dominate the solution i), and S_i (a set of solutions that the solution i dominates). At the end of this procedure, all solutions in the first nondominated front F_1 had their domination count $n_i = 0$. Then for each solution i with $n_i = 0$, it visited each member j of its set S_i and reduced its domination count by one. While doing so, if for any member j the domination count became zero then it was put in a separate list P. These members belonged to the second nondominated front F_2. The above procedure was continued with each member of P and the third front F_3 was identified. This process continued until all fronts were identified. Crowding distance was used to get an estimate of the density of solutions surrounding a particular solution i in the population. Each objective function was normalized before calculating the crowding distance. The results reported showed that the minimum volume of the two bar truss was achieved by MOPSO-CD, whereas the proposed version of TLBO could achieve a minimum stress in the members of truss. The results showed that the minimal cross-sectional area obtained was the smallest among the four methods considered, and the minimal deflections obtained using NSGAII, MOPSO-CD, RM-MEDA, and the TLBO algorithm were equal.

15.1.4 Publications in the Year 2014

Rao and Waghmare (2014) evaluated the performance of the TLBO algorithm against the other optimization algorithms over a set of multiobjective unconstrained and constrained test functions, and the results were compared with other algorithms such as AMGA, clustering MOEA, DECMOSA-SQP, DMOEADD, GDE3, LiuLi algorithm, MOEAD, and MOEADGM. The TLBO algorithm was observed to outperform the other optimization algorithms for the multiobjective unconstrained and constrained benchmark problems.

Rao and More (2014) used TLBO algorithm for optimal selection of design and manufacturing tolerances with an alternative manufacturing process to obtain the optimal solution nearer to the global optimal solution. Three problems were considered and these were: overrunning clutch assembly, knuckle joint assembly with three arms, and a helical spring. Out of these three problems, the problems of overrunning clutch assembly and knuckle joint assembly with three arms were multiobjective optimization problems, and the helical spring problem was a single-objective problem. The comparison of the proposed algorithm was made with the GA, nondominated sorting genetic Algorithm-II (NSGA-II), and multi-objective particle swarm optimization algorithm (MOPSO). It was found that the TLBO algorithm had produced better results when compared to those obtained using GA, NSGA-II, and MOPSO algorithms.

Rao et al. (2014) considered the mathematical models of three important casting processes namely squeeze casting, continuous casting, and die casting for the parameter optimization of respective processes. The TLBO algorithm was used for the parameters optimization of these casting processes. Each process was described with a suitable example which involved respective process parameters. The mathematical model related to the squeeze casting was a multiobjective problem, whereas the model related to the continuous casting was a multiobjective multi-constrained problem, and the problem related to the die casting was a single-objective problem. The mathematical models were previously attempted by GA and SA algorithms. Considerable improvements in results were obtained in all the cases. The detailed literature survey had proved that there was a good scope for the use of advanced optimization techniques like TLBO in the field of parameters optimization of these casting processes. Multiobjective mathematical model was considered in the case of squeeze casting process and initially the objectives were attempted individually where the hardness and tensile strength were improved considerably. A common parameter setting was also obtained for satisfying both these objectives simultaneously. Even though the original problem of continuous casting process was having 17 loss functions but only 10 loss functions were considered for whom the complete information was available. Fair comparison of these 10 loss functions with the results of the previous researchers was justified by considering the 10 complete models of the respective loss functions. The TLBO algorithm had effectively handled the problem and had given significant improvement of above 60 % in total loss function compared to the previous results obtained by SA algorithm. In the case of two loss functions, the undesirability index was almost reduced to zero. For die casting process, the model under consideration was having five input parameters, and it was earlier attempted using GA using 1000 generations. The same model was satisfactorily attempted with only 10 generations, thereby drastically reducing the computational efforts. Moreover, the solution obtained was also believed to be a global optimum solution for the die casting process. Thus, the TLBO algorithm had effectively handled the various mathematical models and proved its capabilities in the field of parameters optimization of casting processes with less computational effort. The TLBO codes were also presented in the Appendix of their work.

Abirami et al. (2014) presented the details of integrated maintenance scheduling for the secure operation. The problem was formulated as a complex optimization problem that affects the unit commitment and economic dispatch schedules. Abirami et al. (2014) used TLBO algorithm as a prime optimization tool and the methodology was tested on standard test systems and it worked well while including generator contingency. Arya and Koshti (2014) presented a load shedding algorithm for alleviating line overloads employing the TLBO algorithm. The buses were selected for load shed based on the sensitivity of severity index with respect to load shed. Load shed was based on the next interval predicted load which could cause emergency situation from thermal limit consideration. Line flow constraints considered next predicted interval and base case loading conditions also. Optimum load shed at the selected buses was obtained for 30-bus and 39-bus standard test systems.

Baykasoğlu et al. (2014) analyzed the performance of TLBO algorithm on combinatorial optimization problems. The authors had provided a detailed literature review about TLBO's applications. The performance of the TLBO algorithm was tested on some combinatorial optimization problems, namely flow shop (FSSP) and job shop scheduling problems (JSSP). It is a well-known fact that scheduling problems are among the most complicated combinatorial optimization problems. Therefore, performance of TLBO algorithm on these problems could give an idea about its possible performance for solving other combinatorial optimization problems. The authors had also provided a comprehensive comparative study along with statistical analyses in order to present the effectiveness of TLBO algorithm on solving scheduling problems. Experimental results showed that the TLBO algorithm has a considerable potential when compared to the best-known heuristic algorithms for the scheduling problems.

Bayram et al. (2014) investigated the applicability of TLBO algorithm in modeling stream dissolved oxygen (DO) prediction. The input parameters selected from a surface water-quality study including 20 indicators for the models were water pH, temperature, electrical conductivity, and hardness, which were measured semimonthly at six monitoring sites selected in an untreated wastewater impacted urban stream during a year due to their direct and indirect effect on DO concentration. The accuracy of TLBO algorithm was compared with those of the ABC algorithm and conventional regression analysis methods. These methods were applied to four different regression forms: quadratic, exponential, linear, and power. There were 144 data for each water-quality indicator, 114 of which were designated for training and the rest for testing patterns in the models. To evaluate the performance of the models, five statistical indices, i.e., sum square error, root mean square error, mean absolute error, average relative error, and determination coefficient, were used. The TLBO method with quadratic form yielded better prediction from among all models with an improvement of nearly 20 %. It was concluded that the equations obtained by employing the TLBO algorithms predicted the stream DO concentration successfully. Therefore, the employment of the TLBO algorithm by water resources and environment managers was encouraged and recommended.

Bouchekara et al. (2014) used TLBO algorithm to solve the optimal power flow problem. In order to show the effectiveness of the method, it was applied to the standard IEEE 30-bus and IEEE 118-bus test systems for different objectives that reflect the performances of the power system. The obtained results and the comparison with other techniques indicated that the TLBO technique provided effective and robust high-quality solution when solving the optimal power flow problem with different complexities. Lin et al. (2014) proposed a direct method to quantify the carbon emissions in turning operations. To determine the coefficients in the quantitative method, real-experimental data were obtained and analyzed in MATLAB. Moreover, a multiobjective TLBO algorithm was proposed and two objectives of minimizing the carbon emissions and the operation time were considered simultaneously. Cutting parameters were optimized and finally the analytic hierarchy process was used to determine the optimal solution which was found to be more environmentally friendly than the cutting parameters determined by the design of experiments method.

Camp and Farshchin (2014) applied a modified TLBO algorithm to fixed geometry space trusses with discrete and continuous design variables. Designs generated by the modified TLBO algorithm were compared with other popular evolutionary optimization methods. In all cases, the objective function was the total weight of the structure subjected to strength and displacement limitations. Designs were evaluated for fitness based on their penalized structural weight, which represented the actual truss weight and the degree to which the design constraints were violated. The computational performance of TLBO designs for several benchmark space truss structures was presented and compared with classical and evolutionary optimization methods. Optimization results indicated that the modified TLBO algorithm could generate improved designs when compared to other population-based techniques and in some cases improved the overall computational efficiency.

Chen et al. (2014) proposed a modified TLBO algorithm for thinning and weighting planar arrays to synthesize the desired antenna array factor. Not only the number of active elements and their corresponding excitation weights were optimized but the peaks of side lobe level, main-lobe width, and current taper ratio were also minimized as objective functions in the multiobjective formulation. The simulation cases demonstrated that the modified TLBO outperformed simulated annealing method and hybrid genetic algorithm through comparisons.

Cheng (2014) introduced TLBO algorithm to select the specific and feasible primers. The specified polymerase chain reaction (PCR) product lengths of 150–300 and 500–800 bp with three melting temperature formulae of Wallace's formula, Bolton and McCarthy's formula and SantaLucia's formula were performed. The author had calculated optimal frequency to estimate the quality of primer selection based on a total of 500 runs for 50 random nucleotide sequences of 'Homo species' retrieved from the National Center for Biotechnology Information. The method was then fairly compared with the genetic algorithm (GA) and memetic algorithm (MA) for primer selection in the literature. The results showed that the method easily found suitable primers corresponding with the setting primer constraints and had preferable performance than the GA and the MA. Furthermore, the method was also compared with the common method Primer. It was concluded that the TLBO algorithm was an interesting primer selection method and a valuable tool for automatic high-throughput analysis. In another work, Chen (2014b) applied TLBO algorithm to screen the primers. The optimal primer frequency (OPF) based on three known melting temperature formulas was estimated by 500 runs for primer design in each different number of generations. The optimal primers were selected from fifty random nucleotide sequences of Homo sapiens. The results indicated that the SantaLucia's formula was better coupled with the method to get higher optimal primer frequency and shorter CPU-time than the Wallace's formula and the Bolton and McCarthy's formula. Through the regression analysis, it was also found that the generations were significantly associated with the optimal primer frequency. The results were helpful for developing the novel TLBO-based computational method to design feasible primers.

Dede (2014) considered several benchmark problems related to truss structures with discrete design variables and showed the efficiency of the TLBO algorithm, and the results were compared with those reported in the literature. It was concluded that the TLBO algorithm could be effectively used in the weight minimization of truss structures. Jordehi (2014) used TLBO algorithm to find the optimal setting of thyristor-controlled series compensators in electric power systems. The experiments were conducted for both *N*–1 and *N*–2 line outage contingencies. The results showed that TLBO performed well in solving this problem. Yang et al. (2014) proposed a new compact TLBO algorithm to combine the strength of the original TLBO and to reduce the memory requirement through a compact structure that utilized an adaptive statistic description to replace the process of a population of solutions. Numerical results on test benchmark functions showed that the new algorithm did not sacrifice the efficiency within the limited hardware resources.

Keesari and Rao (2014) used the TLBO algorithm to solve the job shop scheduling problems to minimize the makespan. The algorithm was tested on 58 job shop scheduling bench mark problems from OR Library, and results were compared with the results obtained using the other algorithms. It was concluded that the TLBO algorithm could be effectively used for job shop scheduling.

Hoseini et al. (2014) modified the teacher phase and the learner phase of the TLBO algorithm. The modified TLBO algorithm was applied to determine the optimal location of automatic voltage regulators (AVRs) in the distribution system. The problem of optimum location of AVRs in a distribution system consisted of multiple objectives such as minimization of energy generation cost, minimization of electrical energy losses, and minimization of voltage deviations. In order to prevent the premature convergence of the TLBO algorithm at the local optimum of the objective function, a cogent strategy to diversify the TLBO population, i.e., mutation was incorporated in the TLBO algorithm. At each step the algorithm mutates vectors by selecting three vectors l_1, l_2, l_3 from the initial population as $l_1, \neq l_2, \neq l_3 \neq i$ ($i = 1, 2,\ldots, N$), where N was the number of population. The best result, worst result, average, standard deviation, and CPU time values obtained using the modified TLBO algorithm were compared with those obtained using GA, PSO and basic TLBO algorithm for different objective functions.

Ghasemi et al. (2014a) applied a modified version of TLBO algorithm and Double Differential Evolution (DDE) algorithm for solving optimal reactive power dispatch (ORPD) problem. The algorithm was applied on IEEE 14-bus, IEEE 30-bus, and IEEE 118-bus power systems for performance assessment and validation purposes. The ORPD problem was used to optimize the active power loss in the transmission network while satisfying equality and inequality constraints at the same time. The results obtained using the algorithm were compared with those obtained using ACO algorithm, DE/best/2/bin algorithm and ABC algorithm. It was observed that the proposed algorithm obtained as lowest value of power loss for all the three test systems (i.e., 0.128978 MW, 0.048596 MW, and 1.139814 MW for IEEE-14 bus, IEEE-30 bus, and IEEE-118 bus test systems, respectively). In another work, Ghasemi et al. (2014b) proposed a novel hybrid algorithm of imperialist competitive algorithm and teaching-learning algorithm for optimal power flow problem with non-smooth cost functions.

Ganguly and Patel (2014) used TLBO algorithm for the global minimization of a loss cost function expressed as a function of three variables *n*, *h,* and *k* in an economic model of *X*-bar chart based on unified approach. A numerical example was solved and the results were found to be better than the earlier published results. Furthermore, the sensitivity analysis using fractional factorial design and analysis of variance were carried out to identify the critical process and cost parameters affecting the economic design.

González-Álvarez et al. (2014) proposed a multiobjective TLBO algorithm for solving the motif discovery problem (MDP). The performance of the MOTLBO algorithm was tested on 12 sequence benchmark dataset and the results were compared with those obtained using other algorithms. In the MDP multiobjective formulation, three conflicting objectives to be maximized were defined: motif length, support, and similarity. The algorithm was used to solve twelve sequence datasets selected from TRANSFAC database as benchmark, choosing instances with different properties (number and size of sequences) which belong to different organisms (drosophila melanogaster 'dm', homo sapiens 'hm,' mus musculus 'mus,' and saccharomyces cerevisiae 'yst') to demonstrate that the proposed algorithm works well in different scenarios. In the proposed algorithm, the teacher phase and the learner phase were exactly similar to those proposed by Rao et al. (2011). However, to be able to solve the multiobjective optimization problems, a dominance-based concept for selecting the teacher was incorporated into the TLBO algorithm. The procedure consists of randomly choosing solutions from the set of nondominated solutions and then the selected solution takes the role of teacher. Dominance concept was also incorporated at the end of the teacher phase, in the learner phase and at the end of learner phase in order to select a better solution when two solutions were compared with each other. However, in all these cases, if the analyzed solutions belong to the same Pareto front, then the dominance concept was not useful and the crowding distance concept was applied as the tie breaker, allowing the user to know which individual provides greater dispersion in the solution set. Besides these concepts, an archive to store the best solutions found by the algorithm was incorporated defining a maximum size of 100 individuals. The results obtained using the TLBO algorithm were compared with those obtained using Differential Evolution with Pareto tournaments (DEPT), multiobjective variable neighborhood search (MO-VNS), nondominated sorting Genetic Algorithm II (NSGA-II), and strength Pareto evolutionary Algorithm 2 (SPEA2).

Krishnasamy and Nanjundappan (2014) integrated the TLBO algorithm with sequential quadratic programming (SQP) to fine tune the better solutions obtained by the TLBO algorithms. To demonstrate the effectiveness of the hybrid TLBO-SQP method, a standard dynamic economic dispatch problem (DEDP) and one practical DEDP with wind power forecasted were tested based on the practical information of wind speed. The main objectives in the dynamic economic dispatch problem were to minimize the total production cost of the power system over a given dispatch period with achieving the required power balance. The dynamic economic dispatch problem was subjected to the constraints of power output and real-power output. The dynamic economic dispatch problem was solved using TLBO-SQP for a 10 unit test system. The results were compared with those obtained using the TLBO algorithm and self-adaptive differential harmony search (CSADHS) algorithm. The minimum generation cost obtained by the TLBO-SQP algorithm for 24 h time duration was \$1018679 as against the cost of \$1018681 and \$1031746 obtained using the CSADHS and TLBO method, respectively. The numerical results revealed that the TLBO-SQP was a suitable method to solve dynamic economic dispatch problem.

Lim and Isa (2014a) proposed to integrate the PSO algorithm with the TLBO algorithm in order to offer an alternative learning strategy in case a particle fails to improve its fitness in the search process. The algorithm was named as the teaching and peer-learning Particle Swarm Optimization (TPLPSO) algorithm. The results obtained using the TPLPSO algorithm were compared with those obtained by using real-coded chemical reaction optimization (RCCRO), group search optimization (GSO), real-coded biography Based Optimization (RCBBO), covariance matrix adaptation evolution strategy (CMAES), generalized generation gap model with generic parent centric recombination (G3PCX) operation, fast evolutionary programming (FEP), and fast evolutionary search (FES). The performance of TPLPSO on the spread spectrum radar polyphase code design problem was compared to the other variants of PSO such as adaptive PSO (APSO), comprehensive learning PSO (CLPSO), constricted PSO, feedback learning PSO with quadratic inertia weight (FLPSO-QIW), Frankenstein PSO (FPSO), fully informed PSO (FIPSO), PSO-time-varying acceleration coefficient (TVAC) with mutation (MPSO-TVAC), random position PSO (RPPSO), PSO with linearly decreasing inertia weights (PSO-LDIW) and unified PSO. The TPLPSO algorithm was proved to be the best optimizer in solving the radar polyphase code design problem, and the TPLPSO was the only algorithm that achieved the best value with the accuracy level of 10^{-1}, while the rest of its competitors achieved the solutions with the accuracy level of 100. The superior performance of the TPLPSO was further verified by the *t*-test. The *h*-values indicated that the searching accuracy of the TPLPSO algorithm was statistically better than the other ten PSO variants.

In order to establish an alternative learning strategy when a particle fails to improve its fitness in the search process, Lim and Isa (2014b) proposed an improved TLBO algorithm adapted to the enhanced framework of PSO known as bidirectional teaching and peer-learning Particle Swarm Optimization (BTPLPSO). The performance of BTPLPSO was tested on 25 benchmark functions and the results were compared with those obtained using other algorithms. Similar to TLBO, the BTPLPSO evolves the particles through two phases, that is, the teaching and peer-learning phases. The number of problems where BTPLPSO significantly outperformed the other variants of PSO were much larger than the number of problems where the former was significantly worse than the latter.

Mandal and Roy (2014) applied the quasi oppositional TLBO (QTLBO) algorithm to solve the multiobjective power flow problem with multiple constraints. The QTLBO algorithm was applied to IEEE 30 bus system, Indian utility 62 bus system, and IEEE 118 bus system. To solve four different single objectives, namely fuel cost minimization, system power loss minimization, voltage stability index minimization, and emission minimization along with three bi-objectives optimization namely minimization of fuel cost and transmission loss; minimization of fuel cost and L-index and minimization of fuel cost and emission and one tri-objective optimization namely fuel cost, minimization of transmission losses, and improvement of voltage stability simultaneously. It was observed that the algorithm had given the best compromising cost, transmission loss, and L-index simultaneously, compared to NSGA and multiobjective Harmony Search algorithm.

Moghadam and Seifi (2014) proposed a fuzzy-TLBO algorithm to solve the optimal reactive power control variable planning (ORPVCP) problem with an objective to minimize the energy loss. The performance of the fuzzy-TLBO algorithm was tested on an IEEE 30 bus test system by considering two different load duration curves. The energy loss obtained using fuzzy-TLBO algorithm was compared with those obtained using different method such as conventional linear programming (LP), fuzzy-LP, modified honey bee mating optimization algorithm (MHBMO) algorithm, and basic TLBO. The best results obtained by the fuzzy-TLBO algorithm were found to be comparable to other algorithms (with

32.5 % reduction in energy loss as compared to the initial state). It was observed that fuzzy-TLBO algorithm could achieve the minimum value of energy loss.

Patel et al. (2014) proposed a simple, efficient, and reliable method to extract all five parameters of a solar cell from a single-illuminated current–voltage (I–V) characteristic using TLBO algorithm. The TLBO algorithm was implemented by developing an interactive numerical simulation using LabVIEW as a programming tool. The effectiveness of the algorithm was validated by applying it to the reported I–V characteristics of different types of solar cells such as silicon, plastic, and dye-sensitized solar cells as well as silicon solar module. The obtained values of parameters by the TLBO algorithm were found to be in very good agreement with the reported values of parameters. The algorithm was also applied to the experimentally measured I–V characteristics of a silicon solar cell and a silicon solar module for the extraction of parameters. It was observed that the TLBO algorithm repeatedly converged to give consistent values of solar cell parameters. It was demonstrated that the program based on TLBO algorithm could be successfully applied to a wide variety of solar cells and modules for the extraction of parameters from a single-illuminated I–V curve with minimal control variables of the algorithm.

Pholdee et al. (2014) integrated the differential operators into the TLBO with a Latin hypercube sampling technique for generation of an initial population in order to improve the flatness of a strip during strip coiling process. The objectives considered were reduction in axial inhomogeneity of the stress distribution and reduction maximum compressive stress. The performance of the hybrid TLBO algorithm was compared with several well-established evolutionary algorithms and it was found that the proposed hybrid approach was powerful for process optimization, especially with a large-scale design problem. The maximum compressive stress and the standard deviation of the stress were decreased by 39.72 and 41.54 %, respectively.

Rao and Patel (2014) proposed a multiobjective improved TLBO algorithm for unconstrained and constrained multiobjective function optimization. The basic TLBO algorithm was improved to enhance its exploration and exploitation capacities by introducing the concept of number of teachers, adaptive teaching factor, tutorial training, and self-motivated learning. The algorithm used a grid-based approach to adaptively assess the nondominated solutions maintained in an external archive. The performance of the algorithm was assessed by implementing it on unconstrained and constrained test problems proposed for the Congress on Evolutionary Computation 2009 (CEC 2009) competition. The performance assessment was done using the inverted generational distance (IGD) measure. However, it may be mentioned here that the parameter of 'number of teachers' used in the modified TLBO algorithm becomes an algorithm-specific parameter and it should have been avoided. Rao and Waghmare (2014) solved the same problems without using any algorithm-specific parameters and produced better results.

Roy and Bhui (2014) applied QTLBO algorithm for solving the nonlinear multiobjective economic emission load dispatch problem of electric power generation with valve-point loading. The QOTLBO was carried out to obtain solution for

6-unit, 10-unit, and 40-unit systems. The objectives were related to the economic dispatch and emission dispatch. The problem was subjected to the constraints of power balance and the capacity. The results obtained using the QTLBO algorithm were compared with those obtained using basic TLBO, GSA, MODE, PDE, NSGA II, and SPEA 2, and it was observed that QTLBO outperformed the other algorithms in terms of both the objectives, i.e., fuel cost and emission in all the considered test systems.

Roy and Sarkar (2014) applied QTLBO to solve the unit commitment problem. The objective was to economically schedule generating units over a short-term planning horizon subjugating to the forecasted demand and other system operating constraints in order to meet the load demand and spinning reserve for each interval. The tests were carried out using a 10-unit system during a scheduling period of 24 h for four different cases. Additionally, the QOTLBO algorithm was also carried out for large-scale power systems viz. 20, 60, 80, and 100 units to prove the scalability of the algorithm. The unit commitment problem was subjected to constraints like power balance, spinning reserve requirement, minimum up time of a unit, minimum down time of a unit, and ramp rate constraints. The test results obtained were compared with those obtained using other algorithms such as evolutionary programming (EP), GA, gravity search algorithm (GSA), IPSO, IQEA-UC, PSO, SA, QEA-UC, and basic TLBO algorithms.

Roy et al. (2014) incorporated oppositional-based learning in the basic TLBO algorithm. The oppositional TLBO algorithm was applied to solve the combined heat and power dispatch problem with bounded feasible operating region. The objective was to find the optimal scheduling of power and heat generation with minimum fuel cost such that both heat and power demands and other constraints were met, while the combined heat and power units were operated in a bounded heat versus power plane. The problem was subjected to the constraints of power production and demand balance and heat production and demand balance. The main idea behind the opposition was the simultaneous consideration of an estimate and its corresponding opposite estimate in order to achieve a better approximation for the current candidate solution. The concept is based on the opposite point and opposite number. In order to verify the effectiveness and efficiency of the proposed TLBO and opposition-based TLBO algorithms for the considered problem, three test systems were presented. The test system 1 consisted of total 7 units out of which four were power only units, one was heat unit and two were cogeneration units. The system power demand and the heat demand were 600 MW and 150 MWth, respectively. The test system 2 consisted of 13 power only units, 6 CHP units, and 5 heat only units. The system power demand and heat demand were 2350 MW and 1250 MWth, respectively. The test system 3 was a large-scale power system with 48 units and it consisted of 26 power only units, 12 CHP units and 10 heat only units. Test system 3 was a large-scale system with more nonlinear elements compared to test systems 1 and 2 and it had more local minima solutions and, therefore, to find the global solution for this case was more difficult. The results obtained using the orthogonal TLBO algorithm for test system 3 were compared with those obtained using other algorithms such as classic PSO (CPSO), time-varing acceleration Coefficient-Particle Swarm Optimization (TVAC-PSO) algorithm, and TLBO. The total fuel cost obtained using CPSO, TVAC-PSO, TLBO, and OTLBO were $119708.8818, $117824.8956, $116739.3640, and 116579.2390, respectively.

Satapathy and Naik (2014) investigated the performance of a new variant of TLBO for global function optimization problems. The performance of the modified TLBO was compared with the state-of-the art forms of PSO, DE, and ABC algorithms. Several advanced variants of PSO, DE, and ABC were considered for the comparison purpose. The suite of benchmark functions were chosen from the competition and special session on real-parameter optimization under IEEE Congress on Evolutionary Computation (CEC) 2005. Statistical hypothesis testing was undertaken to demonstrate the significance of the modified TLBO over other investigated algorithms. Finally, the authors had investigated the data clustering performance of the modified TLBO algorithm over other evolutionary algorithms on a few standard synthetic and artificial datasets. Results revealed that the modified TLBO algorithm performed better than many other algorithms.

Sultana and Roy (2014) applied multiobjective QTLBO algorithm for optimal location of distributed generator in the radial distribution system. The performance of the multiobjective QTLBO algorithm was tested on 33-bus, 69-bus, and 118-bus radial distribution networks. The objectives of distributed generator (DG) placement in radial distribution system were to minimize power losses, to improve voltage profile, and to maximize voltage stability. The optimal location of distributed generator in the radial distribution system was subjected to several operating constraints such as power balance constraint, voltage limit, thermal limit, real-power limit, and reactive power limit. The power loss, voltage profile, and voltage stability index obtained by the proposed algorithm were found to be better.

Assembly line balancing plays a crucial role in modern manufacturing companies in terms of the growth in productivity and reduction in costs. The problem of assigning tasks to consecutive stations in such a way that one or more objectives are optimized subject to the required tasks, processing times, and some specific constraints is called the assembly line balancing problem (ALBP). Depending on production tactics and distinguishing working conditions in practice, assembly line systems show a large diversity. Although a growing number of researchers addressed ALBP over the past fifty years, real-world assembly systems which require practical extensions to be considered simultaneously have not been adequately handled. Tuncel and Aydin (2014) dealt with an industrial assembly system belonging to the class of two-sided line with additional assignment restrictions which are often encountered in practice. The TLBO algorithm was employed to solve the line balancing problem and the computational results were compared in terms of the line efficiency.

Understanding sediment movement in coastal areas is crucial in planning the stability of coastal structures, the recovery of coastal areas, and the formation of new coast. Accretion or erosion profiles formas a result of sediment movement. The characteristics of these profiles depend on the bed slope, wave conditions, and sediment properties. Uzlu et al. (2014a) performed experimental studies in a wave flume with regular waves, considering different values for the wave height (*H0*), wave period (*T*), bed slope (*m*), and mean sediment diameter (*d50*). Accretion profiles developed in these experiments and the geometric parameters of the resulting berms were determined. TLBO and ABC algorithms were applied to regression functions of the data from the physical model. Dimensional and dimensionless equations were found for each parameter. These equations were compared to data from the physical model, to determine the best equation for each parameter and to evaluate the performances of the TLBO and ABC algorithms in the estimation of the berm parameters. Compared to the ABC algorithm, the TLBO algorithm provided better accuracy in estimating the berm parameters. In another work, Uzlu et al. (2014b) applied the ANN (artificial neural network) model with the TLBO algorithm to estimate energy consumption in Turkey. Gross domestic product, population, import, and export data were selected as independent variables in the model. Performances of the ANN–TLBO model and the classical back propagation-trained ANN model (ANN–BP (TLBO) model) were compared using various error criteria to evaluate the model accuracy. Errors of the training and testing datasets showed that the ANN–TLBO model better predicted the energy consumption compared to the ANN–BP model. After determining the best configuration for the ANN–TLBO model, the energy consumption values for Turkey were predicted under three scenarios. The forecasted results were compared between scenarios and with projections by the MENR (Ministry of Energy and Natural Resources). Compared to the MENR projections, all of the analyzed scenarios gave lower estimates of energy consumption and predicted that Turkey's energy consumption would vary between 142.7 and 158.0 Mtoe (million tons of oil equivalent) in 2020.

Wang et al. (2014a) proposed a hybrid TLBO-DE algorithm for chaotic time series prediction. To demonstrate the effectiveness of TLBO-DE approach, it was applied to three typical chaotic nonlinear time series prediction problems. The results obtained using TLBO-DE were compared to those obtained using PSO and basic TLBO based on MSE and it was observed that TLBO-DE performed better than the PSO. In another work, Wang et al. (2014b) introduced a ring neighborhood topology into the original TLBO algorithm to maintain the exploration ability of the population. Different than the traditional method to utilize the global information, the mutation of each learner was restricted within a certain neighboring area so as to fully utilize the whole space and to avoid over-congestion around local optima. Moreover, a mutation operation was presented in order to maintain the diversity of population. To verify the performance of the proposed algorithm, 32 benchmark functions were utilized. Finally, 3 application problems of artificial neural network were examined. The results of 32 benchmark functions and 3 applications of ANN indicated the interesting outcomes of the proposed algorithm.

Due to the interaction among FACTS devices, coordination control of multi-FACTS devices is a hot topic. A multiobjective optimization problem was formulated by Xiao et al. (2014) and a modified TLBO algorithm was presented to coordinate the thyristor-controlled series capacitor (TCSC), static var compensator (SVC), and power angle difference damping characteristics of generators. The optimal parameters of controller were found out to improve the coordination control. A new learner phase was introduced in order to avoid entrapment into local optima. The algorithm was validated and illustrated on IEEE 4-machine 11-bus system.

Disassembly sequence planning (DSP) is a challenging NP-hard combinatorial optimization problem. Xia et al. (2014) presented a simplified teaching-learning-based optimization (STLBO) algorithm for solving DSP problems effectively. The STLBO algorithm inherited the main idea of the teaching-learning-based evolutionary mechanism from the TLBO algorithm, while the realization method for the evolutionary mechanism and the adaptation methods for the algorithm parameters were different. Three new operators were developed and incorporated in the STLBO algorithm to ensure its applicability to DSP problems with complex disassembly precedence constraints: i.e., a feasible solution generator (FSG) used to generate a feasible disassembly sequence, a teaching-phase operator (TPO), and a learning-phase operator (LPO) used to learn and evolve the solutions toward better ones by applying the method of precedence preservation crossover operation. Numerical experiments with case studies on waste product disassembly planning were carried out to demonstrate the effectiveness of the designed operators, and the results exhibited that the developed algorithm performed better than other relevant algorithms under a set of public benchmarks.

Permutation flow shop scheduling (PFSP) is among the most studied scheduling settings. Xie et al. (2014) proposed a hybrid Teaching-Learning-Based Optimization algorithm (HTLBO) which combines the TLBO algorithm for solution evolution and a variable neighborhood search (VNS) for fast solution improvement for PFSP to determine the job sequence with minimization of makespan criterion and minimization of maximum lateness criterion, respectively. To convert the individual to the job permutation, a largest order value (LOV) rule was utilized. Furthermore, an SA algorithm was adopted as the local search method of VNS after the shaking procedure. Experimental comparisons over public PFSP test instances with other competitive algorithms showed the effectiveness of the TLBO algorithm.

Yu et al. (2014) proposed a modified version of TLBO algorithm in which a feedback phase, mutation crossover operation of differential evolution (DE) algorithms, and chaotic perturbation mechanism were incorporated to significantly improve the performance of the algorithm. The feedback phase was used to enhance the learning style of the students and to promote the exploration capacity of the TLBO. The mutation crossover operation of DE was introduced to increase population diversity and to prevent premature convergence. The chaotic perturbation mechanism was used to ensure that the algorithm could escape the local optimal. Simulation results based on ten unconstrained benchmark problems and five constrained engineering design problems showed that the proposed version of TLBO algorithm was better than, or at least comparable to, other state-of-the-art algorithms. Yu et al. (2014) commented that the claim made by Črepinšek et al. (2012) in another study that Waghmare (2013) used different success rates was unsuitable. Furthermore, the comparisons of evolutionary algorithms conducted by Veček et al. (2014) (with Črepinšek as a co-author) attempted to cast the TLBO algorithm in a poor light, although this attempt may also be seen as not meaningful as the findings of Veček et al (2014) were simply comparisons of the basic TLBO algorithm with different modified versions of DE and did not consider comparisons with other important algorithms.

Zou et al. (2014) proposed a modified TLBO algorithm with dynamic group strategy (DGSTLBO). The DGSTLBO enables each learner to learn from the mean of his corresponding group, rather than the mean of the class, in the teacher phase. Furthermore, each learner employs the random learning strategy or the quantum-behaved learning strategy in his corresponding group in the learner phase. Regrouping occurs dynamically after a certain number of generations, helping to maintain the diversity of the population, and discourage premature convergence. The effectiveness and feasibility of the DGSTLBO experiments were conducted on 18 benchmark functions with 50 dimensions and the mean and standard deviation of solutions obtained for 50 independent runs with the termination criteria set to 100,000 function evaluations. The results of DGSTLBO algorithm were compared with those obtained using self-adaptive Differential Evolution (SaDE), local version of Particle Swarm Optimization with constriction factor (PSO-cf-Local), fitness - distance ratio-based particle swarm optimization (FDR-PSO), and TLBO. The results showed that DGSTLBO algorithm performed better for functions F_1, F_2, F_3, F_4, F_6, F_7, F_8, F_9, F_{11}, F_{12}, F_{14}, F_{15}, F_{16}, and F_{17}. Thus, it was concluded that the DGSTLBO algorithm was a suitable method for global optimization problems.

15.1.5 Publications in the Year 2015

Agrawal et al. (2015) proposed TLBO algorithm-based iris recognition system in which feature extraction phase of iris recognition system was optimized using TLBO algorithm. The process of feature extraction was performed by texture feature extraction Gabor wavelet transform technique. The TLBO algorithm was then applied on these features. The TLBO algorithm acquired the feature of iris image as a student and generated the optimized feature template as a teacher. The proposed algorithm was compared with the other iris recognition methods such as standard iris recognition system and Genetic Algorithm optimized iris recognition system. Experimental results when applied to CASIA dataset showed superior performance of the TLBO algorithm with better recognition rate.

Barisal (2015) presented TLBO algorithm and its application to automatic load frequency control of multisource power system having thermal, hydro, and gas power plants. In this extensive study, the algorithm was applied in multiarea and multisource realistic power system without and with DC link between two areas in order to tune the PID controller which was used for automatic generation control (AGC). The potential and effectiveness of the proposed algorithm was compared with that of DE and optimal output feedback controller tuning performance for the same power systems. The dynamic performance of the controller was investigated by different cost functions like integral of absolute error (IAE), integral of squared error (ISE), integral of time-weighted squared error (ITSE), integral of time multiplied absolute error (ITAE), and the robustness of the optimized controller was verified by its response toward changing in load and system parameters. It was found that the dynamic performance of the controller was better and also the proposed system was more robust and stable to wide changes in system loading, parameters, size and locations of step load perturbation and different cost functions.

Boudjefdjouf et al. (2015) combined the time-domain reflectometry response extracted from vector network analyzer measurements and the TLBO technique and applied to the diagnosis of wiring networks. The approach consisted of two steps. In the first step, propagation along the cables was modeled using the forward model. In the second step, TLBO was used to solve the inverse problem to deduce physical information about the defects in the wiring network. The approach was successfully tested on several cases and for different configurations. Comparisons of the time domain reflectometry/TLBO approach results with the measurements revealed the high potential of the approach for wiring network diagnosis.

In order to decrease the computation cost and improve the global performance of the original TLBO algorithm, the area copying operator of the producer–scrounger (PS) model was introduced into TLBO for global optimization problems by Chen et al. (2015a). The algorithm was tested on different kinds of benchmark problems and the results indicated that the proposed algorithm has competitive performance to some other algorithms in terms of accuracy, convergence speed, and success rate. In another work, Chen et al. (2015b) proposed a variant of TLBO algorithm with multiclasses cooperation and simulated annealing operator (SAMCCTLBO). To take full advantage of microteaching, the population was divided into several subclasses, the mean of all learners in teacher phase of original TLBO was replaced by the mean solutions of different subclasses, the modification might make the mean solutions improved quickly. The effectiveness of the algorithm was tested on several benchmark functions and the results demonstrated that SAMCCTLBO has some good performances when compared with some other evolutionary algorithms.

Cheng et al. (2015) proposed TLBOMPD (TLBO-based mutagenic primer design) in order to improve the mutagenic primer design to search for more feasible mutagenic primers and provide the latest available restriction enzymes. The method maintained the original Wallace's formula for the calculation of melting temperature, more accurate calculation formulas of GC-based melting temperature, and thermodynamic melting temperature were especially introduced. Mutagenic matrix was also reserved to increase the efficiency of judging whether a hypothetical mutagenic primer can bring available restriction enzymes for recognizing the target SNP. Furthermore, the core of SNP-RFLPing version 2 was used to enhance the mining work for restriction enzymes based on the latest REBASE. Twenty-five SNPs with mismatch PCR-RFLP screened from 288 SNPs in human SLC6A4 gene were used to appraise the TLBOMPD. Also, the computational results were compared with those of the GAMPD.

Cho and Kim (2015) introduced a new design method using a hybrid optimization algorithm for the electromagnet used in maglev transportation vehicles. Maglev system typically uses electromagnetic suspension, which is more advantageous than electro dynamic suspension. However, the structural constraints must be considered in the optimal design of an electromagnet for an electromagnetically suspended vehicle. A hybrid optimization algorithm based on the TLBO algorithm and clonal selection was used to design an electromagnet satisfying the structural constraints. The proposed method was verified by MATLAB simulations which showed that the method was more efficient than the conventional methods.

Das and Padhy (2015) forecasted a financial derivatives instrument (the commodity futures contract index) using techniques based on recently developed machine learning techniques. The authors had developed a hybrid method that combined a support vector machine (SVM) with TLBO algorithm. The viability and efficiency of this hybrid model was assessed by forecasting the daily closing prices of the COMDEX commodity futures index, traded in the Multi Commodity Exchange of India Limited. The experimental results showed that the proposed model was effective and performed better than the PSO + SVM hybrid and standard SVM models.

Dede and Ayvaz (2015) used TLBO algorithm for the size and shape optimization of structures. The cross-sectional areas of the bar element and the nodal coordinates of the structural system were the design variables for size and shape optimization, respectively. Displacement, allowable stress and the Euler buckling stress were taken as the constraints for the problem considered. Some truss structures were designed using the TLBO algorithm to show the efficiency of the TLBO algorithm. The results obtained were compared with those reported in the literature. It was concluded that the TLBO algorithm can be effectively used in combined size and shape optimization of the structures.

Dokeroglu (2015) proposed a set of TLBO-based hybrid algorithms to solve the challenging combinatorial optimization problem of quadratic assignment. Individuals were trained with recombination operators and later a robust Tabu search engine processed them. The performances of sequential and parallel TLBO-based hybrid algorithms were compared with those of state-of the-art metaheuristics in terms of the best solution and computational effort. It was shown experimentally that the performance of the proposed algorithms was competitive with the best reported algorithms for the solution of the quadratic assignment problem with which many real-life problems can be modeled. Durai et al. (2015) provided a framework using TLBO algorithm for the computation of coefficients for quadratic and cubic cost functions, valve-point loading, piece-wise quadratic cost, and emission functions. The effectiveness of TLBO was demonstrated on 5 standard test systems and a practical Indian utility system, involving varying degree of complexity. The TLBO algorithm yielded better results than the benchmark least error square (LES) method and other evolutionary algorithms. The economic deviation was also tested with existing systems.

Ghasemi et al. (2015) investigated the possibility of using TLBO algorithm as a solution for the optimal power flow (OPF) problems using Lévy mutation strategy for optimal settings of OPF problem control variables. The performance was studied and evaluated on the standard IEEE 30-bus and IEEE 57-bus test systems with different objective functions and was compared with the other methods reported in the literature. At the end, the results which were extracted from implemented simulations confirmed Lévy mutation TLBO (LTLBO) as an effective solution for the OPF problem.

Govardhan and Roy (2015) presented the individual and collective impact of three distributed energy resources (DERs), namely, wind power generator as a renewable energy source, plug-in electric vehicles (PEVs) and emergency demand response program (EDRP) on unit commitment. An inconsistent nature of wind speed and wind power was characterized by the Weibull probability distribution function considering overestimation and underestimation cost model of the stochastic wind power. The extensive economic analysis of unit commitment with DERs was carried out to attain the least total cost of the entire system. The TLBO algorithm was employed to solve the unit commitment problem considering IEEE standard 10 unit test system, and it was found that the combined effect of wind power generator, plug-in electric vehicles, and emergency demand response program on unit commitment significantly lessened the total cost of the system.

Huang et al. (2015) proposed a new hybrid algorithm named teaching-learning-based cuckoo search (TLCS) for the constrained optimization problems. The TLCS modifieed the cuckoo search (CS) based on the TLBO and then was applied for constrained engineering design problems. Experimental results on several well-known constrained engineering design problems demonstrated the effectiveness, efficiency, and robustness of the TLCS. Moreover, the TLCS obtained some solutions better than those previously reported in the literature.

Hosseinpour and Bastaee (2015) presented a multiobjective optimal location of on-load tap changers (OLTCs) in distribution systems at spirit of distributed generators (DGs) based on TLBO coupled with SA algorithm (SA-TLBO). In the suggested algorithm, teacher and learner phases were modified by SA. The algorithm utilized several teachers and considered the teachers as external cache to store the found Pareto optimal solutions during the search process. The approach allowed the decision maker to take one of the Pareto optimal solutions (by trade-off) for different applications. The performance of the SA-TLBO algorithm on a 70-bus distribution network in comparison with other methods such as GA, PSO, SA, and TLBO was remarkable.

The complex nature of the petrochemical industries necessitates an efficient decision on a large number of factors so as to optimally operate a plant. Production planning is an integral part of the petrochemical industry and requires the optimal selection of processes, production levels, and products to maximize its profit. Previously an MILP formulation was proposed for guiding the petrochemical industry development in Saudi Arabia. Kadambur and Kotecha (2015) stated the limitations of this formulation and proposed an alternate elitist TLBO algorithm-based strategy to overcome them. The benefits of this strategy included the determination of better production plans that lead to higher profits and were demonstrated on eight case studies in the literature. The strategy is generic and can be applied to determine production plans of multiple levels in various industries.

Kumar et al. (2015) focused on machining aspects of CFRP (epoxy) composites using single-point cutting tool. The optimal setting, i.e., the most favorable combination of process parameters (spindle speed, feed rate, depth of cut, and fiber orientation angle) were derived in view of multiple and conflicting requirements of machining performance yields viz. material removal rate, surface roughness of the turned product, and the cutting force. The study initially derived mathematical models (objective functions) using statistics of nonlinear regression for correlating various process parameters with respect to the output responses. In the next phase, the study utilized TLBO algorithm in order to determine the optimal machining conditions for achieving satisfactory machining performances. Application potential of TLBO algorithm was compared to that of GA. It was observed that the exploration of TLBO algorithm appeared more fruitful in contrast to GA in the context of this experimental research focused on machining of CFRP composites.

Kurada et al. (2015) persuaded a novel automatic clustering algorithm (AutoTLBO) with a credible prospect by coalescing automatic assignment of k value in partitioned clustering algorithms and cluster validations into TLBO. The objectives of the algorithm were thoroughly tested over microarray datasets. The investigation results that endorse AutoTLBO were impeccable in obtaining optimal number of clusters, co-expressed cluster profiles, and gene patterns. The work was further extended by inputting the AutoTLBO algorithm outcomes into benchmarked bioinformatics tools to attain optimal gene functional enrichment scores. The concessions from these tools indicated excellent implications and significant results, justifying that the outcomes of AutoTLBO were incredible. Thus, both these rendezvous investigations gave an impression that AutoTLBO arises as an impending colonizer in this hybrid approach.

Li et al. (2015) proposed a discrete TLBO (DTLBO) for solving the flowshop rescheduling problem. Five types of disruption events, namely machine breakdown, new job arrival, cancelation of jobs, job processing variation, and job release variation were considered simultaneously. The proposed algorithm aimed to minimize two objectives, i.e., the maximal completion time and the instability performance. Four discretisation operators were developed for the teaching phase and learning phase to enable the TLBO algorithm to solve rescheduling problems. In addition, a modified iterated greedy-based local search was embedded to enhance the searching ability of the proposed algorithm. Furthermore, four types of DTLBO algorithms were developed to make detailed comparisons with different parameters. Experimental comparisons on 90 realistic flowshop rescheduling instances with other efficient algorithms indicated that the proposed algorithm was competitive in terms of its searching quality, robustness, and efficiency.

Omidvar et al. (2015) focused on selecting optimal factors combination that cause maximum bending angle in laser bending of AA6061-T6. For this purpose, an L25 Taguchi orthogonal design (four factors-five levels) was used to design the experiments. The process main factors were laser power, spot diameter, pulse duration, and scanning speed and the main response was the bending angle. To correlate relationship between process factors and the bending angle, a radial basis function neural network (RBFNN) was utilized. Then the developed RBFNN model was used as an objective function for maximizing deformation through TLBO algorithm. Results indicated that the laser power of 3.6 kW, spot diameter of 2 mm, pulse duration of 0.9 ms, and scanning speed of 2 mm/s led to maximal bending angle about 28.7°. The optimal results were verified by confirmatory experiments.

Sahu et al. (2015) dealt with the design of a novel fuzzy proportional–integral–derivative (PID) controller for automatic generation control (AGC) of a two unequal area interconnected thermal system. The TLBO algorithm was applied to obtain the parameters of the fuzzy-PID controller. The design problem was formulated as an optimization problem and TLBO was employed to optimize the parameters of the fuzzy-PID controller. The superiority of the approach was demonstrated by comparing the results with some of the recently published approaches such as Lozi map-based chaotic optimization algorithm (LCOA), GA, pattern search (PS) and SA-based PID controller for the same system under study employing the same objective function. It was observed that the TLBO algorithm optimized fuzzy-PID controller provided better dynamic performance in terms of settling time, overshoot and undershoot in frequency and tie-line power deviation as compared to LCOA, GA, PS, and SA-based PID controllers. Further, robustness of the system was studied by varying all the system parameters from −50 to +50 % in step of 25 %. Analysis also revealed that the TLBO optimized fuzzy-PID controller gains were quite robust and need not be reset for wide variation in system parameters.

Tiwary et al. (2015) described a technique for optimizing inspection and repair-based availability of distribution systems. Optimum duration between two inspections was obtained for each feeder section with respect to cost function and subject to satisfaction of availability at each load point. The TLBO algorithm was used for availability optimization. The developed algorithm was implemented on radial and meshed distribution systems. The results obtained were compared with those obtained with differential evolution.

Xu et al. (2015) proposed an effective TLBO algorithm to solve the flexible job shop scheduling problem. The TLBO algorithm was incorporated with a special encoding scheme to represent solutions and a decoding method was employed to transfer a solution to a feasible schedule in the fuzzy sense. A bi-phase crossover scheme based on the teaching-learning mechanism and special local search operators was also incorporated into the search framework of the TLBO to balance the exploration and exploitation capabilities. The objective of the problem was to determine both the assignment of machines for all the operations and the sequence of the operations on all the machines to minimize the maximum completion time. The results obtained using TLBO algorithm were compared to those obtained using EDA, ABC, CGA, DIGA, PEGA, and PSO + SA algorithms. For each instance, the TLBO was run 20 times independently. The maximum generations were set as 300 which were much smaller than the other algorithms. The comparative results demonstrated the effectiveness and efficiency of the proposed TLBO algorithm in solving the flexible job shop scheduling.

An integrated approach for real-time model-based state of charge (SOC) estimation of Lithium-ion batteries was proposed by Zhang et al. (2015). Firstly, an autoregression model was adopted to reproduce the battery terminal behavior, combined with a nonlinear complementary model to capture the hysteresis effect. The model parameters, including linear parameters and nonlinear parameters, were optimized off-line using a hybrid optimization method that combined the TLBO method and the least square method. Second, using the trained model, two real-time model-based SOC estimation methods were presented, one based on the real-time battery open circuit voltage (OCV) regression model achieved through weighted recursive least square method, and the other based on the state estimation using the extended Kalman filter method (EKF). To tackle the problem caused by the flat OCV-versus-SOC segments when the OCV-based SOC estimation method was adopted, a method combining the coulombic counting and the OCV-based method was proposed. Finally, modeling results and SOC estimation results were presented and analyzed using the data collected from LiFePo4 battery cell.

Rao and More (2015) carried out single objective as well as multiobjective design optimization of heat pipe using TLBO algorithm. Two examples of heat pipe were presented. The results of application of TLBO algorithm for the design optimization of heat pipe were compared with the NPGA (Niched Pareto Genetic Algorithm), GEM (Grenade Explosion Method) and GEO (Generalized External optimization). It was found that the TLBO algorithm had produced better results as compared to those obtained using NPGA, GEM, and GEO algorithms. The convergence behavior of the TLBO algorithm to a near global solution was observed to be more effective than GEM, NPGA, and GEO results.

Črepinšek et al. (2015) tried to justify their previous work (Črepinšek et al. 2012) which was commented upon by Waghmare (2013). However, the justification made by them is not convincing and many of the statements made by them on the work of Waghmare (2013) are unhealthy in spirit. There was no such inexact replication of computational experiments by Waghmare (2013) and the criticism made by Črepinšek et al. (2015) on the capability of the reviewers is uncalled for. The authors had incorrectly opined that there is insufficient evidence that TLBO is better than the other nature-inspired algorithms on large-scale problems. However, Waghmare (2013) had already proved the better performance of the algorithm for optimization problems with 100 dimensions size. It was not only Waghmare (2013) but researchers like Baykasoğlu et al. (2014), Rao and Patel (2012, 2013a), Satapathy et al. (2013), Satapathy and Naik (2014), Zou et al. (2014) and many others had already proved that the TLBO algorithm is suitable for high dimensional (i.e., large scale) problems. Waghmare (2013) showed that TLBO algorithm can perform well on low dimensional problems also by properly tuning the common control parameters of the algorithm. Thus, the opinion of Črepinšek et al. (2015) on this aspect is incorrect and they could not prove their point. Rao and Patel (2012) explained the concepts of TLBO algorithm including the concept of elitism. The paper of Rao and Patel (2012) was published in March 2012 whereas the paper of Črepinšek et al. (2012) was published in May 2012. It was not the fault of Rao and Patel (2012) if Črepinšek et al. (2012) had not read their paper which is freely available since 15th March 2012 on the journal's website before getting their own paper published on 28th May 2012.

A statement made by Črepinšek et al. (2015) that "Waghmare is trying to justify that there is no need to explain all the steps of an algorithm using some other examples… disagree with Waghmare's explanation that this is acceptable behaviour simply because it has happened before" is incorrect as Waghmare (2013) had not made such comments in his paper. Also, the concept of calculation of function evaluations required for duplicate removal is quite unusual and one will not come across such concept in the widely accepted algorithms like GA, PSO, ABC, DE, ACO, etc. Hence, it is not a big point to discuss about in actual practice and probably no more redundant discussion is needed in this regard. Furthermore, the authors had mentioned about few "true" parameter-less algorithms and claimed that TLBO is not a parameter-less algorithm. But it should be realized that the so-called algorithms mentioned by Črepinšek et al. (2015) also require values of parameters (for example, parameter-less GA of Harik and Lobo (1999) requires fixing the crossover probability of 0.5 and selection operator of 4 and runs multiple populations in a cascade-like manner; the PLES proposed by Papa (2008) and Papa et al. (2012) requires the parameters to set virtually, according to the complexity of the problem and according to the statistical properties of the solutions found). Hence, it does not mean that the "true" parameter-less algorithms do not require the parameters. The TLBO algorithm has never claimed that it is a parameter-less algorithm. What it has claimed is that it requires only the common parameters (like population size and number of generations) to tune and it does not require algorithm-specific parameters unlike the other algorithms such as GA, PSO, ABC, DE, ACO, SFL, etc. The common control parameters are required by all the algorithms and no algorithm is exceptional in this issue (and how the parameters are tuned is another issue). Also, another statement made by Črepinšek et al. (2015) that "Instead of setting an algorithm-specific control parameter—the probability of mutation/crossover, the same effect is obtained in Waghmare (2013) by properly setting 'common' parameters—population size and number of generations… instead of setting one 'algorithm-specific' control parameter, it is necessary to set two 'common' control parameters" is totally incorrect. It is to be understood by Črepinšek et al. (2015) that the GA algorithm requires algorithm-specific parameter (s) in addition to the common parameters and TLBO requires only the common parameters. It may be mentioned here that Rao (2016) had developed another algorithm-specific parameter-less algorithm named as "Jaya Algorithm".

Rao and Waghmare (2015a) presented the performance of (TLBO) algorithm to obtain the optimum geometrical dimensions of a robot gripper. Five objectives, namely the difference between the maximum and minimum gripping forces, the force transmission ratio, the shift transmission ratio, length of all the elements of the gripper, and the effort of the gripper mechanism were considered as objective functions. The problem had seven design variables, including six variables as dimensions of the gripper and one variable as the angle between two links of the gripper. Three robot grippers were optimized and the computational results showed that the TLBO algorithm is better or competitive to other optimization algorithms reported in the literature for the considered problem.

Rao and Waghmare (2015b) presented the performance of the TLBO algorithm to obtain the optimum set of design and operating parameters for a smooth flat plate solar air heater (SFPSAH). Maximization of thermal efficiency was considered as an objective function for the thermal performance of SFPSAH. The number of glass plates, irradiance, and the Reynolds number were considered as the design parameters and wind velocity, tilt angle, ambient temperature, and emissivity of the plate were considered as the operating parameters to obtain the thermal performance of the SFPSAH using the TLBO algorithm. The computational results had shown that the TLBO algorithm was better or competitive to other optimization algorithms reported in the literature for the considered problem.

Rao and Waghmare (2015c) investigated the performance of the TLBO algorithm for multiobjective design optimization of a plate-fin heat sink equipped with flow-through and impingement-flow air cooling system. The Pareto front of multiobjective design of plate-fin heat sink was effectively located and had a good distribution of points along the curve. The results obtained using the TLBO algorithm were compared with the other optimization methods available in the literature for the considered optimization problems. The optimal results of plate-fin heat sink showed that the TLBO algorithm is better or competitive to the other optimization algorithms reported in the literature for the considered problem. The dynamic heat dissipation performance of the plate-fin heat sink was also investigated using the finite element software-ANSYS-12.1. From the dynamic heat dissipation analysis, it was concluded that the plate-fin heat sink with flow-through air cooling system, though has a larger size, is better than the plate-fin heat sink with impingement-flow cooling system.

In addition to the above works, so many research papers have been published in various international journals and conference proceedings using the TLBO algorithm or its modifications. Many researchers are using the TLBO algorithm or its modifications to solve the problems related to different fields of engineering and science. The number of research papers is continuously increasing at a faster rate.

References

Abirami, M., Ganesan, S., Subramanian, S., Anandhakumar R., 2014. Source and transmission line maintenance outage scheduling in a power system using teaching learning based optimization algorithm. Applied Soft Computing 21, 72–83.

Agrawal, S., Sharma, S., Silakari, S., 2015. Teaching learning based optimization (TLBO) based improved iris recognition system. Advances in Intelligent Systems and Computing 330, 735–740.

Arya, L.D., Koshti, A., 2014, Anticipatory load shedding for line overload alleviation using teaching learning based optimization (TLBO). International Journal of Electrical Power & Energy Systems 63, 862–877.

Barisal, A.K., 2015. Comparative performance analysis of teaching learning based optimization for automatic load frequency control of multi-source power systems. Electrical Power and Energy Systems 66, 67–77.

Bayram, A., Uzlu, E., Kankal, M., Dede, T., 2014. Modeling stream dissolved oxygen concentration using teaching–learning based optimization algorithm. Environmental Earth Sciences. doi:10.1007/s12665-014-3876-3.

Baykasoğlu, A., Hamzadayi, A., Köse, S.Y., 2014. Testing the performance of teaching-learning based optimization (TLBO) algorithm on combinatorial problems: Flowshop and job shop scheduling cases. Information Sciences 276(20), 204–218.

Bouchekara, H.R.E.H., Abido, M.A., Boucherma, M., 2014. Optimal power flow using teaching-learning-based optimization technique. Electric Power Systems Research 114, 49–59.

Boudjefdjouf, H., Mehasni, R., Orlandi, A., Bouchekara, H.R.E.H., de Paulis, F., Smail, M.K., 2015. Diagnosis of multiple wiring faults using time-domain reflectometry and teaching–learning-based optimization. Electromagnetics 35, 10–24.

Camp, C.V., Farshchin, M., 2014. Design of space trusses using modified teaching-learning-based optimization. Engineering Structures 62–63, 87-97.

Chen, X., Luo, Z., He, X., Zhu, L., 2014. Thinning and weighting of planar arrays by modified teaching learning based optimization algorithm. Journal of Electromagnetic Waves and Applications 28(15), 1924–1934.

Chen, D., Zou, F., Wang, J., Yuan, W., 2015a. A teaching–learning-based optimization algorithm with producer–scrounger model for global optimization. Soft Computing 19, 745–762.

Chen, D., Zou, F., Wang, J., Yuan, W., 2015b. SAMCCTLBO: a multi-class cooperative teaching–learning-based optimization algorithm with simulated annealing. Soft Computing. doi 10.1007/s00500-015-1613-9.

Cheng, Y-H., 2014. Computational intelligence-based polymerase chain reaction primer selection based on a novel teaching-learning-based optimization. IET Nanobiotechnology 8(4), 238–246.

Cheng, Y-H., et al., 2015. A novel teaching-learning-based optimization for improved mutagenic primer design in mismatch PCR-RFLP SNP genotyping. IEEE/ACM Transactions on Computational Biology and Bioinformatics.

Cho, J.H., Kim, Y.T., 2015. Optimal design of electromagnet for Maglev vehicles using hybrid optimization algorithm. Soft Computing 19, 901–907.

Črepinšek, M., Liu, S-H., Mernik, L., 2012. A note on teaching learning based optimization algorithm. Information Sciences 212, 79–93.

Črepinšek, M., Liu, S-H., Mernik, L., Mernik, M., 2015. Is a comparison of results meaningful from the inexact replications of computational experiments? Soft Computing. doi:10.1007/s00500-014-1493-4.

Das, S. P., Padhy, S., 2015. A novel hybrid model using teaching–learning-based optimization and a support vector machine for commodity futures index forecasting. International Journal of Machine Learning and Cybernetics, 10.1007/s13042-015-0359-0.fs.

Dede, T., 2013. Optimum design of grillage structures to LRFD-AISC with teaching-learning based optimization. Structure and Multidisciplinary Optimization 48(5), 955–964.

Dede, T., 2014. Application of teaching-learning-based-optimization algorithm for the discrete optimization of truss structures. KSCE Journal of Civil Engineering 18(6), 1759–1767.

Dede, T., Ayvaz, Y., 2015.Combined size and shape optimization of structures with a new metaheuristic algorithm. Applied Soft Computing 28, 250–258.

Degertekin, S.O., Hayalioglu, M.S., 2013. Sizing truss structures using teaching learning based optimization. Computers & Structures, 119(1), 177–188.

Dokeroglu, T., 2015. Hybrid teaching-learning-based optimization algorithm for the quadratic assignment problem. Computers & Industrial Engineering 85, 86–101.

Durai, S., Subramanian, S., Ganesan, S., 2015. Improved parameters for economic dispatch problems by teaching-learning-based optimization. International Journal of Electrical Power & Energy Systems 67, 11–24.

Ganguly, A., Patel, S.K., 2014. A teaching-learning-based optimization approach for economic design of X-bar control chart. Applied Soft Computing 24, 643–653.

Ghasemi, M., Ghanbarian, M.M., Ghavidel, S., Rahmani, S., Moghaddam, E.M., 2014a. Modified teaching learning algorithm and double differential evolution algorithm for optimal reactive power dispatch problem: A comparative study. Information Sciences 278, 231–249.

Ghasemi, M., Ghavidel, S., Gitizadeh, M., Akbari, E., 2015. An improved teaching–learning-based optimization algorithm using levy mutation strategy for non-smooth optimal power flow. Electrical Power and Energy Systems 65, 375–384.

Ghasemi, M., Ghavidel, S., Rahmani, S., Roosta, A., Falah, H., 2014b. A novel hybrid algorithm of imperialist competitive algorithm and teaching learning algorithm for optimal power flow problem with non-smooth cost functions. Engineering Applications of Artificial Intelligence 29, 54–69.

García, J.A.M., Mena, A.J.G., 2013. Optimal distributed generation location and size using a modified teaching-learning-based optimization algorithm. International Journal of Electrical Power & Energy Systems 50, 65–75.

González-Álvarez, D.L., Vega-Rodríguez, M.A., Gómez-Pulido, J.A., Sánchez-Pérez, J.M., 2014. Predicting DNA motifs by using evolutionary multiobjective optimization. IEEE Transactions on Systems, Man and Cybernetics Part C: Applications and Reviews, 1–13.

Govardhan, M., Roy, R., 2015. Ecomomic analysis of unit commitment with distributed energy resources. International Journal of Electrical Power & Energy Systems 71, 1–14.

Harik, G.R., Lobo, F., 1999. A parameter-less genetic algorithm, in: Technical report, University of Illinois at Urbana-Champaign.

Hoseini, M., Hosseinpour, H., Bastaee, B., 2014.A new multi objective optimization approach in distribution systems. Optimization Letters 8, 181–199.

Hosseinpour, H., Bastaee, B., 2015. Optimal placement of on-load tap changers in distribution networks using SA-TLBO method. International Journal of Electrical Power & Energy Systems 64, 1119–1128.

Huang, J., Gao, L., Li, X., 2015. A teaching–learning-based cuckoo search for constrained engineering design problems. Advances in Global Optimization 95, 375–386.

Jain, N.K., Jain, V.K., Deb, K., 2007. Optimization of process parameters of mechanical type advanced machining processes using genetic algorithm. International Journal of Machine Tools and Manufacture, 47, 900–919.

Jiang, X., Zhou, J., 2013. Hybrid DE-TLBO algorithm for solving short term hydro-thermal optimal scheduling with incommensurable Objectives, in: Proceedings of IEEE 32nd Chinese Control Conference, 26–28 July, 2474–2479.

Jordehi, A.R., 2014. Optimal setting of TCSCs in power systems using teaching–learning-based optimisation algorithm. Neural Computing and Applications. doi:10.1007/s00521-014-1791-x.

Kadambur, R., Kotecha, P., 2015. Multi-level production planning in a petrochemical industry using elitist teaching–learning-based-optimization. Expert Systems with Applications 42, 628–641.

Karaboga, D., Akay, B., 2009. A comparative study of Artificial Bee Colony algorithm. Applied Mathematics and Computation 214(1), 108–132.

Keesari, H.S., Rao, R.V., 2014. Optimization of job shop scheduling problems using teaching-learning-based optimization algorithm. OPSEARCH 51(4), 545–561.

Krishnanand, K.R., Hasani, S.M.F, Panigrahi, B.K., Panda S.K., 2013. Optimal power flow solution using self–evolving brain–storming inclusive teaching–learning–based algorithm. in: Proceeding of International Conference on Swarm Intelligence, Lecture Notes in Computer Science 7928, 338–345.

Krishnasamy, U., Nanjundappan, D., 2014. A refined teaching-learning based optimization algorithm for dynamic economic dispatch of integrated multiple fuel and wind power plants. Mathematical Problems in Engineering, 2014, 1–14.

Kumar, A., Kumar, V. R., Datta, S., Mahapatra, S.S., 2015. Parametric appraisal and optimization in machining of CFRP composites by using TLBO (teaching–learning based optimization algorithm). Journal of Intelligent Manufacturing. doi 10.1007/s10845-015-1050-8.

Kundu, S., Biswas, S., Das, S., Bose, D., 2012. Selective teaching-learning based niching technique with local diversification strategy. Swarm, Evolutionary and Memetic Computing, Lecture Notes in Computer Science 7677, 160–168.

Kurada, R.R., Pavan, K.K., Rao, A.A., 2015. Automatic teaching–learning-based optimization: A novel clustering method for gene functional enrichments. Computational Intelligence Techniques for Comparative Genomics, SpringerBriefs in Applied Sciences and Technology, 17–35.

Li, G., Niu, P., s, W., Liu, Y., 2013. Model NOx emissions by least squares support vector machine with tuning based on ameliorated teaching–learning-based optimization. Chemometrics and Intelligent Laboratory Systems 126, 11–20.

Li, J., Pan, Q., Mao, K., 2015. A discrete teaching-learning-based optimisation algorithm for realistic flowshop rescheduling problems. Engineering Applications of Artificial Intelligence 37, 279–292.

Lim, W.H., Isa, N.A.M., 2014a. Teaching and peer-learning particle swarm optimization. Applied Soft Computing 18, 39–58.

Lim, W.H., Isa, N.A.M., 2014b. Bidirectional teaching and peer-learning particle swarm optimization. Information Sciences 280, 111–134.

Lin, W., Yu, D.Y., Wang, S., Zhang, C., Zhang, S., Tian, H., Luo M., Liu S., 2014. Multiobjective teaching–learning-based optimization algorithm for reducing carbon emissions and operation time in turning operations. Engineering Optimization, http://dx.doi.org/10.1080/0305215X.2014.928818.

Mandal, B., Roy, P.K., 2013. Optimal reactive power dispatch using quasi oppositional teaching learning based optimization. Electrical Power and Energy Systems 53, 123–134.

Mandal, B., Roy, P.K., 2014. Multiobjective optimal power flow using quasi-oppositional teaching learning based optimization. Applied Soft Computing 21, 590–606.

Mardaneh, M., Golestaneh, F., 2014. Harmonic optimization of diode-clamped multilevel inverter using teaching-learning-based optimization algorithm. IETE Journal of Research 59(1), 9–16.

Medina, M.A., Coello, C.A.C., Ramirez, J.M., 2013. Reactive Power Handling by a Multiobjective Teaching Learning Optimizer Based on Decomposition. IEEE Transactions on Power Systems 28(4), 3629–3637.

Mernik, M., Liu, S-H., Karaboga, D., Črepinšek, M., 2015. On clarifying misconceptions when comparing variants of the Artificial Bee Colony Algorithm by offering a new implementation. Information Sciences 291, 115–127.

Moghadam, A., Seifi, A.R., 2014. Fuzzy-TLBO optimal reactive power control variables planning for energy loss minimization. Energy Conversion and Management 77, 208–215.

Mohapatra, A., Panigrahi, B.K., Singh, B., Bansal, R., 2012. Optimal placement of capacitors in distribution networks using modified teaching learning based optimization algorithm. Swarm, Evolutionary and Memetic Computing, Lecture Notes in Computer Science 7677, 398–405.

Niknam, T., Rasoul, A.A., Narimani, M.R., 2012a.A new multi objective optimization approach based on TLBO for location of automatic voltage regulators in distribution systems. Engineering Applications of Artificial Intelligence 25, 1577–1588.

Niknam, T., Rasoul, A.A., Narimani, M.R., 2012b. An efficient scenario-based stochastic programming framework for multi-objective optimal micro-grid operation. Applied Energy 99, 455–470.

Niknam, T., Rasoul, A.A., Aghaei, J., 2013. A new modified teaching-learning algorithm for reserve constrained dynamic economic dispatch. IEEE Transactions on Power Systems 28(2), 749–763.

Omidvar, M., Fard, R.K., Sohrabpoor, H., Teimouri, R., 2015. Selection of laser bending process parameters for maximal deformation angle through neural network and teaching–learning-based optimization algorithm. Soft Computing 19, 609–620.

Papa, G., 2008. Parameter-less evolutionary search, in: Proceedings of genetic and evolutionary computation conference, 1133–1134.

Papa, G., Vukašinović, V., Korošec, P., 2012. Guided restarting local search for production planning. Engineering Applications of Artificial Intelligence 25, 242–253.

Patel, S.J., Panchal, A.K., Kheraj, V., 2014. Extraction of solar cell parameters from a single current–voltage characteristic using teaching learning based optimization algorithm. Applied Energy 119, 384–393.

Pawar, P.J., Rao, R.V., 2013a. Parameter optimization of machining processes using teaching–learning-based optimization algorithm. International Journal of Advanced Manufacturing Technology 67(5–8), 995–1006.

Pawar, P.J., Rao, R.V., 2013b. Erratum to: Parameter optimization of machining processes using teaching-learning-based optimization algorithm. International Journal of Advanced Manufacturing Technology 67, 1955.

Pholdee, N., Park, W.W., Kim, D.K., Im Y T, Bureerat S, Kwon H C, Chun M S (2014) Efficient hybrid evolutionary algorithm for optimization of a strip coiling process. Engineering Optimization 47(4), 521–532.

Rajasekhar, A., Rani, R., Ramya, K., Abraham, A., 2012. Elitist teaching learning opposition based algorithm for global optimization, in: Proceeding of IEEE International Conference on Systems, Man, and Cybernetics. Seoul, 1124–1129.

Rao, R.V., 2011. Advanced Modeling and Optimization of Manufacturing Processes: International Research and Development. London: Springer-Verlag.

Rao, R.V., 2016. Jaya: A simple and new optimization algorithm for solving constrained and unconstrained problems. International Journal of Industrial Engineering Computations 7(1), 19–34.

Rao, R.V., Kalyankar, V.D., 2013a. Parameter optimization of modern machining processes using teaching–learning-based optimization algorithm. Engineering Applications of Artificial Intelligence 26, 524–531.

Rao, R.V., Kalyankar, V.D., 2013b. Multi-pass turning process parameter optimization using teaching–learning-based optimization algorithm. Scientia Iranica Transactions E: Industrial Engineering 20(3), 967–974.

Rao, R.V., Kalyankar, V.D., Waghmare, G., 2014. Parameters optimization of selected casting processes using teaching–learning-based optimization algorithm. Applied Mathematical Modelling 38, 5592–5608.

Rao, R.V., More, K.C., 2014. Advanced optimal tolerance design of machine elements using teachinglearning-based optimization algorithm Production & Manufacturing Research: An Open Access Journal 2(1), 71–94.

Rao, R.V., More, K.C., 2015. Optimal design of the heat pipe using TLBO (teaching-learning-based optimization) algorithm. Energy 80, 535–544.

Rao, R.V., Patel, V., 2012. An elitist teaching-learning-based optimization algorithm for solving complex constrained optimization problems. International Journal of Industrial Engineering Computations 3(4), 535–560.

Rao, R.V., Patel, V., 2013a. Comparative performance of an elitist teaching-learning-based optimization algorithm for solving unconstrained optimization problems. International Journal of Industrial Engineering Computations 4(1), 29–50.

Rao, R.V., Patel, V., 2013b. Multiobjective optimization of heat exchangers using a modified teaching-learning-based optimization algorithm. Applied Mathematical Modelling 37, 1147–1162.

Rao, R.V., Patel, V., 2013c. Multiobjective optimization of two stage thermoelectric cooler using a modified teaching–learning-based optimization algorithm. Engineering Applications of Artificial Intelligence 26, 430–445.

Rao, R.V., Patel, V., 2013d. An improved teaching-learning-based optimization algorithm for solving unconstrained optimization problems. Scientia Iranica Transactions D: Computer Science & Engineering and Electrical Engineering 20(3), 710–720.

Rao, R.V., Patel, V., 2014. A multiobjective improved teaching-learning based optimization algorithm for unconstrained and constrained optimization problems. International Journal of Industrial Engineering Computations 5(1), 1–22.

Rao, R.V., Pawar, P.J., 2010. Parameter optimization of a multi-pass milling process using non-traditional optimization algorithms. Applied Soft Computing 10(2), 445–456.

Rao, R.V., Savsani, V.J., 2011a. Mechanical Design Optimization using Advanced Optimization Techniques. New York: Springer-Verlag.

Rao, R.V., Savsani, V.J., 2011b. Multiobjective design optimization oif a robot gripper using TLBO technique. Proceedings of the Second Indo-Russian Joint Workshop on Computational Intelligence, Modern Heuristics in Automation and Robotics, Novisibirsk State Technical University, Russia, 10–13 September, 184–188.

Rao, R.V., Savsani, V.J., Vakharia, D.P., 2011. Teaching–learning-based optimization: A novel method for constrained mechanical design optimization problems. Computer-Aided Design 43, 303–315.

Rao, R.V., Savsani, V.J., Vakharia, D.P., 2012a. Teaching–Learning-Based Optimization: An optimization method for continuous non-linear large scale problems. Information Sciences 183, 1–15.

Rao, R.V., Savsani, V.J., Balic, J., 2012b. Teaching–learning-based optimization algorithm for unconstrained and constrained real-parameter optimization problems. Engineering Optimization 44(12), 1447–1462.

Rao, R.V., Waghmare, G.G., 2013. Solving Composite Test Functions Using Teaching-Learning-Based Optimization Algorithm. in: Proceedings of the International Conference on Frontiers of Intelligent Computing: Theory and Applications (FICTA), Advances in Intelligent Systems and Computing 199, 395–403.

Rao, R.V., Waghmare, G.G., 2014. A comparative study of a teaching–learning-based optimization algorithm on multiobjective unconstrained and constrained functions. Journal of King Saud University–Computer and Information Sciences 26, 332–346.

Rao, R.V., Waghmare G., 2015a. Design optimization of robot grippers using teaching-learning based optimization algorithm. Advanced Robotics 29(6), 431–447.

Rao, R.V., Waghmare, G., 2015b. Optimization of thermal performance of a smooth flat-plate solar air heater using teaching–learning-based optimization algorithm. Cogent Engineering 2 (1), 1–28.

Rao, R.V., Waghmare, G.G., 2015c. Multiobjective design optimization of a plate-fin heat sink using a teaching-learning-based optimization algorithm. Applied Thermal Engineering 76, 521–529.

Rasoul, A.A., Niknam, T., Roosta, A., Malekpour, A.R., Zarea, M., 2012. Probabilistic multiobjective wind-thermal economic emission dispatch based on point estimated method. Energy 37, 322–335.

Roy, P.K., Sur, A., Pradhan, D.K., 2013. Optimal short-term hydro-thermal scheduling using quasi-oppositional teaching learning based optimization. Engineering Applications of Artificial Intelligence 26, 2516–2524.

Roy, P.K., Bhui, S., 2014. Multiobjective quasi-oppositional teaching learning based optimization for economic emission load dispatch problem. Electrical Power and Energy Systems 53, 937–948.

Roy, P.K., Paul, C., Sultana, S., 2014. Oppositional teaching learning based optimization approach for combined heat and power dispatch. Electrical Power and Energy Systems 57, 392–403.

Roy, P.K., Sarkar, R., 2014. Solution of unit commitment problem using quasi-oppositional teaching learning based algorithm. Electrical Power and Energy Systems 60, 96–106.

Sahu, B.K., Pati, S., Mohanty, P.K., Panda, S., 2015. Teaching–learning based optimization algorithm based fuzzy-PID controller for automatic generation control of multi-area power system. Applied Soft Computing 27, 240–249.

Satapathy, S.C., Naik, A., 2012. Improved teaching learning based optimization for global function optimization. Decision Science Letters 2, 23–34.

Satapathy, S.C., Naik, A., Parvathi, K., 2013. A teaching learning based optimization based on orthogonal design for solving global optimization problems. Springer Plus 2 (130), 1–12.

Satapathy, S.C., Naik, A., 2014. Modified teaching-learning-based optimization algorithm for global numerical optimization—A comparative study. Swarm and Evolutionary Computation 16, 28–37.

Shabanpour-Haghighi, A., Seifi, A.R., Niknam, T., 2014a. A modified teaching learning based optimization for multiobjective optimal power flow problem. Energy Conversion and Management 77, 597–607.

Shabanpour-Haghighi, A., Seifi, A.R., Niknam, T., 2014b. A modified teaching learning based optimization for multiobjective optimal power flow problem. Energy Conversion and Management 77, 597–607.

Singh, M., Panigrahi, B.K., Abhyankar, A.R., 2013. Optimal coordination of directional over-current relays usingteaching-learning-based optimization (TLBO) algorithm. International Journal of Electrical Power & Energy Systems 50, 33–41.

Singh, R., Verma, H.K., 2014. Teaching-learning-based optimization algorithm for parameter identification in the design of IIR filters. Journal of The Institution of Engineers (India): Series B 94(4), 285–294.

Sultana, S., Roy, P.K., 2014. Multiobjective quasi-oppositional teaching learning based optimization for optimal location of distributed generator in radial distribution systems. Electrical Power and Energy Systems 63, 534–545.

Tiwary, A., Arya, L.D., Arya, R., Choube, S.C., 2015. Inspection–repair based availability optimization of distribution systems using teaching learning based optimization. Journal of The Institute of Engineers (India): Series B. doi:10.1007/s40031-015-0196-2.

Tang, D., Zhao, J., Li, H., 2013. An improved teaching-learning-based optimization algorithm with memetic method for global optimization.International Journal of Advancements in Computing Technology 5(9), 942–949.

Theja, B.S., Rajasekhar, A., 2013. An optimal design of coordinated PI based PSS with TCSC controller using modified teaching learning based optimization, in: Proceedings of World Congress on Nature and Biologically Inspired Computing, 99–106.

Togan, V., 2012. Design of planar steel frames using teaching-learning based optimization. Engineering Structures 34, 225–234.

Tuncel, G., Aydin, D., 2014. Two-sided assembly line balancing using teaching-learning-based optimization algorithm. Computers & Industrial Engineering 74, 291–299.

Tuo, S., Yong, L., Zhou, T., 2013. An improved harmony search based on teaching–learning strategy for unconstrained optimization problems. Mathematical Problems in Engineering 2013, 1–29, http://dx.doi.org/10.1155/2013/413565.

Uzlu, E., Komurcu, M.I., Kankal, M., Dede, T., Ozturk, H.T., 2014a. Prediction of berm geometry using a set of laboratory tests combined with teaching–learning-based optimization and artificial bee colony algorithms. Applied Ocean Research 48, 103–113.

Uzlu, E., Kankal, M., Akpinar, A., Dede, T., 2014b. Estimates of energy consumption in Turkey using neural networks with the teaching–learning-based optimization algorithm. Energy 75, 295–303.

Veček, N., Mernik, M., Črepinšek, M., 2014. A chess rating system for evolutionary algorithms: A new method for the comparison and ranking of evolutionary algorithms. Information Sciences 277, 656–679.

Waghmare, G., 2013. Comments on "A Note on Teaching Learning Based Optimization Algorithm". Information Sciences 229, 159–169.

Wang, L., Zou, F., Hei, X., Yang, D., Chen, D., Jiang, Q., Cao, Z., 2014a. A hybridization of teaching–learning-based optimization and differential evolution for chaotic time series prediction. Neural Computing and Applications 25(6), 1407–1422.

Wang, L., Zou, F., Hei, X., Yang, D., Chen, D., Jiang, Q., 2014b. An improved teaching-learning-based optimization with neighborhood search for applications of ANN. Neurocomputing, 143(2), 231–247.

Xia, K., Gao, L., Li, W., Chao, K.M., 2014. Disassembly sequence planning using a simplified teaching–learning-based optimization algorithm. Advanced Engineering Informatics 28, 518–527.

Xiao, L., Zhu, Q., Li, C., Cao, Y., Tan, Y., Li L., 2014. Application of modified teaching-learning algorithm in coordination optimization of TCSC and SVC. Pattern Recognition, Communications in Computer and Information Sciences 483, 44–53.

Xie, Z., Zhang, C., Shao, X., Lin, W., Zhu, H., 2014. An effective hybrid teaching–learning-based optimization algorithm for permutation flow shop scheduling problem. Advances in Engineering Software 77, 35–47.

Xu, Y., Wang, L., Wang, S., Liu, M., 2015. An effective teaching–learning-based optimization algorithm for the flexible job-shop scheduling problem with fuzzy processing time. Neurocomputing 148, 26–268.

Yildiz, A.R., 2013. Optimization of multi-pass turning operations using hybrid teaching learning-based approach. International Journal of Advanced Manufacturing Technology 66, 1319–1326.

Yang, Z., Li, K., Guo, Y., 2014. A new compact teaching–learning-based optimization method. Intelligent Computing Methodologies, Lecture Notes in Computer Science 8589, 717–726.

Yu, K., Wang, X., Wang, Z., 2014. An improved teaching-learning-based optimization algorithm for numerical and engineering optimization problems. Journal of Intelligent Manufacturing, doi:10.1007/s10845-014-0918-3.

Zhang, C., Li, K., Pei, L., Zhu, C., 2015. An integrated approach for real-time model-based state-of-charge estimation of lithium-ion batteries. Journal of Power Sources 283(1), 24–36.

Zou, F., Wang, L., Hei, X., Chen, D., Wang, B., 2013a. Multiobjective optimization using teaching–learning-based optimization algorithm. Engineering. Applications of Artificial Intelligence 26, 1291–1300.

Zou, F., Wang, L., Hei, X., Jiang, Q., Yang, D., 2013b. Teaching-learning-based optimization algorithm in dynamic environments. Swarm, Evolutionary and Memetic Computing, Lecture Notes in Computer Science 8297, 389–400.

Zou, F., Wang, L., Hei, X., Chen, D., Yang, D., 2014. Teaching-learning-based optimization with dynamic group strategy for global optimization. Information Sciences 273, 112–131.

Epilogue

The TLBO algorithm has carved a niche for itself in the field of advanced optimization and many researchers have understood the potential of the algorithm. The algorithm-specific parameterless concept of the algorithm is one of the attracting features of the algorithm in addition to its *simplicity*, *robustness,* and the *ability to provide global or near-global optimum solutions* in *comparatively less number of function evaluations*. Few researchers have made modifications to the basic TLBO algorithm and proved the effectiveness of the modified versions. However, to make the implementation of TLBO algorithm simpler, it is desirable to maintain the algorithm-specific parameterless concept in the modified versions also. Otherwise, the user will be faced with the burden of tuning the algorithm-specific parameters in addition to the common control parameters. The logic of the TLBO algorithm (shown in flowcharts of Chap. 2) is such that the algorithm can be used with equal ease for maximization as well as minimization problems. Single objective as well as multiobjective optimization problems can be easily attempted. The working of the TLBO algorithm demonstrated step-by-step by means of examples in Chap. 2 is expected to be useful to beginners to understand the basics of the algorithm. One can understand the *ease* of the algorithm after reading through the steps.

The potential and effectiveness of the TLBO algorithm are demonstrated in this book by reporting the results of application on the complex composite functions, constrained and unconstrained single objective, as well as multiobjective benchmark functions and practical design problems. The performance of the algorithm is compared with other well-known evolutionary algorithms and swarm intelligence-based algorithms. In the majority of cases the results obtained by the TLBO algorithm and its variants are found superior or competitive to other optimization algorithms such as GA, EP, ES, SA, ACO, PSO, DE, ABC, BBO, GSA, SFL, NSAGA-II, SPEA, MOEA, etc.

The purpose of this epilogue is not to claim that the TLBO algorithm is the 'best' algorithm among all the optimization algorithms available in the literature. In fact, there may not be any such 'best' algorithm existing! The TLBO algorithm is relatively a new algorithm and has strong potential to solve optimization problems. If the algorithm is found to have certain limitations, then the efforts of the researchers should be to find the ways to overcome the limitations and to further

R.V. Rao, *Teaching Learning Based Optimization Algorithm*,
DOI 10.1007/978-3-319-22732-0

strengthen the algorithm. The efforts should not be in the form of destructive criticism. What can be said with more confidence at present about the TLBO algorithm is that it is simple to apply, it has no algorithm-specific parameters, and it provides the optimum results in comparatively less number of function evaluations. There is no need to make any undue criticism on these aspects, as the algorithm has established itself and has set itself distinct. Researchers are encouraged to make improvements to the TLBO algorithm so that the algorithm will become much more powerful with much improved performance. It is hoped that researchers belonging to different disciplines of engineering and sciences (physical, life, and social) will find the TLBO algorithm as a powerful tool to optimize the systems and processes.

Appendix
TLBO and ETLBO Codes for Multiobjective Unconstrained and Constrained Optimization Problems

The TLBO and ETLBO codes for multiobjective unconstrained and constrained optimization problems are given below. The user has to create separate MATLAB files but the files are to be saved in a single folder. For example, UF1solution.m and UF1.m are to be saved in a single folder in the case of unconstrained optimization problems and CF1solution.m and CF1.m are to be saved in a single folder in the case of constrained optimization problems. These codes may be used for reference and the user may define the objective function(s), design variables, and their ranges as per his own requirements. The codes are developed for the UF1 and CF1 functions of Chap. 4. A priori approach of solving the multiobjective optimization problems is used in these codes by assigning equal weightages to the objectives.

R.V. Rao, *Teaching Learning Based Optimization Algorithm*,
DOI 10.1007/978-3-319-22732-0

A.1 TLBO Code for the Multiobjective Unconstrained Function

```
Multiobjective unconstrained function code
%%%%%%%%%%%%%%%%%%%%%%%%%%%%%%%%%%%%%%%%%%% UF1
%%%%%%%%%%%%%%%%%%%%%%%%%%%%%%%%%%%%%%%%%%%
Function File: UF1solution.m

clear all
clc
ps=5000;   % no of students
nd=1;
d=30; % no of design variables or no of subjects which are taken by students
ng=30;    % no of generations
ni=0;    % no of iterations
ub=ones(1,nd)*1;  %upper boundary to the variables
lb=zeros(1,nd);   %lower boundary to the variables
r=1;
truea=1;
k=1;
range=repmat((ub-lb),ps,1);
lower=repmat(lb,ps,1);

fxnewl=zeros(ps,1);
xnewl=rand(ps,nd).*range+lower; %initial random solution
for i=1:ps          % evaluation of objective function
  x=xnewl(i,1:nd);
  [f]=UF1(x,d);
   fxnewl(i,1)=f;
end
  true=1;
while(true)
  ni=ni+1;
  k=k+1;

  %%%%%%%%%%% teacher phase %%%%%%%%

md=mean(xnewl);                     % mean of the total students
indext=find(fxnewl==min(fxnewl)); % capturing the teacher position using objective function values
bt=indext;
bestteacher=zeros(1,nd);
tf=round(1+rand);
for j=1:nd
   bestteacher(1,j)=xnewl(bt(1,1),j);
end
diff=zeros(1,nd);
for j=1:nd
  diff(1,j)=rand*(bestteacher(1,j)-tf*md(1,j));
end
xnew=zeros(ps,nd);
for i=1:ps
  for j=1:nd     %%% improving the initial solution in teacher phase
    xnew(i,j)=xnewl(i,j)+diff(1,j);
    if(xnew(i,j)>ub(1,j))  % for values goes beyond upper limit
      xnew(i,j)=ub(1,j);
    elseif(xnew(i,j)<lb(1,j)) % for values goes beyond lower limit
      xnew(i,j)=lb(1,j);
    end
```

```
  end
end
fxnew=zeros(ps,1);
for i=1:ps          % evaluation of objective function
   x=xnew(i,1:nd);
   [f]=UF1(x,d);
    fxnew(i,1)=f;
end
for i=1:ps  %% applying greedy selection process
   if(fxnew(i,1)> fxnewl(i,1)) % if old value is better, then keep it as it is
      fxnew(i,1)=fxnewl(i,1);
      xnew(i,:)=xnewl(i,:);
   end
end

%%%%%%%%%%%%%%%  learner phase  %%%%%%%%%%%%%%%
indexl=find(fxnew==min(fxnew)); % capturing the best learner position using objective function value
bl=indexl;
bestl=zeros(1,nd);
for j=1:nd
    bestl(1,j)=xnew(bl(1,1),j);
end
for i=1:ps          %% improving the level other learners using best learner
   for j=1:nd
      xnewl(i,j)=xnew(i,j)+rand*(bestl(1,j)-xnew(i,j));
   end
end
for i=1:ps          % evaluation of objective function
   x=xnewl(i,1:nd);
   [f]=UF1(x,d);
    fxnewl(i,1)=f;
end

for i=1:ps  %% applying greedy selection process
   if(fxnewl(i,1)> fxnew(i,1)) % if old value is better, then keep it as it is
      fxnewl(i,1)=fxnew(i,1);
      xnewl(i,:)=xnew(i,:);
   end
end
ham=find(fxnewl==min(fxnewl));
globalmin=fxnewl(ham(1,1));
   if(ni==ng)
      ni=0;
      true=0;
   else
      true=1;
   end
%    disp ('num of iterations')
%    disp (ni)
%    disp(' global min')
%    disp (globalmin)
 fprintf('no of runs =%d num of iterations =%d Global min=%g\n',r,ni,globalmin);

end

%%%%%%%%%%%%%%%%%%%%%%%%%%%% Function File: UF1.m
%%%%%%%%%%%%%%%%%%%%%%%%%%%%%%%%
function [f]=UF1(x,d)
sumodd=0;
sumeven=0;
```

```
j=2;
d=30;
% xj=rand;

while (j<=d)
    xj=rand;
%     xjodd=sin((6*pi*xj)+((j+1)*pi/d));
    sumodd=sumodd+(xj-sin((6*pi*x)+((j+1)*pi/d)))^2;
%     xjeven=sin((6*pi*xj)+(j*pi/d));
    sumeven=sumeven+(xj-sin((6*pi*x)+(j*pi/d)))^2;
    j=j+2;
end
J1=29;
J2=30;
f1=x+2*(sumodd)/abs(J1);
f2=1-sqrt(x)+2*(sumeven)/abs(J2);
F=zeros(200,1);
for y=1:200
  F(y,1)=0.5*(f1)/1.18+0.5*(f2)/0.17;
end
f=min(F);
```

A.2 TLBO Code for the Multiobjective Constrained Function

```
Multiobjective constrained function code
%%%%%%%%%%%%%%%%%%%%%%%%%%%%%%%%%%%%%%%%%%% CF1
%%%%%%%%%%%%%%%%%%%%%%%%%%%%%%%%%%%%%%%%%%%%
Function File: CF1solution.m
clear all
clc
ps= 500;   % no of students
nd=1;
d=10; % no of design variables or no of subjects which are taken by students
ng=300;   % no of generations
ni=0;   % no of iterations
ub=ones(1,nd)*1;   %upper boundary to the variables
lb=zeros(1,nd);   %lower boundary to the variables
r=1;
truea=1;
k=1;
% while(truea)

range=repmat((ub-lb),ps,1);
lower=repmat(lb,ps,1);

fxnewl=zeros(ps,1);
xnewl=rand(ps,nd).*range+lower; %initial random solution
% for i=1:ps
%   for j=2:d
%       xnewl(i,j)=sin((6*pi*x(i,1))+(j*pi/d));
%   end
% end
s=1;
for i=1:ps          % evaluation of objective function
   x=xnewl(i,1:nd);
   [f,xnew1]=CF1(x,d,range,lower);
    fxnewl(i,1)=f;
    xnewl(i,1:nd)=xnew1;
end
```

```
  true=1;
while(true)
   ni=ni+1;
   k=k+1;

   %%%%%%%%%%% teacher phase %%%%%%%%

md=mean(xnewl);                    % mean of the total students
indext=find(fxnewl==min(fxnewl)); % capturing the teacher position using objective function values
bt=indext;
bestteacher=zeros(1,nd);
tf=round(1+rand);
for j=1:nd
    bestteacher(1,j)=xnewl(bt(1,1),j);
end
diff=zeros(1,nd);
for j=1:nd
   diff(1,j)=rand*(bestteacher(1,j)-tf*md(1,j));
end
xnew=zeros(ps,nd);
for i=1:ps
  for j=1:nd    %%% improving the initial solution in teacher phase
     xnew(i,j)=xnewl(i,j)+diff(1,j);
     if(xnew(i,j)>ub(1,j))  % for values goes beyound upper limit
       xnew(i,j)=ub(1,j);
     elseif(xnew(i,j)<lb(1,j)) % for values goes beyound lower limit
        xnew(i,j)=lb(1,j);
     end
  end
end
fxnew=zeros(ps,1);
for i=1:ps          % evaluation of objective function
   x=xnew(i,1:nd);
   [f,xnew1]=CF1(x,d,range,lower);
    fxnew(i,1)=f;
   xnew(i,1:nd)=xnew1;
end
for i=1:ps  %% applying greedy selection process
   if(fxnew(i,1)> fxnewl(i,1)) % if old value is better then keep it as it is
      fxnew(i,1)=fxnewl(i,1);
      xnew(i,:)=xnewl(i,:);
   end
end

%%%%%%%%%%%%%%%% learner phase %%%%%%%%%%%%%%%%
indexl=find(fxnew==min(fxnew)); % capturing the best learner position using objective function value
bl=indexl;
bestl=zeros(1,nd);
for j=1:nd
    bestl(1,j)=xnew(bl(1,1),j);
end
for i=1:ps          %% improving the level other learners using best learner
   for j=1:nd
      xnewl(i,j)=xnew(i,j)+rand*(bestl(1,j)-xnew(i,j));
   end
end
for i=1:ps          % evaluation of objective function
   x=xnewl(i,1:nd);
   [f,xnew1]=CF1(x,d,range,lower);
    fxnewl(i,1)=f;
```

```
    xnewl(i,1:nd)=xnew1;
end
for i=1:ps  %% applying greedy selection process
   if(fxnewl(i,1)> fxnew(i,1)) % if old value is better, then keep it as it is
      fxnewl(i,1)=fxnew(i,1);
      xnewl(i,:)=xnew(i,:);
   end
end
ham=find(fxnewl==min(fxnewl));
globalmin=fxnewl(ham(1,1));
   if(ni==ng)
      ni=0;
      true=0;
   else
      true=1;
   end
   fprintf('no of runs =%d num of iterations =%d Global min=%g\n',r,ni,globalmin);
  y(s)=globalmin;
  s=ni+1;


end

%%%%%%%%%%%%%%%%%%%%%%%%%%%%%% Function File: CF1.m
%%%%%%%%%%%%%%%%%%%%%%%%%%%%%%%%%%%%%
function [f,xnew]=CF1(x,d,range,lower)
sumodd=0;
sumeven=0;
j=2;
xj=rand;
while (j<=d)
%    xj=rand;
   xjodd=xj^(0.5*(1+(3*(j+1-2)/(d-2))));
   sumodd=sumodd+(xjodd-x^(0.5*(1+(3*(j+1-2)/(d-2)))))^2;
   xjeven=xj^(0.5*(1+(3*(j-2)/(d-2))));
   sumeven=sumeven+(xjeven-x^(0.5*(1+(3*(j-2)/(d-2)))))^2;
   j=j+2;
end
J1=9;
J2=10;
f1=x+2*(sumodd)/abs(J1);
f2=1-sqrt(x)+2*(sumeven)/abs(J2);
a=1;
N=10;
Q=(f1+f2-a*abs(sin(N*pi*(f1-f2+1)))-1);
while (Q<0)
      xnew=rand(1,1).*range(1,1)+lower(1,1);
      x=xnew;
      sumodd=0;
      sumeven=0;
      j=2;
      while (j<=d)
         xjodd=xj^(0.5*(1+(3*(j+1-2)/(d-2))));
         sumodd=sumodd+(xjodd-xnew^(0.5*(1+(3*(j+1-2)/(d-2)))))^2;
         xjeven=xj^(0.5*(1+(3*(j-2)/(d-2))));
         sumeven=sumeven+(xjeven-xnew^(0.5*(1+(3*(j-2)/(d-2)))))^2;
         j=j+2;
      end
      f1=x+2*(sumodd)/abs(J1);
      f2=1-sqrt(x)+2*(sumeven)/abs(J2);
```

```
        Q=(f1+f2-a*abs(sin(N*pi*(f1-f2+1)))-1);
    end
    xnew=x;
       y=unidrnd(200);
       f=0.5*(f1)/1.07+0.5*(f2)/0.86;
```

A.3 ETLBO Code for the Multiobjective Unconstrained Function

Multiobjective unconstrained function elitist TLBO code
%% UF1
%%
Function File: UF1solution.m

```
clear all
clc
ps=5000;   % no of students
nd=1;
d=30; % no of design variables or no of subjects which are taken by students
ng=30;    % no of generations
ni=0;     % no of iterations
ub=ones(1,nd)*1;  %upper boundary to the veriables
lb=zeros(1,nd);   %lower boundary to the veriables
r=1;
truea=1;
k=1;
range=repmat((ub-lb),ps,1);
lower=repmat(lb,ps,1);
fxnewl=zeros(ps,1);
xnewl=rand(ps,nd).*range+lower; %intial random solution
elite=12;

for i=1:ps          % evaluation of objective function
   x=xnewl(i,1:nd);
   [f]=UF1(x,d);
    fxnewl(i,1)=f;
end
  true=1;
while(true)
   ni=ni+1;
   k=k+1;
   for i = 1: elite
 xelite(i,:)=xnewl(i,1:nd);
 felite(i,:)=fxnewl(i);


end
   %%%%%%%%%%%% teacher phase %%%%%%%%

md=mean(xnewl);                  % mean of the total students
indext=find(fxnewl==min(fxnewl)); % capturing the teacher position using objective function values
bt=indext;
bestteacher=zeros(1,nd);
tf=round(1+rand);
for j=1:nd
    bestteacher(1,j)=xnewl(bt(1,1),j);
end
diff=zeros(1,nd);
for j=1:nd
   diff(1,j)=rand*(bestteacher(1,j)-tf*md(1,j));
```

```
end
xnew=zeros(ps,nd);
for i=1:ps
  for j=1:nd     %%% improving the initial solution in teacher phase
     xnew(i,j)=xnewl(i,j)+diff(1,j);
     if(xnew(i,j)>ub(1,j))  % for values goes beyound upper limit
       xnew(i,j)=ub(1,j);
     elseif(xnew(i,j)<lb(1,j)) % for values goes beyound lower limit
        xnew(i,j)=lb(1,j);
     end
  end
end
fxnew=zeros(ps,1);
for i=1:ps          % evaluation of objective function
   x=xnew(i,1:nd);
   [f]=UF1(x,d);
    fxnew(i,1)=f;
end
for i=1:ps  %% applying greedy selection process
   if(fxnew(i,1)> fxnewl(i,1)) % if old value is better then keep it as it is
      fxnew(i,1)=fxnewl(i,1);
      xnew(i,:)=xnewl(i,:);
   end
end

%%%%%%%%%%%%%%  learner phase %%%%%%%%%%%%%%
indexl=find(fxnew==min(fxnew)); % capturing the best learner position using objective function valuesr
bl=indexl;
bestl=zeros(1,nd);
for j=1:nd
    bestl(1,j)=xnew(bl(1,1),j);
end
for i=1:ps          %% improoving the level other learners using best learner
   for j=1:nd
      xnewl(i,j)=xnew(i,j)+rand*(bestl(1,j)-xnew(i,j));
   end
end
for i=1:ps          % evaluation of objective function
   x=xnewl(i,1:nd);
   [f]=UF1(x,d);
    fxnewl(i,1)=f;
end

for i=1:ps  %% applying greedy selection process
   if(fxnewl(i,1)> fxnew(i,1)) % if old value is better then keep it as it is
      fxnewl(i,1)=fxnew(i,1);
      xnewl(i,:)=xnew(i,:);
   end
end
for i = ps+1-elite:ps
   elitecount=1;
 xnew1(i,:)=xelite(elitecount,:);

   elitecount=1+elitecount;


end
xnew1=x;
[f]=UF1(x,d);
fnew1=f;
ham=find(fxnewl==min(fxnewl));
```

```
globalmin=fxnewl(ham(1,1));
   if(ni==ng)
      ni=0;
      true=0;
   else
      true=1;
   end
fprintf('no of runs =%d num of iterations =%d Global min=%g\n',r,ni,globalmin);

end


%%%%%%%%%%%%%%%%%%%%%%%%%%% Function File: UF1.m
%%%%%%%%%%%%%%%%%%%%%%%%%%%%%%%
function [f]=UF1(x,d)
sumodd=0;
sumeven=0;
j=2;
d=30;
while (j<=d)
      xj=rand;
      sumodd=sumodd+(xj-sin((6*pi*x)+((j+1)*pi/d)))^2;
      sumeven=sumeven+(xj-sin((6*pi*x)+(j*pi/d)))^2;
      j=j+2;
end
J1=29;
J2=30;
f1=x+2*(sumodd)/abs(J1);
f2=1-sqrt(x)+2*(sumeven)/abs(J2);
F=zeros(200,1);
for y=1:200
   F(y,1)=0.5*(f1)/1.18+0.5*(f2)/0.17;
end
f=min(F);
```

A.4 ETLBO Code for the Multiobjective Constrained Function

```
Multiobjective constrained elitist function code
%%%%%%%%%%%%%%%%%%%%%%%%%%%%%%%%%%%%%%%%%%%% CF1
%%%%%%%%%%%%%%%%%%%%%%%%%%%%%%%%%%%%%%%
Function File: CF1solution.m
clear all
clc
ps= 500;    % no of students
nd=1;
d=10; % no of design variables or no of subjects which are taken by students
ng=300;    % no of generations
ni=0;    % no of iterations
ub=ones(1,nd)*1;   %upper boundary to the veriables
lb=zeros(1,nd);   %lower boundary to the veriables
r=1;
truea=1;
k=1;
elite=16;

range=repmat((ub-lb),ps,1);
lower=repmat(lb,ps,1);

fxnewl=zeros(ps,1);
```

```
xnewl=rand(ps,nd).*range+lower; %intial random solution
s=1;
for i=1:ps        % evaluation of objective function
   x=xnewl(i,1:nd);
   [f,xnew1]=CF1(x,d,range,lower);
    fxnewl(i,1)=f;
    xnewl(i,1:nd)=xnew1;
end
  true=1;
while(true)
   ni=ni+1;
   k=k+1;
   for i = 1: elite
 xelite(i,:)=xnewl(i,1:nd);
 felite(i,:)=fxnewl(i);


end
   %%%%%%%%%%%% teacher phase %%%%%%%%

md=mean(xnewl);               % mean of the total students
indext=find(fxnewl==min(fxnewl)); % capturing the teacher position using objective function values
bt=indext;
bestteacher=zeros(1,nd);
for j=1:nd
    bestteacher(1,j)=xnewl(bt(1,1),j);
end
diff=zeros(1,nd);
for j=1:nd
   diff(1,j)=rand*(bestteacher(1,j)-md(1,j));
end
xnew=zeros(ps,nd);
for i=1:ps
  for j=1:nd    %%% improving the initial solution in teacher phase
     xnew(i,j)=xnewl(i,j)+diff(1,j);
     if(xnew(i,j)>ub(1,j))  % for values goes beyound upper limit
       xnew(i,j)=ub(1,j);
     elseif(xnew(i,j)<lb(1,j)) % for values goes beyound lower limit
       xnew(i,j)=lb(1,j);
     end
  end
end
fxnew=zeros(ps,1);
for i=1:ps        % evaluation of objective function
   x=xnew(i,1:nd);
   [f,xnew1]=CF1(x,d,range,lower);
    fxnew(i,1)=f;
    xnew(i,1:nd)=xnew1;
end
for i=1:ps  %% applying greedy selection process
   if(fxnew(i,1)> fxnewl(i,1)) % if old value is better then keep it as it is
      fxnew(i,1)=fxnewl(i,1);
      xnew(i,:)=xnewl(i,:);
   end
end

%%%%%%%%%%%%%%%  learner phase  %%%%%%%%%%%%%%%
indexl=find(fxnew==min(fxnew)); % capturing the best learner position using objective function valuesr
bl=indexl;
bestl=zeros(1,nd);
```

```
for j=1:nd
    bestl(1,j)=xnew(bl(1,1),j);
end
for i=1:ps          %% improoving the level other learners using best learner
   for j=1:nd
      xnewl(i,j)=xnew(i,j)+rand*(bestl(1,j)-xnew(i,j));
   end
end
for i=1:ps         % evaluation of objective function
   x=xnewl(i,1:nd);
   [f,xnew1]=CF1(x,d,range,lower);
    fxnewl(i,1)=f;
    xnewl(i,1:nd)=xnew1;
end
for i=1:ps  %% applying greedy selection process
   if(fxnewl(i,1)> fxnew(i,1)) % if old value is better then keep it as it is
      fxnewl(i,1)=fxnew(i,1);
      xnewl(i,:)=xnew(i,:);
   end
end
for i = ps+1-elite:ps
   elitecount=1;
 xnew1(i,:)=xelite(elitecount,:);

   elitecount=1+elitecount;

end
xnew1=x;
[f,xnew1]=CF1(x,d,range,lower);
fnew1=f;
ham=find(fxnewl==min(fxnewl));
globalmin=fxnewl(ham(1,1));
   if(ni==ng)
      ni=0;
      true=0;
   else
      true=1;
   end
   fprintf('no of runs =%d num of iterations =%d Global min=%g\n',r,ni,globalmin);
   y(s)=globalmin;
   s=ni+1;


end

 %%%%%%%%%%%%%%%%%%%%%%%%%%%%%%%% Function File: CF1.m
%%%%%%%%%%%%%%%%%%%%%%%%%%%%%%%%%%%%%%
function [f,xnew]=CF1(x,d,range,lower)
sumodd=0;
sumeven=0;
j=2;
xj=rand;
while (j<=d)
   xjodd=xj^(0.5*(1+(3*(j+1-2)/(d-2))));
   sumodd=sumodd+(xjodd-x^(0.5*(1+(3*(j+1-2)/(d-2)))))^2;
   xjeven=xj^(0.5*(1+(3*(j-2)/(d-2))));
   sumeven=sumeven+(xjeven-x^(0.5*(1+(3*(j-2)/(d-2)))))^2;
   j=j+2;
end
J1=9;
```

```
J2=10;
f1=x+2*(sumodd)/abs(J1);
f2=1-sqrt(x)+2*(sumeven)/abs(J2);
a=1;
N=10;
Q=(f1+f2-a*abs(sin(N*pi*(f1-f2+1)))-1);
while (Q<0)
    xnew=rand(1,1).*range(1,1)+lower(1,1);
    x=xnew;
    sumodd=0;
    sumeven=0;
    j=2;
    while (j<=d)
       xjodd=xj^(0.5*(1+(3*(j+1-2)/(d-2))));
       sumodd=sumodd+(xjodd-xnew^(0.5*(1+(3*(j+1-2)/(d-2)))))^2;
       xjeven=xj^(0.5*(1+(3*(j-2)/(d-2))));
       sumeven=sumeven+(xjeven-xnew^(0.5*(1+(3*(j-2)/(d-2)))))^2;
       j=j+2;
    end
    f1=x+2*(sumodd)/abs(J1);
    f2=1-sqrt(x)+2*(sumeven)/abs(J2);
    Q=(f1+f2-a*abs(sin(N*pi*(f1-f2+1)))-1);
end
xnew=x;
  y=unidrnd(200);
  f=0.5*(f1)/1.07+0.5*(f2)/0.86;
```

In addition to the above, the readers may also refer to Rao and Patel (2012) and Rao et al. (2014) for more codes. A tutorial on TLBO algorithm for solving the unconstrained and constrained optimization problems is available in Rao (2016).

References

Rao, R.V., 2016. Review of applications of TLBO algorithm and a tutorial for beginners to solve the unconstrained and constrained optimization problems. Decision Science Letters 5, 1–30.

Rao, R.V., Kalyankar, V.D., Waghmare, G., 2014. Parameters optimization of selected casting processes using teaching-learning-based optimization algorithm. Applied Mathematical Modelling 38, 5592–5608.

Rao, R.V., Patel, V., 2012. An elitist teaching-learning-based optimization algorithm for solving complex constrained optimization problems. International Journal of Industrial Engineering Computations 3(4), 535–560.

Index

R.V. Rao, *Teaching Learning Based Optimization Algorithm*,
DOI 10.1007/978-3-319-22732-0

The manufacturer's authorised representative in the EU is Springer Nature Customer Service Centre GmbH, Europaplatz 3, 69115 Heidelberg, Germany. If you have any concerns regarding our products, please contact ProductSafety@springernature.com

Printed and bound by CPI Group (UK) Ltd, Croydon, CR0 4YY
01/10/2026
02225975-0011